Python数据可视化

数据类型、库与实践

熊 斌 孙文丽 编著

中国铁道出版社有限公司
CHINA RAILWAY PUBLISHING HOUSE CO., LTD.

内 容 简 介

本书循序渐进地深入讲解了使用Python语言实现数据可视化的核心知识，并通过具体实例的实现过程演练了数据分析的方法和流程，旨在帮助有一定Python语言基础的读者较为系统地了解和熟悉各类数据类型、库在可视化实践中的具体应用，完善基于Python语言的数据爬取、分析和可视化的应用链条，顺利迈入应用实践。

图书在版编目（CIP）数据

Python数据可视化：数据类型、库与实践/熊斌，孙文丽编著.—北京：中国铁道出版社有限公司，2021.7
ISBN 978-7-113-27565-5

Ⅰ.①P… Ⅱ.①熊… ②孙… Ⅲ.①软件工具-程序设计 Ⅳ.①TP311.56

中国版本图书馆CIP数据核字（2020）第262840号

书　　名：Python 数据可视化：数据类型、库与实践
Python SHUJU KESHIHUA: SHUJU LEIXING KU YU SHIJIAN
作　　者：熊　斌　孙文丽

责任编辑：荆　波　　**编辑部电话**：（010）51873026　　**邮箱**：the-tradeoff@qq.com
封面设计：MXK DESIGN STUDIO
责任校对：焦桂荣
责任印制：赵星辰

出版发行：中国铁道出版社有限公司（100054，北京市西城区右安门西街 8 号）
印　　刷：国铁印务有限公司
版　　次：2021 年 7 月第 1 版　2021 年 7 月第 1 次印刷
开　　本：787 mm×1 092 mm 1/16　**印张**：20.25　**字数**：506 千
书　　号：ISBN 978-7-113-27565-5
定　　价：79.00 元

版权所有　侵权必究

凡购买铁道版图书，如有印制质量问题，请与本社读者服务部联系调换。电话：（010）51873174

打击盗版举报电话：（010）63549461

前　言

■ 写作目的

互联网的普及让我们迎来了大数据时代，各行各业都在产生大量的数据；如何更好地用好这些数据，让它为我们服务；数据挖掘、清洗和分析应运而生，数据整理完毕后如何以更直观、更有效的方式呈现出来，这也是本书的写作目的。

随着数据运营技术的不断成熟，现在数据可视化工具也开始增多，数据可视化的目的是让用户更好地使用数据。当数据变得可视化之后有以下两个好处：

第一，直接给出用户想要的数据并且以最简单的视图呈现出来；

第二，根据数据不断变化的现实特点，动态显示数据，不但给出数据的整体直观印象，更给出一段时间内的发展变化轨迹。

随着我国大数据战略规划的提出和推广，在未来一段时间内，数据可视化技能将成为数据分析从业人员的必备技能，甚至会衍生出专业的数据可视化岗位，未来可期。

■ 图书特色

（1）内容较全面

详细讲解了在Python语言框架下数据可视化所需要的开发技术，循序渐进地讲解了这些技术的使用方法和技巧，力求帮助读者尽快建立起Python数据可视化的开发框架和思维。

（2）实例驱动

在图书讲解方式上，笔者采用了理论加实例的模式，通过对实例的分析，能够实现对所讲知识点的横向切入和纵向比较，不但能让读者有更多的实践演练机会，还可以从不同的方位展现一个知识点的用法。

（3）全程二维码讲解视频

为了能更加紧密地贴合实践，笔者专门制作了本书的讲解视频，书中每一个二级目录下都有一个二维码，扫描二维码可以观看本节内容的讲解视频，既包括对正文内容的讲解，也包括对实例的实践操作。

■ 读者对象

本书主要面向以下读者对象：

- 专业数据分析人员；
- 数据库工程师和管理员；

- 数据可视化研发工程师。

■ 整体下载包

为了方便不同网络环境下的读者学习，也为了提升图书的附加价值，笔者将书中 55 个扫码视频和源代码整理成整体下载包，读者可以通过封底二维码和下载链接学习。

备用网盘下载地址：https://pan.baidu.com/s/1oIX50mwN0TITuhKZJUMqcA

提取码：y2sr

■ 致谢与批评

感谢中国铁道出版社有限公司各位编辑的大力支持，正是各位编辑的求实、耐心和效率，才使得本书能够在这么短的时间内得以出版。另外，也十分感谢我的家人给予的巨大支持。毕竟笔者水平有限，书中存在疏漏之处在所难免，诚请读者提出宝贵意见或建议，以便修订并使之更臻完善。

最后感谢您购买本书，希望本书能成为您数据可视化编程路上的领航者，祝您阅读愉快！

熊　斌

2021 年 5 月

目　录

第 1 章

数据可视化基础

数据可视化使数据变得更容易理解，其内在价值更直观。数据可视化软件正帮助越来越多的企业从浩如烟海的复杂数据中理出头绪，化繁为简，变成看得见的财富，从而实现更有效的决策过程。Python 是一门面向对象的程序开发语言，被公认为是开发数据可视化程序的最佳语言。本章将详细介绍 Python 语言的基础知识和常见的数据可视化技术，为读者步入本书后面知识的学习打下基础。

1.1 什么是数据可视化

数据可视化是一个非常重要的概念，是指将一些文字形式的数据集用图形、图像的形式表示，并利用数据分析和开发工具发现其中未知信息的处理过程。

↑扫码看视频（本节视频课程时间：4 分 41 秒）

1.1.1 数据可视化介绍

数据可视化 Data Visualization 和信息可视化 Infographics 是两个相近的专业领域名词。狭义上的数据可视化是指将数据用统计图表的方式呈现出来，而信息图形（信息可视化）则是将非数字的信息进行可视化。前者用于传递信息，后者用于表现抽象或复杂的概念、技术和信息。

因为在现实中，信息包含了数字信息和非数字信息，所以从广义上讲，数据可视化是信息可视化中的一类。从原词的解释来讲，数据可视化重点突出的是“可视化”，而信息可视化的重点则是“图示化”。从整体而言，可视化是数据、信息以及科学等多个领域图示化技术的统称。

数据可视化的主要目的是借助于图形化的手段，清晰直观地传达与沟通信息。但是，这并不意味着数据可视化为实现其功能用途而枯燥乏味，或者为了看上去绚丽多彩而显得极其复杂。为了有效地传达思想概念，美学形式与功能需要齐头并进，通过直观地传达关键的方面与特征，从而实现对于相当稀疏而又复杂的数据集的深入洞察。然而，很多设计人员往往并不能很好地把握设计与功能之间的平衡，做出了华而不实的数据可视化形式，无法达到其主要目的，也就是传达与沟通信息。

数据可视化与信息图形、信息可视化、科学可视化以及统计图形密切相关。当前，在研究、教学和开发领域，数据可视化仍是一个极为活跃而又关键的方面。“数据可视化”这个术语实现了成熟的科学可视化领域与较年轻的信息可视化领域的统一。

1.1.2 数据可视化的起源

一般认为，数据可视化起源于统计学诞生的时代，但是真正追溯其根源，可以把时间往前推 10 个世纪。让我们一起来看一幅 10 世纪的数据可视化作品，如图 1-1 所示。这幅作品应该是目前能找到的，时代最久远的数据可视化作品了，是由一位天文学家创作。在这幅作品中，包含了很多现代统计图形元素：坐标轴，网格，时间序列。

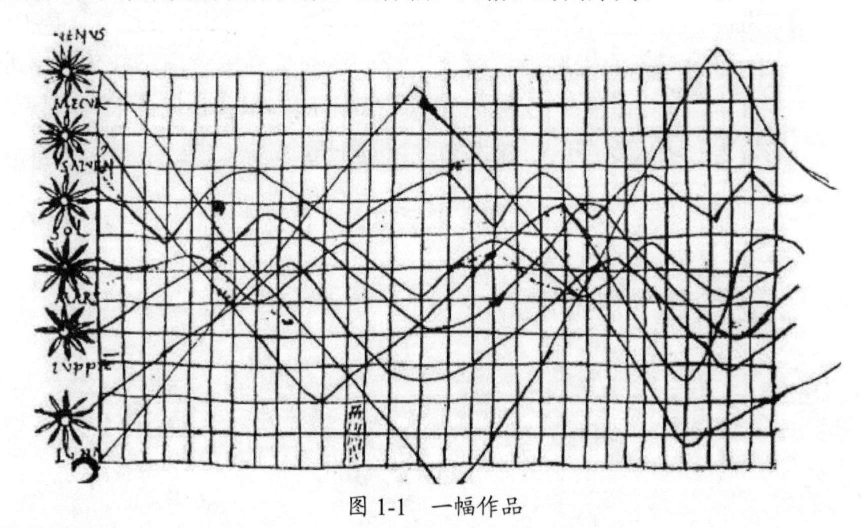

图 1-1　一幅作品

在欧洲文艺复兴时期（14 世纪 ~17 世纪）出现了很多科学家和艺术家，出现了各种测量技术，例如著名的“笛卡儿”几何和坐标系理论，费马和帕斯卡发展出了概率论，英国人开始了人口统计学研究。这些科学和艺术的发展，为数据可视化正式打开了大门。

在 18 世纪，微积分、物理、化学、数学都开始蓬勃发展，统计学也开始出现了萌芽。数据的价值开始被人们重视起来，人口、商业等经验数据开始被系统地收集整理并记录下来，各种图表和图形也开始诞生。例如在 1702 年，著名天文学家哈雷计算并绘制出了哈雷彗星的轨道图，在地图的网格上用等值线标注了磁偏角，如图 1-2 所示。苏格兰工程师 William Playfair（1759—1823）创造了今天我们习以为常的几种基本数据可视化图形，例如折线图、条形图和饼图，如图 1-3 所示。

19 世纪是现代图形学的开始，科学技术的迅速发展，使得工业革命从英国扩散到欧洲大陆和北美。随着社会对数据的积累和应用的需求，现代数据可视化，统计图形和主题图的主要表达方式，在这几十年间基本上都出现了。在这个时期内，数据可视化的重要发展包括：统计图形方面：散点图、直方图、极坐标图形和时间序列图等当代统计图形的常用形式都已出现。主题图方面：主题地图和地图集成为这个年代展示数据信息的一种常用方式，应用领域涵盖社会、经济、疾病、自然等各个主题。

19 世纪中晚期，可以说是数据可视化的黄金时期，出现了大量的经典作品，人们对数据价值的理解在图表的帮助下逐渐普及开来。欧洲的官方统计机构也普遍建立起来，高斯和拉普拉斯奠定了统计理论的基础。

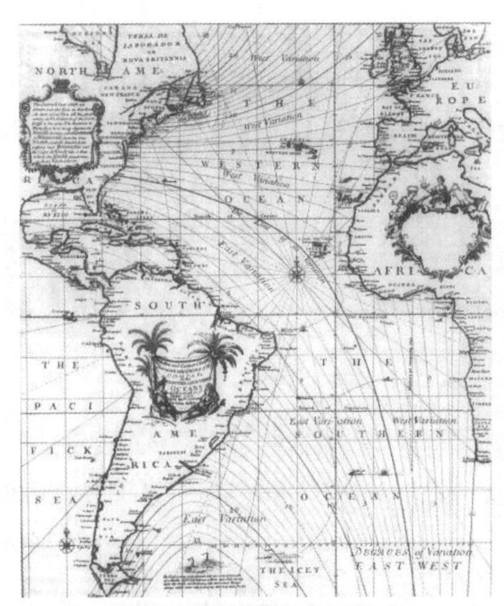

图 1-2　哈雷彗星的轨道图

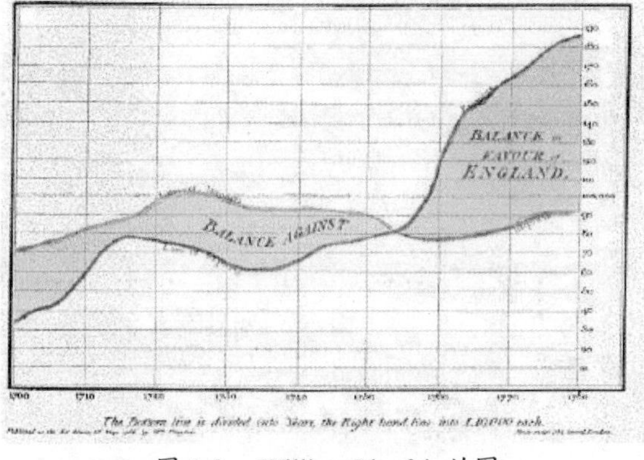

图 1-3　William Playfair 的图

时间进入到 20 世纪下半段，随着计算机技术的兴起，数据统计处理变得越来越高效。理论层面，数理统计也把数据分析变成了科学。工业和科学发展导致的对数据处理的迫切需求把这门科学逐渐运用到各行各业。统计的各个应用分支不断建立起来，处理着各自行业面对的数据问题。在应用当中，图形表达占据了重要的地位，相对于参数估计假设检验，明快、直观的图形形式更容易被人接受。

进入 21 世纪以来，计算机技术获得了长足的进展，计算机图形学，高分辨率、高色深还原度的屏幕应用越来越广泛，数据可视化的需求也正在变得越来越强烈，数据可视化进入了一个黄金时代。

1.1.3　数据可视化的意义

如果我们使用百度搜索“数据可视化”，大概能查到 42 700 000 个词条，并且这个数据会随着时间的推移稳步增加。“数据可视化”是当今数据分析领域中发展最快也是最引人注目的领域。也许有人会说，数据可视化就是画图，看不出来研究的价值在哪里。也有很多人认为，数据可视化就是把数据从冰冷的数字或文字转换成图形，这样看起来更丰富一些，酷炫一点儿。其实不然，一个好的数据可视化，不仅可以给人们带来视觉上的冲击，而且还能够揭示蕴含在数据中的规律和道理。

数据可视化的意义是帮助人们更好地分析数据，信息的质量很大程度上依赖于其表达方式。对数字罗列组成的数据中所包含的意义进行分析，使分析结果可视化。其实数据可视化的本质就是视觉对话。数据可视化将技术与艺术完美结合，借助图形化的手段，清晰有效地传达与沟通信息。一方面，数据赋予可视化以价值；另一方面，可视化增加数据的灵性，两者相辅相成，帮助企业从信息中提取知识、从知识中获取价值。

严格来说，数据可视化的意义和优势如下：

（1）传递速度快

人脑对视觉信息的处理要比书面信息快 10 倍。使用图表来总结复杂的数据，可以确保对关系的理解要比那些混乱的报告或电子表格更快。

（2）数据显示的多维性

在可视化分析下，数据将每一维的值分类、排序、组合和显示，这样就可以看到表示对象或事件的数据的多个属性或变量。

（3）更直观的展示信息

大数据可视化报告使我们能够用一些简单的图形就能体现那些复杂的信息，甚至单个图形也能做到。决策者可以轻松地解释各种不同的数据源。丰富又有意义的图形有助于让忙碌的主管和业务伙伴了解问题和未解决的计划。

（4）大脑记忆能力的限制

实际上我们在观察物体的时候，我们的大脑和计算机一样有长期的记忆（memory 硬盘）和短期的记忆（cache 内存）。只有我们记下文字、诗歌、物体，一遍一遍地在短期记忆出现了之后，它们才可能进入长期记忆。

1.2 数据可视化的相关技术

一直以来，数据可视化就是一个处于不断演变之中的概念，其边界在不断扩大。因此，最好是对其加以宽泛的定义。数据可视化指的是技术上较为高级的技术方法，而这些技术方法允许利用图形、图像处理、计算机视觉以及用户界面，通过表达、建模以及对立体、表面、属性以及动画的显示，对数据加以可视化解释。与立体建模之类的特殊技术方法相比，数据可视化所涵盖的技术方法要广泛得多。在本节的内容中，将详细讲解实现数据可视化的相关技术，为读者的学习做好指引方向。

↑扫码看视频（本节视频课程时间：3 分 8 秒）

1.2.1 数据采集和数据挖掘

数据采集（有时缩写为 DAQ 或 DAS），又称为“数据获取”或“数据收集”，是指对现实世界进行采样，以便产生可供计算机处理的数据的过程。例如，在现实中常见的网络爬虫技术就属于数据采集模块。通常，数据采集过程中包括为获得所需信息，对于信号和波形进行采集并对它们加以处理的过程。数据采集系统的组成元件当中包括用于将测量参数转换为电信号的传感器，而这些电信号则是由数据采集硬件来负责获取的。

数据挖掘是指对大量数据加以分类整理并挑选出相关信息的过程。数据挖掘通常为商业智能组织和金融分析师普遍采用；不过，在科学领域，数据挖掘也越来越多地用于从现代实验与观察方法所产生的庞大数据集之中提取信息。数据挖掘被描述为“从数据之中提取隐含的、先前未知的、潜在有用信息的非凡过程”，以及“从大型数据集或数据库之中提取有用信息的科学”。与企业资源规划相关的数据挖掘是指对大型交易数据集进行统计分析和逻辑分析，从中寻找出可能有助于决策制定工作的模式的过程。

注意：数据可视化分析和数据挖掘的异同

数据可视化分析和数据挖掘的目标都是从数据中获取信息与知识，但是手段不同。数据可视化分析是将数据呈现给用户以易于感知的图形符号，让用户交互地理解数据。数据挖掘是通过计算机自动或者半自动地获取数据隐含的知识，并将获取的知识直接给予用户。也就

是说，数据可视化可以看到交互界面，更适合于探索性地分析数据。而数据挖掘面对的是一堆活生生但“黑不溜秋”的数据，需要像挖矿一样从中发现金子。

1.2.2 数据分析

数据分析是指为了提取有用信息和形成结论而对数据加以详细研究和概括总结的过程。数据分析与数据挖掘密切相关，但数据挖掘往往倾向于关注较大型的数据集，较少侧重于推理，且常常采用的是最初为另外一种不同目的而采集的数据。在统计学领域，有些人将数据分析划分为描述性统计分析、探索性数据分析以及验证性数据分析；其中，探索性数据分析侧重于在数据之中发现新的特征，而验证性数据分析则侧重于已有假设的证实或证伪。

数据分析包括两种类型，具体说明如下：

（1）探索性数据分析：是指为了形成值得假设的检验而对数据进行分析的一种方法，是对传统统计学假设检验手段的补充。该方法由美国著名统计学家约翰·图基命名；

（2）定性数据分析：又称为“定性资料分析”、“定性研究”或者“质性研究资料分析”，是指对诸如词语、照片、观察结果之类的非数值型数据（或者说资料）的分析。

2010 年后数据可视化工具基本以表格，图形（Chart），地图等可视化元素为主，数据可进行过滤、钻取、数据联动、跳转、高亮等分析手段做动态分析。可视化工具可以提供多样的数据展现形式、多样的图形渲染形式、丰富的人机交互方式、支持商业逻辑的动态脚本引擎等。

1.3 Python 语言和数据可视化

Python 语言是一门全能型的编程语言，不但可以开发网络爬虫程序收集数据，而且也可以开发出可视化的数据分析图。在本节的内容中，将简要介绍 Python 语言的优势和实现数据可视化的相关技术。

↑扫码看视频（本节视频课程时间：2 分 0 秒）

1.3.1 TIOBE 编程语言的地位

根据 IEEE 发布的《2018 年最热门的编程语言》（2018 list of top programming languages），Python 在整体排名中位居榜首。而在 TIOBE 编程语言社区排行榜中，Python 语言的排名十分靠前，仅次于传统的“霸主”语言 Java 和 C。表 1-1 展示了 2020 年 6 月 TIOBE 编程语言使用率统计信息 。

表 1-1

2020 年排名	语言	2020 年占有率 (%)
1	C	14.32
2	Java	11.23
3	Python	11.03
4	C++	7.14

注意：TIOBE 编程语言社区排行榜是编程世界的一个重要参考数据，是编程语言流行趋势的一个重要指标，此榜单每月更新一次，这份排行榜排名基于互联网上有经验的程序员、课程和第三方厂商的数据。

1.3.2 Python 语言的优点

Python 语言为什么这么火呢？主要有如下两个原因。

（1）简单

无论是对于广大学习者还是程序员，简单就代表着一切，代表着最大的吸引力。既然都能实现同样的目的，人们有什么理由不去选择更加简单的开发语言呢？例如，在运行 Python 程序时，只需要简单地输入 Python 代码后即可运行，而不需要像其他语言（例如 C 或 C++）那样，需要经过编译和连接等中间步骤。Python 可以立即执行程序，这样便形成了一种交互式编程体验和不同情况下快速调整的能力，往往在修改代码后能立即看到程序改变后的效果。

（2）功能强大

Python 语言可以被用来作为批处理语言，写一些简单工具，处理一些数据，作为其他软件的接口调试等。Python 语言可以用来作为函数语言，进行人工智能程序的开发，具有 Lisp 语言的大部分功能。Python 语言可以用来作为过程语言，进行我们常见的应用程序开发，可以和 VB 等语言一样应用。 Python 语言可以用来作为面向对象语言，具有大部分面向对象语言的特征，经常作为大型应用软件的原型开发，然后再用 C++ 语言改写，而有些应用软件则是直接使用 Python 开发的。

1.3.3 Python 数据可视化的相关技术

Python 语言的最大优势是可以使用内置库和第三方库实现数据可视化，这样可以节省开发时间，提高开发效率。下面列出了常见的实现数据可视化功能的相关库。

1. 数据采集和挖掘

数据采集和挖掘是指从现实中获取数据信息，特别是网络中的数据信息。在 Python 语言中提供了多个库实现数据采集和挖掘功能，具体说明如下：

（1）请求库：实现 HTTP 请求操作，具体名称与解释如下表 1-2 所示。

表 1-2

名称	解　释
urllib	一系列用于操作 URL 的功能
requests	基于 urllib 编写的，阻塞式 HTTP 请求库，发出一个请求，一直等待服务器响应后，程序才能进行下一步处理
selenium	自动化测试工具。一个调用浏览器的 driver，通过这个库你可以直接调用浏览器完成某些操作，比如输入验证码
aiohttp	基于 asyncio 实现的 HTTP 框架。异步操作借助于 async/await 关键字，使用异步库进行数据抓取，可以极大地提高效率

（2）解析库：从网页中提取信息，具体名称与解释如下表 1-3 所示。

表 1-3

名称	解释
beautifulsoup	HTML 和 XML 的解析，从网页中提取信息，同时拥有强大的 API 和多样解析方式
pyquery	jQuery 的 Python 实现，能够以 jQuery 的语法来操作解析 HTML 文档，易用性和解析速度都很好
lxml	支持 HTML 和 XML 的解析，支持 XPath 解析方式，而且解析效率非常高
tesserocr	一个 OCR 库，在遇到验证码（图形验证码为主）的时候，可直接用 OCR 进行识别

（3）爬虫框架，具体名称与解释如下表 1-4 所示。

表 1-4

名称	解释
Scrapy	很强大的爬虫框架，可以满足简单的页面爬取（比如可以明确获知 url pattern 的情况）。用这个框架可以轻松爬下来如亚马逊商品信息之类的数据。但是对于稍微复杂一点儿的页面，如微博的页面信息，这个框架就满足不了需求了
Crawley	高速爬取对应网站的内容，支持关系和非关系数据库，数据可以导出为 JSON、XML 等
Portia	可视化爬取网页内容
newspaper	提取新闻、文章以及内容分析
python-goose	用 Java 语言编写的文章提取工具
cola	一个分布式爬虫框架。项目整体设计有点糟，模块间耦合度较高

2. 数据可视化

通过数据采集和挖掘会得到海量的数据，巨量的数据呈现在我们面前，这些毫无规律的数据是没有任何意义的。数据可视化的功能是让这些数据变得更加有价值。在 Python 语言中提供了多个库实现数据可视化功能，具体说明如表 1-5 所示。

表 1-5

名称	解释
Matplotlib	是一个最基础的 Python 可视化库，作图风格接近 MATLAB，所以称为 matplotlib。一般都是从 matplotlib 上手 Python 数据可视化，然后开始做纵向与横向拓展
Seaborn	是一个基于 matplotlib 的高级可视化效果库，针对的点主要是数据挖掘和机器学习中的变量特征选取，Seaborn 可以用短小的代码去绘制、描述更多维度数据的可视化效果图
Plotly	是做交互可视化的一把利器，同时支持 Python 和 R 语言，并且实现了在线导入数据做可视化并保存内容在云端 server 的功能。做演示的时候，只需要在本地的 jupyter notebook 与 plotly server 建立通信，即可调用已经做好的可视化内容作为展示。Plotly 同时有 freemium 和 premium 两种账户，免费账户已经可以满足基本需要
Pandas	是建立在 Matplotlib 之上的，当用 Pandas 中的 df.plot() 时，用得其实是别人用 Matplotlib 写的代码。因此，这些图在美化方面是相似的，自定义图时用的语法也都非常相似
Pygal	和其他常用的绘图包一样，也是用图形框架语法来构建图像的。由于绘图目标比较简单，因此这是一个相对简单的绘图包
Echart	一个纯 Javascript 实现的数据可视化库，百度的产品，常应用于软件产品开发或网页的统计图表模块。可在 Web 端高度定制可视化图表，图表种类多，动态可视化效果好，各类图表、各类形式都完全开源免费。能处理较大数据量，3D 绘图也毫不逊色，结合百度地图使用会更加出色

第 2 章

常用的数据采集技术

进行数据可视化处理的前提是有要处理的数据，这些数据是从哪儿来的呢？一些是个人、机构或组织提供的，一些需要大家从网络中收集归纳。在本章的内容中，我们首先讲解了处理不同类型数据的原理，然后通过网络爬虫技术和第三方爬虫框架 Scarpy 从实践层面对数据采集技术进行了阐述，为读者步入本书后面知识的学习打下基础。

2.1 处理网络数据

互联网改变了人们的生活方式，在网络中保存了大量的数据。例如在网页中用 HTML 和 XML 存储的数据，还有用 HTTP 传输的数据等。在本节的内容中，将详细讲解使用 Python 网络库处理网络数据的知识。

↑扫码看视频（本节视频课程时间：18 分 12 秒）

2.1.1 解析 HTML 和 XML 数据

在互联网应用中，经常使用 HTML 和 XML 技术保存或展示数据信息。在 Python 中提供了多个库解析 HTML 和 XML 数据，具体说明如下：

1. 使用内置库解析 XML 数据

在 Python 应用程序中，有两种常见的 XML 编程接口，分别是 SAX 和 DOM。所以与之对应的是，Python 语言有两种解析 XML 文件的方法，分别是 SAX 和 DOM 方法。其中库 XML 由表 2-1 所示的核心模块构成。

表 2-1

模块名称	说　明
xml.etree.ElementTree	提供了处理 ElementTree 的成员，是一个轻量级的 XML 处理器
xml.dom	用于定义 DOM 的 API，提供了处理 DOM 标记的成员
xml.dom.minidom	提供了处理最小 DOM 的成员
xml.dom.pulldom	提供了构建部分 DOM 树的成员
xml.sax	提供了处理 SAX2 的基类和方法成员
xml.parsers.expat	绑定了 Expat 解析器的功能，能够使用注册处理器处理不同 XML 的文档部分

例如在下面的实例代码中，演示了使用库 xml.etree.ElementTree 读取 XML 文件的过程。

其中 XML 文件 test.xml 的具体实现代码如下所示：

```
<students>
    <student name='赵敏'     sex='男' age='35'/>
    <student name='周芷若'   sex='女' age='38'/>
    <student name='小昭'     sex='女' age='22'/>
</students>
```

例如在下面的实例代码中，演示了使用 SAX 方法解析 XML 文件的过程。其中实例文件 movies.xml 是一个基本的 XML 文件，在里面保存了一些和电影有关的资料信息。文件 movies.xml 的具体实现代码如下所示：

```
<collection shelf="Root">
<movie title="深入敌后">
<type>War, Thriller</type>
<format>DVD</format>
<year>2003</year>
<rating>三星</rating>
<stars>10</stars>
<description>战争故事</description>
</movie>
<movie title="变形金刚">
<type>Anime, Science Fiction</type>
<format>DVD</format>
<year>1989</year>
<rating>五星</rating>
<stars>8</stars>
<description>科幻片</description>
</movie>
<movie title="枪神">
<type>Anime, Action</type>
<format>DVD</format>
<episodes>4</episodes>
<rating>四星</rating>
<stars>10</stars>
<description>警匪片</description>
</movie>
<movie title="伊什塔尔">
<type>Comedy</type>
<format>VHS</format>
<rating>五星</rating>
<stars>2</stars>
<description>希腊神话</description>
</movie>
</collection>
```

实例文件 sax.py 的功能是解析文件 movies.xml 的内容，具体实现代码如下所示：

```
import xml.sax
class MovieHandler( xml.sax.ContentHandler ):
def __init__(self):
      self.CurrentData = ""
      self.type = ""
      self.format = ""
      self.year = ""
      self.rating = ""
      self.stars = ""
      self.description = ""
   # 元素开始调用
def startElement(self, tag, attributes):
      self.CurrentData = tag
if tag == "movie":
print ("*****Movie*****")
title = attributes["title"]
```

```
print ("Title:", title)
   # 元素结束调用
def endElement(self, tag):
      if self.CurrentData == "type":                   # 处理 XML 中的 type 元素
print ("Type:", self.type)
      elif self.CurrentData == "format":               # 处理 XML 中的 format 元素
print ("Format:", self.format)
      elif self.CurrentData == "year":                 # 处理 XML 中的 year 元素
print ("Year:", self.year)
      elif self.CurrentData == "rating":               # 处理 XML 中的 rating 元素
print ("Rating:", self.rating)
      elif self.CurrentData == "stars":                # 处理 XML 中的 stars 元素
print ("Stars:", self.stars)
      elif self.CurrentData == "description":          # 处理 XML 中的 description 元素
print ("Description:", self.description)
      self.CurrentData = ""
   # 读取字符时调用
def characters(self, content):
if self.CurrentData == "type":
         self.type = content
elif self.CurrentData == "format":
         self.format = content
elif self.CurrentData == "year":
         self.year = content
elif self.CurrentData == "rating":
         self.rating = content
elif self.CurrentData == "stars":
         self.stars = content
elif self.CurrentData == "description":
         self.description = content
if ( __name__ == "__main__"):
   # 创建一个 XMLReader
parser = xml.sax.make_parser()
   # turn off namespaces
parser.setFeature(xml.sax.handler.feature_namespaces, 0)
   # 重写 ContentHandler
   Handler = MovieHandler()
parser.setContentHandler( Handler )
parser.parse("movies.xml")
```

执行后的效果如图 2-1 所示。

```
*****Movie*****
Title: 深入敌后
Type: War, Thriller
Format: DVD
Year: 2003
Rating: 三星
Stars: 10
Description: 战争故事
*****Movie*****
Title: 变形金刚
Type: Anime, Science Fiction
Format: DVD
Year: 1989
Rating: 五星
Stars: 8
Description: 科幻片
*****Movie*****
Title: 枪神
Type: Anime, Action
Format: DVD
Rating: 四星
Stars: 10
Description: 警匪片
```

图 2-1　执行效果

例如在下面的实例文件 dom.py 中，演示了使用 DOM 方法解析 XML 文件的过程。实例文件 dom.py 的功能是解析文件 movies.xml 的内容，具体实现代码如下所示：

```
from xml.dom.minidom import parse
import xml.dom.minidom
# 使用 minidom.parse 解析器 XML 文件
DOMTree = xml.dom.minidom.parse("movies.xml")
collection = DOMTree.documentElement
if collection.hasAttribute("shelf"):
   print ("根元素是 : %s" % collection.getAttribute("shelf"))

# 在集合中获取所有电影
movies = collection.getElementsByTagName("movie")

# 打印每部电影的详细信息
for movie in movies:
   print ("*****Movie*****")
   if movie.hasAttribute("title"):
      print ("Title 电影名 : %s" % movie.getAttribute("title"))

   type = movie.getElementsByTagName('type')[0]
   print ("Type 电影类型 : %s" % type.childNodes[0].data)
   format = movie.getElementsByTagName('format')[0]
   print ("Format 电影格式 : %s" % format.childNodes[0].data)
   rating = movie.getElementsByTagName('rating')[0]
   print ("Rating 电影评分 : %s" % rating.childNodes[0].data)
   description = movie.getElementsByTagName('description')[0]
   print ("Description 电影简介 : %s" % description.childNodes[0].data)
```

执行后会输出：

```
根元素是 : Root
*****Movie*****
Title 电影名 : 深入敌后
Type 电影类型 : War, Thriller
Format 电影格式 : DVD
Rating 电影评分 : 三星
Description 电影简介 : 战争故事
*****Movie*****
Title 电影名 : 变形金刚
Type 电影类型 : Anime, Science Fiction
Format 电影格式 : DVD
Rating 电影评分 : 五星
Description 电影简介 : 科幻片
*****Movie*****
Title 电影名 : 枪神
Type 电影类型 : Anime, Action
Format 电影格式 : DVD
Rating 电影评分 : 四星
Description 电影简介 : 警匪片
*****Movie*****
Title 电影名 : 伊什塔尔
Type 电影类型 : Comedy
Format 电影格式 : VHS
Rating 电影评分 : 五星
Description 电影简介 : 希腊神话
```

2. 使用库 Beautiful Soup 解析网络数据

Beautiful Soup 是一个重要的 Python 库，其功能是将 HTML 和 XML 文件的标签信息解析成树形结构，然后提取 HTML 或 XML 文件中指定标签属性对应的数据。库 Beautiful Soup 经常被用在爬虫项目中，通过使用库 Beautiful Soup 可以极大地提高开发效率。

Beautiful Soup 3 目前已经停止开发，其官方推荐使用 Beautiful Soup 4，本书讲解的是 Beautiful Soup 4。开发者可以使用如下两种命令安装库 Beautiful Soup：

```
pip install beautifulsoup4
easy_install beautifulsoup4
```

在安装 Beautiful Soup 4 后还需要安装文件解析器，Beautiful Soup 不但支持 Python 标准库中的 HTML 解析器，而且还支持第三方的解析器（例如 lxml）。根据开发者所用操作系统的不同，可以使用如下命令来安装 lxml：

```
$ apt-get install Python-lxml
$ easy_install lxml
$ pip install lxml
```

例如在下面的实例文件 bs01.py 中，演示了使用库 Beautiful Soup 解析 HTML 代码的过程。

```
from bs4 import BeautifulSoup
html_doc = """
<html><head><title>The Dormouse's story</title></head>
<body>
<p class="title"><b> 睡鼠的故事 </b></p>

<p class="story"> 在很久以前有三个可爱的小熊宝宝，名字分别是
<a href="http://example.com/elsie" class="sister" id="link1">Elsie</a>,
<a href="http://example.com/lacie" class="sister" id="link2">Lacie</a> 和
<a href="http://example.com/tillie" class="sister" id="link3">Tillie</a>;
and they lived at the bottom of a well.</p>

<p class="story">...</p>
"""
soup = BeautifulSoup(html_doc,"lxml")
print(soup)
```

通过上述代码，解析了 html_doc 中的 HTML 代码，执行后会输出如下解析结果：

```
<html><head><title>The Dormouse's story</title></head>
<body>
<p class="title"><b> 睡鼠的故事 </b></p>
<p class="story"> 在很久以前有三个可爱的小熊宝宝，名字分别是
<a class="sister" href="http://example.com/elsie" id="link1">Elsie</a>,
<a class="sister" href="http://example.com/lacie" id="link2">Lacie</a> 和
<a class="sister" href="http://example.com/tillie" id="link3">Tillie</a>;
and they lived at the bottom of a well.</p>
<p class="story">...</p>
</body></html>
```

在下面的实例文件 bs02.py 中，演示了使用库 Beautiful Soup 解析指定 HTML 标签的过程。

```
from bs4 import BeautifulSoup

html = '''
<html><head><title> 睡鼠的故事 </title></head>
<body>
<p class="title"><b> 睡鼠的故事 </b></p>

<p class="story">0 在很久以前有三个可爱的小熊宝宝，名字分别是
<a href="http://example.com/elsie" class="sister" id="link1">Elsie</a>,
<a href="http://example.com/lacie" class="sister" id="link2">Lacie</a> 和
<a href="http://example.com/tillie" class="sister" id="link3">Tillie</a>;
它们快乐的住在大森林里 </p>
<p class="story">...</p>
'''
soup = BeautifulSoup(html,'lxml')
print(soup.title)
```

```
print(soup.title.name)
print(soup.title.string)
print(soup.title.parent.name)
print(soup.p)
print(soup.p["class"])
print(soup.a)
print(soup.find_all('a'))
print(soup.find(id='link3'))
```

执行后将输出指定标签的信息：

```
<title>睡鼠的故事</title>
title
睡鼠的故事
head
<p class="title"><b>睡鼠的故事</b></p>
['title']
<a class="sister" href="http://example.com/elsie" id="link1">Elsie</a>
[<a class="sister" href="http://example.com/elsie" id="link1">Elsie</a>, <a class="sister" href="http://example.com/lacie" id="link2">Lacie</a>, <a class="sister" href="http://example.com/tillie" id="link3">Tillie</a>]
<a class="sister" href="http://example.com/tillie" id="link3">Tillie</a>
```

3. 使用库 bleach 过滤数据

在使用 Python 开发 Web 程序时，开发者面临一个十分重要的安全性问题：跨站脚本注入攻击（黑客利用网站漏洞从用户一端盗取重要信息）。为了解决跨站脚本注入攻击漏洞，最常用的做法是设置一个访问白名单，设置只显示指定的 HTML 标签和属性。在现实应用中，最常用的 HTML 过滤库是 Bleach，能够实现基于白名单的 HTML 清理和文本链接模块。

我们可以使用如下两种命令安装库 bleach：

```
pip install bleach
easy_install bleach
```

例如在下面的实例文件 guolv.py 中，演示了使用方法 bleach.clean() 过滤处理 HTML 标签的过程。

```
import bleach
# tag 参数示例
print(bleach.clean(
     u'<b><i>例子 1</i></b>',
     tags=['b'],
))

# attributes 为 list 示例
print(bleach.clean(
     u'<p class="foo" style="color: red; font-weight: bold;">例子 2</p>',
     tags=['p'],
     attributes=['style'],
     styles=['color'],
))
# attributes 为 dict 示例
attrs = {
     '*': ['class'],
     'a': ['href', 'rel'],
     'img': ['alt'],
}
print(bleach.clean(
    u'<img alt="an example" width=500>例子 3',
    tags=['img'],
    attributes=attrs
 ))
```

```
# attributes 为 function 示例
def allow_h(tag, name, value):
     return name[0] == 'h'
print(bleach.clean(
    u'<a href="http://example.com" title="link">例子 4</a>',
    tags=['a'],
    attributes=allow_h,
 ))


# style 参数示例
tags = ['p', 'em', 'strong']
attrs = {
     '*': ['style']
 }
styles = ['color', 'font-weight']
print(bleach.clean(
     u'<p style="font-weight: heavy;">例子 5</p>',
     tags=tags,
     attributes=attrs,
     styles=styles
))
# protocol 参数示例
print(bleach.clean(
     '<a href="smb://more_text">例子 6</a>',
     protocols=['http', 'https', 'smb']
))
print(bleach.clean(
     '<a href="smb://more_text">例子 7</a>',
     protocols=bleach.ALLOWED_PROTOCOLS + ['smb']
))

#strip 参数示例
print(bleach.clean('<span>例子 8</span>'))
print(bleach.clean('<b><span>例子 9</span></b>', tags=['b']))

print(bleach.clean('<span>例子 10</span>', strip=True))
print(bleach.clean('<b><span>例子 11</span></b>', tags=['b'], strip=True))

# strip_comments 参数示例
html = 'my<!-- commented --> html'
print(bleach.clean(html))
print(bleach.clean(html, strip_comments=False))
```

执行后会输出：

```
<b>&lt;i&gt;例子 1&lt;/i&gt;</b>
<p style="color: red;">例子 2</p>
<img alt="an example">例子 3
<a href="http://example.com">例子 4</a>
<p style="font-weight: heavy;">例子 5</p>
<a href="smb://more_text">例子 6</a>
<a href="smb://more_text">例子 7</a>
&lt;span&gt;例子 8&lt;/span&gt;
<b>&lt;span&gt;例子 9&lt;/span&gt;</b>
例子 10
<b>例子 11</b>
my html
my<!-- commented --> html
```

4. 使用库 html5lib 解析网络数据

在 Python 程序中，可以使用库 html5lib 解析 HTML 文件。库 html5lib 是用纯 Python 语言

编写实现的，其解析方式与浏览器相同。在本章前面讲解的库 Beautiful Soup，使用的是 lxml 解析器，而库 html5lib 是 Beautiful Soup 支持的另一种解析器。

我们可以使用如下两种命令安装库 html5lib：

```
pip install html5lib
easy_install html5lib
```

例如在下面的实例文件 ht501.py 中，演示了使用 html5lib 解析 HTML 文件的过程。

```
from bs4 import BeautifulSoup
html_doc = """
<html><head><title> 睡鼠的故事 </title></head>
<body>
<p class="title"><b> 睡鼠的故事 </b></p>

<p class="story"> 在很久以前有三个可爱的小熊宝宝，名字分别是
<a href="http://example.com/elsie" class="sister" id="link1">Elsie</a>,
<a href="http://example.com/lacie" class="sister" id="link2">Lacie</a> 和
<a href="http://example.com/tillie" class="sister" id="link3">Tillie</a>;
它们快乐的生活在大森林里 .</p>

<p class="story">...</p>
"""
soup = BeautifulSoup(html_doc,"html5lib")
print(soup)
```

执行后会输出：

```
<html><head><title> 睡鼠的故事 </title></head>
<body>
<p class="title"><b> 睡鼠的故事 </b></p>

<p class="story"> 在很久以前有三个可爱的小熊宝宝，名字分别是
<a class="sister" href="http://example.com/elsie" id="link1">Elsie</a>,
<a class="sister" href="http://example.com/lacie" id="link2">Lacie</a> 和
<a class="sister" href="http://example.com/tillie" id="link3">Tillie</a>;
它们快乐的生活在大森林里 .</p>

<p class="story">...</p>
</body></html>
```

5. 使用库 MarkupSafe 解析数据

在 Python 程序中，使用库 MarkupSafe 可以将具有特殊含义的字符替换掉，这样可以减轻注入攻击（把用户输入、提交的数据当作代码来执行），能够将不受信任的用户输入的信息安全地显示在页面上。

我们可以使用如下两种命令安装库 MarkupSafe：

```
pip install MarkupSafe
easy_install MarkupSafe
```

例如在下面的实例文件 mark01.py 中，演示了使用库 MarkupSafe 构建安全 HTML 的过程。

```
from markupsafe import Markup, escape
# 实现支持 HTML 字符串的 Unicode 子类
print(escape("<script>alert(document.cookie);</script>"))
tmpl = Markup("<em>%s</em>")
print(tmpl % "Peter > Lustig")

# 可以通过重写 __html__ 功能自定义等效 HTML 标记
class Foo(object):
  def __html__(self):
   return '<strong>Nice</strong>'
```

```
print(escape(Foo()))
print(Markup(Foo()))
```

执行后会输出：

```
&lt;script&gt;alert(document.cookie);&lt;/script&gt;
<em>Peter &gt; Lustig</em>
<strong>Nice</strong>
<strong>Nice</strong>
```

2.1.2 处理 HTTP 数据

HTTP（HyperText Transfer Protocol）是互联网上应用最为广泛的一种网络协议，所有的 WWW 文件都必须遵守这个标准。在网络应用中，大多数数据是通过 HTTP 传递的。在 Ptyhon 语言中提供了多个库来处理 HTTP 数据，接下来我们详细讲解这些库的用法。

1. 使用 Python 内置库 http 处理网络数据

在 Python 语言中，使用内置库 http 实现了对 HTTP 协议的封装。在库 http 中主要包含如表 2-2 所示的核心模块。

表 2-2

模块名称	说　明
http.client	底层的 HTTP 协议客户端，可以为 urllib.request 模块所用
http.server	提供了处理 socketserver 模块的功能类
http.cookies	提供了在 HTTP 传输过程中处理 Cookies 应用的功能类
http.cookiejar	提供了实现 Cookies 持久化支持的功能类

在 http.client 模块中，主要包括如下两个处理客户端应用的类。

- HTTPConnection：基于 HTTP 协议的访问客户端。
- HTTPResponse：基于 HTTP 协议的服务端回应。

例如在下面的实例文件 fang.py 中，演示了使用 http.client.HTTPConnection 对象访问指定网站的过程。

```
from http.client import HTTPConnection                    # 导入内置模块
# 基于 HTTP 协议的访问客户端
mc = HTTPConnection('www.baidu.com:80')
mc.request('GET','/')                                      # 设置 GET 请求方法
res = mc.getresponse()                                     # 获取访问的网页
print(res.status,res.reason)                               # 打印输出响应的状态
print(res.read().decode('utf-8'))                          # 显示获取的内容
```

在上述实例代码中只是实现了一个基本的访问实例，首先实例化 http.client.HTTPConnection 对指定请求的方法为 GET，然后使用 getresponse() 方法获取访问的网页，并打印输出响应的状态。执行效果如图 2-2 所示。

在现实应用中，有时需要通过 HTTP 协议以客户端的形式访问多种服务，例如下载服务器中的数据，或同一个基于 REST 的 API 进行交互。通过使用 urllib.request 模块，可以实现简单的客户端访问任务，例如要发送一个简单的 HTTP GET 请求到远端服务器上，只需通过下面的实例文件 fang1.py 即可实现。

```
200 OK
<!DOCTYPE html><!--STATUS OK-->
<html>
<head>
    <meta http-equiv="content-type" content="text/html;charset=utf-8">
    <meta http-equiv="X-UA-Compatible" content="IE=Edge">
    <link rel="dns-prefetch" href="//s1.bdstatic.com"/>
    <link rel="dns-prefetch" href="//t1.baidu.com"/>
    <link rel="dns-prefetch" href="//t2.baidu.com"/>
    <link rel="dns-prefetch" href="//t3.baidu.com"/>
    <link rel="dns-prefetch" href="//t10.baidu.com"/>
    <link rel="dns-prefetch" href="//t11.baidu.com"/>
    <link rel="dns-prefetch" href="//t12.baidu.com"/>
    <link rel="dns-prefetch" href="//b1.bdstatic.com"/>
    <title>百度一下，你就知道</title>
    <link href="http://s1.bdstatic.com/r/www/cache/static/home/css/index.css" rel="stylesheet" type="text/css" />
    <!--[if lte IE 8]><style index="index" >#content{height:480px\9}#m{top:260px\9}</style><![endif]-->
    <!--[if IE 8]><style index="index" >#u1 a.mnav,#u1 a.mnav:visited{font-family:simsun}</style><![endif]-->
    <script>var hashMatch = document.location.href.match(/#+(.*wd=[^&].+)/);if (hashMatch && hashMatch[0] && hashMatch[1])
    <script>function h(obj){obj.style.behavior='url(#default#homepage)';var a = obj.setHomePage('//www.baidu.com/');}</scri
```

图 2-2　执行效果

```
from urllib import request, parse

url = 'http://httpbin.org/get'

# Dictionary of query parameters (if any)
parms = {
   'name1' : 'value1',
   'name2' : 'value2'
}

# Encode the query string
querystring = parse.urlencode(parms)

# Make a GET request and read the response
u = request.urlopen(url+'?' + querystring)
resp = u.read()

import json
from pprint import pprint

json_resp = json.loads(resp.decode('utf-8'))
pprint(json_resp)
```

执行后会输出：

```
{'args': {'name1': 'value1', 'name2': 'value2'},
 'headers': {'Accept-Encoding': 'identity',
             'Connection': 'close',
             'Host': 'httpbin.org',
             'User-Agent': 'Python-urllib/3.6'},
 'origin': '27.211.158.101',
 'url': 'http://httpbin.org/get?name1=value1&name2=value2'}
```

2. 使用库 requests 处理网络数据

库 Requests 是使用 Python 语言对库 urllib 的封装升级，使用库 Requests 会比使用 urllib 更加方便。我们可以使用如下两种命令安装库 requests：

```
pip install requests
easy_install requests
```

例如在下面的实例文件 Requests01.py 中，演示了使用库 requests 返回指定 URL 地址请求的过程。

```
import requests
```

```
r = requests.get(url='http://www.toppr.net')  # 最基本的 GET 请求
print(r.status_code)  # 获取返回状态
r = requests.get(url='http://www.toppr.net', params={'wd': 'python'})
# 带参数的 GET 请求
print(r.url)
print(r.text)  # 打印解码后的返回数据
```

在上述代码中，创建了一个名为 r 的 Response 对象，我们可以从这个对象中获取所有想要的信息。执行后会输出：

```
200
http://www.toppr.net/?wd=python
<!DOCTYPE html PUBLIC "-//W3C//DTD XHTML 1.0 Transitional//EN" "http://www.w3.org/TR/xhtml1/DTD/xhtml1-transitional.dtd">
<html xmlns="http://www.w3.org/1999/xhtml">
<head>
<meta http-equiv="X-UA-Compatible" content="IE=edge">
<meta http-equiv="Content-Type" content="text/html; charset=gbk" />
<title>门户 -  Powered by Discuz!</title>

<meta name="keywords" content="门户 " />
<meta name="description" content="门户 " />
<meta name="generator" content="Discuz! X3.2" />
#省略后面的结果
```

上述实例只是演示了 get 接口的用法，其实其他接口的用法也十分简单，代码如下：

```
requests.get('https://github.com/timeline.json') #GET 请求
requests.post("http://httpbin.org/post") #POST 请求
requests.put("http://httpbin.org/put") #PUT 请求
requests.delete("http://httpbin.org/delete") #DELETE 请求
requests.head("http://httpbin.org/get") #HEAD 请求
requests.options("http://httpbin.org/get") #OPTIONS 请求
```

例如想查询 http://httpbin.org/get 页面的具体参数，需要在 url 中加上这个参数。假如我们想看有没有 Host=httpbin.org 这条数据，url 形式应该是 http://httpbin.org/get?Host=httpbin.org。例如在下面的实例文件 Requests02.py 中，提交的数据是往这个地址传送 data 中的数据。

```
import requests
url = 'http://httpbin.org/get'
data = {
    'name': 'python',
    'age': '25'
}
response = requests.get(url, params=data)
print(response.url)
print(response.text)
```

执行后会输出：

```
http://httpbin.org/get?name=zhangsan&age=25
{
  "args": {
    "age": "25",
    "name": "python "
  },
  "headers": {
    "Accept": "*/*",
    "Accept-Encoding": "gzip, deflate",
    "Connection": "close",
    "Host": "httpbin.org",
    "User-Agent": "python-requests/2.12.4"
  },
  "origin": "39.71.61.153",
```

```
    "url": "http://httpbin.org/get?name=python&age=25"
}
```

3. 使用库 urllib3 处理网络数据

在 Python 应用中，库 urllib3 提供了一个线程安全的连接池，能够以 post 方式传输文件。我们可以使用如下两种命令安装库 urllib3：

```
pip install urllib3
easy_install urllib3
```

在下面的实例文件 urllib303.py 中，演示了使用库 urllib3 中的 post() 方法创建请求的过程。

```
import urllib3
http = urllib3.PoolManager()
# 信息头
header = {
            'User-Agent': 'Mozilla/5.0 (Windows NT 2.1; Win64; x64) AppleWebKit/537.36 (KHTML, like Gecko) Chrome/63.0.3239.108 Safari/537.36'
}
r = http.request('POST',
                     'http://httpbin.org/post',
                     fields={'hello':'Python'},
                     headers=header)
print(r.data.decode())

# 对于 POST 和 PUT 请求 (request)，需要手动对传入数据进行编码，然后加在 URL 之后：
encode_arg = urllib.parse.urlencode({'arg': '我的信息：'})
print(encode_arg.encode())
r = http.request('POST',
                 'http://httpbin.org/post?'+encode_arg,
                 headers=header)
# unicode 解码
print(r.data.decode('unicode_escape'))
```

执行后会输出：

```
{
  "args": {},
  "data": "",
  "files": {},
  "form": {
    "hello": "Python"
  },
  "headers": {
    "Accept-Encoding": "identity",
    "Connection": "close",
    "Content-Length": "129",
    "Content-Type": "multipart/form-data; boundary=b33b20053e6444ee947a6b7b3f4572b2",
    "Host": "httpbin.org",
    "User-Agent": "Mozilla/5.0 (Windows NT 2.1; Win64; x64) AppleWebKit/537.36 (KHTML, like Gecko) Chrome/63.0.3239.108 Safari/537.36"
  },
  "json": null,
  "origin": "39.71.61.153",
  "url": "http://httpbin.org/post"
}

b'arg=%E6%88%91%E7%9A%84'
{
  "args": {
    "arg": "我的"
  },
```

```
    "data": "",
    "files": {},
    "form": {},
    "headers": {
      "Accept-Encoding": "identity",
      "Connection": "close",
      "Content-Length": "0",
      "Host": "httpbin.org",
       "User-Agent": "Mozilla/5.0 (Windows NT 2.1; Win64; x64) AppleWebKit/537.36
(KHTML, like Gecko) Chrome/63.0.3239.108 Safari/537.36"
    },
    "json": null,
    "origin": "39.71.61.153",
    "url": "http://httpbin.org/post?arg= 我的信息 "
  }
```

例如在下面的实例文件 urllib305.py 中，演示了使用库 urllib3 获取远程 CSV 数据的过程。

```
import urllib3
# 两个文件的源 url
url1 = 'http://earthquake.usgs.gov/earthquakes/feed/v1.0/summary/all_week.csv'
url2 = 'http://earthquake.usgs.gov/earthquakes/feed/v1.0/summary/all_month.csv'
# 开始创建一个 HTTP 连接池
http = urllib3.PoolManager()
# 请求第一个文件并将结果写入文件：
response = http.request('GET', url1)
with open('all_week.csv', 'wb') as f:
    f.write(response.data)
# 请求第二个文件并将结果写入 CSV 文件：
response = http.request('GET', url2)
with open('all_month.csv', 'wb') as f:
    f.write(response.data)

# 最后释放这个 HTTP 连接的占用资源：
response.release_conn()
```

执行后会将这两个远程 CSV 文件下载保存到本地，如图 2-3 所示。

在下面的实例文件 urllib302.py 中，演示了使用库 urllib3 抓取显示某资讯类网站头条新闻的方法。

实例文件代码如下：

图 2-3　下载保存到本地的 CSV 文件

```
from bs4 import BeautifulSoup
import urllib3

def get_html(url):
    try:
        userAgent = 'Mozilla/5.0 (Windows; U; Windows NT 2.1; en-US; rv:1.9.1.6)
Gecko/20091201 Firefox/3.5.6'
        http = urllib3.PoolManager(timeout=2)
        response = http.request('get', url, headers={'User_Agent': userAgent})
        html = response.data
        return html
    except Exception as e:
        print(e)
        return None

def get_soup(url):
    if not url:
        return None
    try:
        soup = BeautifulSoup(get_html(url))
    except Exception as e:
        print(e)
```

```
            return None
        return soup

    def get_ele(soup, selector):
        try:
            ele = soup.select(selector)
            return ele
        except Exception as e:
            print(e)
        return None

    def main():
        url = 'http://www.ifeng.com/'
        soup = get_soup(url)
        ele = get_ele(soup, '#headLineDefault > ul > ul:nth-of-type(1) > li.topNews
> h1 > a')
        headline = ele[0].text.strip()
        print(headline)

    if __name__ == '__main__':
        main()
```

因为头条新闻是随着时间的推移发生变化的，所以每次的执行效果可能不一样。

2.1.3　处理 URL 数据

URL（Uniform Resoure Locator）的中文含义是统一资源定位器，也就是我们平常说的WWW 网址。在 Python 语言中提供了多个处理 URL 数据的库，例如内置库 urllib 和第三方库 furl。

1. 使用内置库 urllib

在 Python 程序中，可以内置库 urllib 处理 URL 请求。urllib 非常简单，在内置库 urllib 中主要包括如表 2-3 所示的核心模块。

表 2-3

模块名称	说　明
urllib.request	用于打开指定的 URL 网址
urllib.error	用于处理 URL 访问异常
urllib.parse	用于解析指定的 URL
urllib.robotparser	用于解析 robots.txt 文件，robots 是 Web 网站跟爬虫之间的协议，可以用 txt 格式的文本方式告诉对应的爬虫被允许的权限

注意：因为 urllib.error 和 urllib.robotparser 用得比较少，也比较简单，所以在本书中不再介绍这两个子模块的用法。

（1）使用 urllib.request 模块

在 urllib.request 模块中定义了打开指定 URL 的方法和类，甚至可以实现身份验证、URL 重定向和 Cookies 存储等功能。例如在下面的实例文件 url.py 中，演示了使用方法 urlopen() 在百度搜索关键词中得到第一页链接的过程。

```
    from urllib.request import urlopen              # 导入 Python 的内置模块
    from urllib.parse import urlencode              # 导入 Python 的内置模块
    import re                                       # 导入 Python 的内置模块
    ##wd = input('输入一个要搜索的关键字: ')
    wd= 'www.toppr.net'                             # 初始化变量 wd
    wd = urlencode({'wd':wd})                       # 对 URL 进行编码
```

```
url = 'http://www.baidu.com/s?' + wd          # 初始化 url 变量
page = urlopen(url).read()                    # 打开变量 url 的网页并读取内容
# 定义变量 content，对网页进行编码处理，并实现特殊字符处理
content = (page.decode('utf-8')).replace("\n","").replace("\t","")
title = re.findall(r'<h3 class="t".*?h3>', content)
# 正则表达式处理
title = [item[item.find('href =')+6:item.find('target=')] for item in title]
# 正则表达式处理
title = [item.replace(' ','').replace('"','') for item in title]
# 正则表达式处理
for item in title:                            # 遍历 title
    print(item)                               # 打印显示遍历值
```

在上述实例代码中，使用方法 urlencode() 对搜索的关键字“www.toppr.net”进行 URL 编码，在拼接到百度的网址后，使用 urlopen() 方法发出访问请求并取得结果，最后通过将结果进行解码和正则搜索与字符串处理后输出。如果将程序中的注释去除而把其后一句注释删掉，就可以在运行时自主输入搜索的关键词。执行效果如图 2-4 所示。

```
http://www.baidu.com/link?url=hm6N8CdYPCSxsCsreajusLxba8mRVPAgc1D_WBhkYb7
http://www.baidu.com/link?url=Nlf7T18nlQ0pke8pH8CIzg0V_wjqTKRtQ2NXLs-wUzvLHM0UknbUflsJT3DLE2G0m6JW5GlRoBx-GbF6epS7sa
http://www.baidu.com/link?url=cbZcgLHZSTFBp6tFwWGwTuVq6xE3FcjM_d-cIH5qRNrkkXaETLwKKj9n9Rhlvvi8
http://www.baidu.com/link?url=0AHbz_vI3wIC_ocpmRc3jzcjJeu3gDeImuXcGfKu1zKGtaZ50-KR-HfGchsHSyGY
http://www.baidu.com/link?url=h591VC_3X6t7hm6eptcTS0dxFe5c4Z7XznyLzpqkJ1Z6a01WptFh4IS37h6LhzIC
```

图 2-4　执行效果

注意：urllib.response 模块是 urllib 使用的响应类，定义了和 urllib.request 模块类似接口、方法和类，包括 read() 和 readline()。为了节省本书篇幅，下面不再进行讲解。

（2）使用 urllib.parse 模块

在 Python 程序中，urllib.parse 模块提供了一些用于处理 URL 字符串的功能。这些功能主要是通过下面所讲的三种方法实现的。

① 方法 urlpasrse.urlparse()

方法 urlparse() 的功能是将 URL 字符串拆分成前面描述的一些主要组件，其语法结构具体如下：

```
urlparse (urlstr, defProtSch=None, allowFrag=None)
```

方法 urlparse() 将 urlstr 解析成一个 6 元组（prot_sch, net_loc, path, params, query, frag）。如果在 urlstr 中没有提供默认的网络协议或下载方案，defProtSch 会指定一个默认的网络协议。allowFrag 用于标识一个 URL 是否允许使用片段。例如下面是一个给定 URL 经 urlparse() 后的输出。

```
>>> urlparse.urlparse('http://www.python.org/doc/FAQ.html')
('http', 'www.python.org', '/doc/FAQ.html', '', '', '')
```

② 方法 urlparse.urlunparse()

方法 urlunparse() 的功能与方法 urlpase() 完全相反，能够将经 urlparse() 处理的 URL 生成 urltup 这个 6 元组（prot_sch, net_loc, path, params, query, frag）拼接成 URL 并返回。可以用如下所示的方式表示其等价性：

```
urlunparse(urlparse(urlstr)) ≡ urlstr
```

下面是使用 urlunpase() 的语法，具体解释不再赘述。

```
urlunparse(urltup)
```

③ 方法 urlparse.urljoin()

在处理多个相关的 URL 时需要使用 urljoin() 方法的功能，例如在一个 Web 页中可能会产生一系列页面的 URL。方法 urljoin() 的语法格式如下所示：

```
urljoin (baseurl, newurl, allowFrag=None)
```

方法 urljoin() 能够取得根域名，并将其根路径（net_loc 及其前面的完整路径，但是不包括末端的文件）与 newurl 连接起来。例如下面的演示过程：

```
>>> urlparse.urljoin('http://www.python.org/doc/FAQ.html',
... 'current/lib/lib.htm')
'http://www.python.org/doc/current/lib/lib.html'
```

假设有一个身份验证（登录名和密码）的 Web 站点，通过验证的最简单方法是在 URL 中使用登录信息进行访问，例如 http://username:passwd@www.python.org。但是这种方法的问题是它不具有可编程性。通过使用 urllib 可以很好地解决这个问题，假设合法的登录信息是：

```
LOGIN = 'admin'
PASSWD = "admin"
URL = 'http://localhost'
REALM = 'Secure AAA'
```

此时便可以通过下面的实例文件 pa.py，实现使用 urllib 实现 HTTP 身份验证的过程。

```
import urllib.request, urllib.error, urllib.parse

① LOGIN = 'admin'
PASSWD = "admin"
URL = 'http://localhost'
② REALM = 'Secure AAA'

③ def handler_version(url):
    hdlr = urllib.request.HTTPBasicAuthHandler()
    hdlr.add_password(REALM,
        urllib.parse.urlparse(url)[1], LOGIN, PASSWD)
    opener = urllib.request.build_opener(hdlr)
    urllib.request.install_opener(opener)
④     return url

⑤ def request_version(url):
    import base64
    req = urllib.request.Request(url)
    b64str = base64.b64encode(
        bytes('%s:%s' % (LOGIN, PASSWD), 'utf-8'))[:-1]
    req.add_header("Authorization", "Basic %s" % b64str)
⑥     return req

⑦ for funcType in ('handler', 'request'):
    print('*** Using %s:' % funcType.upper())
    url = eval('%s_version' % funcType)(URL)
    f = urllib.request.urlopen(url)
    print(str(f.readline(), 'utf-8'))
⑧ f.close()
```

我们对上述代码作一下简要解析。

①～②实现普通的初始化功能，设置合法的登录验证信息。

③～④定义函数 handler_version()，添加验证信息后建立一个 URL 开启器，安装该开启器以便所有已打开的 URL 都能用到这些验证信息。

⑤～⑥定义函数 request_version() 创建一个 Request 对象，并在 HTTP 请求中添加简单的 base64 编码的验证头信息。在 for 循环里调用 urlopen() 时，该请求用来替换其中的 URL 字符串。

⑦～⑧分别打开了给定的 URL，通过验证后会显示服务器返回的 HTML 页面的第一行（转储了其他行）。如果验证信息无效会返回一个 HTTP 错误（并且不会有 HTML）。

2. 使用库 furl 处理数据

在 Python 应用中，库 furl 是一个快速处理 URL 应用的小型 Python 库，可以方便开发者以更加优雅的方式操作 URL 地址。我们可以使用如下命令安装 furl：

```
pip install furl
```

例如在下面的实例文件 url02.py 中，演示了使用库 furl 处理 URL 参数的过程。

```
from furl import furl
f= furl('http://www.baidu.com/?bid=12331')
# 打印参数
print(f.args)
# 增加参数
f.args['haha']='123'
print(f.args)
# 修改参数
f.args['haha']='124'
print(f.args)
# 删除参数
del f.args['haha']
print(f.args)
```

执行后会输出：

```
{'bid': '12331'}
{'bid': '12331', 'haha': '123'}
{'bid': '12331', 'haha': '124'}
{'bid': '12331'}
```

2.2 网络爬虫技术

在本章上一节的内容中，讲解了处理网络数据的基本知识，接下来将要讲解的爬虫技术也属于处理网络数据范畴。在本节的内容中，将详细讲解开发网络爬虫程序的知识，为读者步入本书后面知识的学习打下基础。

↑扫码看视频（本节视频课程时间：18 分 26 秒）

2.2.1 网络爬虫基础

对于网络爬虫这一概念，大家可以将其理解为在网络中爬行的一只小蜘蛛，如果将互联网比作一张大网，那么可以将爬虫看作是在这张网上爬行的蜘蛛，如果它遇到你喜欢的资源，那么这只小蜘蛛就会把这些信息抓取下来为己所用。

我们在浏览网页时可能会看到许多好看的图片，例如在打开网页 http://image.baidu.com/ 时会看到很多张图片以及百度搜索框。我们用肉眼看到的网页实际是由 HTML 代码构成的，爬虫的功能是分析和过滤这些 HTML 代码，然后将有用的资源（例如图片和文字等）抓取出来。在现实应用中，被抓取出来的爬虫数据十分重要，通常是进行数据分析的原始资料。

在使用爬虫抓取网络数据时，必须要有一个目标的 URL 才可以获取数据。网络爬虫从一个或若干初始网页的 URL 开始，在抓取网页的过程中，不断从当前页面上抽取新的 URL，并将 URL 放入队列中。当满足系统设置的停止条件时，爬虫会停止抓取操作。为了使用爬取到的数据，系统需要存储被爬虫抓取到的网页，然后进行一系列的数据分析和过滤工作。

在现实应用中，网络爬虫获取网络数据的流程可以分为 4 个步骤。

（1）模拟浏览器发送请求

在客户端使用 HTTP 技术向目标 Web 页面发起请求，即发送一个 Request。在 Request 中包含请求头和请求体等，是访问目标 Web 页面的前提。Request 请求方式有一个缺陷，即不能执行 JS 和 CSS 代码。

（2）获取响应内容

如果目标 Web 服务器能够正常响应，在客户端会得到一个 Response 响应，Response 内容包含 HTML、JSON、图片和视频等类型的信息。

（3）解析内容

解析目标网页的内容，既可以使用正则表达式提高解析效率，也可以使用第三方解析库提高解析效率。常用的第三方解析库有 Beautifulsoup 和 pyquery 等。

（4）保存数据

在现实应用中，通常将爬取的数据保存到数据库（例如 MySQL、Mongdb、Redis 等）或不同格式的文件（例如 CSV、JSON 等）中，这样可以为下一步的数据分析工作做好准备。

2.2.2 使用 Beautiful Soup 爬取网络数据

Beautiful Soup 是一个著名的 Python 库，功能是从 HTML 或 XML 文件中提取数据。在现实应用中，通常在爬虫项目中使用 Beautiful Soup 获取网页数据。在作者写作本书时，Beautiful Soup 的最新版本是 Beautiful Soup 4。

1. 安装 Beautiful Soup

在新版的 Debain 或 Ubuntu 系统中，可以通过如下系统软件包管理命令进行安装：

```
$ apt-get install Python-bs4
```

因为 Beautiful Soup 4 通过 PyPi 发布的，所以可以通过 easy_install 或 pip 来安装。库 Beautiful Soup 4 对应的包的名字是 beautifulsoup4，我们可以通过如下命令进行安装：

```
easy_install beautifulsoup4
pip install beautifulsoup4
```

在安装 Beautiful Soup 后还需要安装解析器，Beautiful Soup 不但支持 Python 标准库中的 HTML 解析器，而且还支持一些第三方的解析器，例如 lxml。根据操作系统不同，可以选择如下命令来安装 lxml：

```
apt-get install Python-lxml
easy_install lxml
pip install lxml
```

另一个可以供 Beautiful Soup 使用的解析器是 html5lib，这是一款用纯 Python 语言实现的解析器。html5lib 的解析方式与浏览器相同 , 可以选择如下命令来安装 html5lib。

```
apt-get install Python-html5lib
easy_install html5lib
pip install html5lib
```

2. 使用 Beautiful Soup

在下面的实例中，演示了使用 BeautifulSoup 解析一个网页的过程。实例文件 bs03.py 的功能是解析某出版社官方网站的主页，具体实现代码如下所示：

```
import urllib.request
from bs4 import BeautifulSoup

url = "http://www.tup.×××××.edu.cn/index.html"
page = urllib.request.urlopen(url)
soup = BeautifulSoup(page,"html.parser")
print(soup)
```

执行后会显示解析后获取的网页代码，如图 2-5 所示。

```
C:\ProgramData\Anaconda3\python.exe E:/123/codes--data/1/1-2/bs03.py

<!DOCTYPE html>

<html lang="en">
<head><meta charset="utf-8"/><title>
  ××大学出版社
</title><meta content="text/html; charset=utf-8" http-equiv="Content-Type"/><meta content="IE=8" http-equiv="X-UA-Compatible"/><meta
<script src="js/xslb/jquery-1.4.2.min.js" type="text/javascript"></script>
</head>
<body>
<div class="nav">
<div class="logo">
<div class="logo1">
<ul>
<li class="logozc"><a href="member/register.aspx">注册</a></li>
<li>
<div class="logofh">
            +</div>
          <a href="member/dl.aspx">登录</a></li>
</ul>
</div>
<div class="logo2">
<div class="logo21">
<img alt="" src="img/logo.jpg"/></div>
<div class="logo22">
<div class="logo22-1 bxs">
          书问搜索</div>
<div class="logo22-2 bxs">
          站内搜索</div>
```

图 2-5　解析结果的部分解图

在下面的实例中，演示了使用 BeautifulSoup 解析并获取网页信息的过程。

实例文件 bs04.py 的功能是，解析并获取 ×× 大学出版社官网中最新的 15 本 Python 图书信息，并打印输出这 15 本 Python 图书的基本信息。文件 bs04.py 的具体实现代码如下所示。

```
from urllib.request import urlopen
from bs4 import BeautifulSoup
html=urlopen("http://www.tup.×××××.edu.cn/booksCenter/booklist.html?keyword=python&keytm=8E323D219188916A8F")
bsObj=BeautifulSoup(html,"html.parser")

nameList=bsObj.findAll("ul",{"class":"b_list"})
for name in nameList:
        print(name.get_text()))
```

通过上述代码，可以解析 ×× 大学出版社官网中关键字为“python”的图书信息，只是抓取了第一个分页的 15 本图书的书名。执行后会输出：

```
Python 3.8 从入门到精通（视频 ... 王英英 9787302552116 定价：89 元
Python 机器学习及实践 梁佩莹 9787302539735 定价：79 元
Python 案例教程 钱毅湘、熊福松 9787302550587 定价：39 元
```

```
Python 程序设计 黄蔚　熊福松　9787302550235 定价：59 元
青少年学 Python 编程（配套视频 ... 龙豪杰 9787302552123 定价：59 元
Python 机器学习算法与应用 邓立国 9787302548997 定价：69 元
Python 3.8 从零开始学 刘宇宙、刘艳 9787302552147 定价：79.80 元
深度学习：从 Python 到 TensorFlo... 叶虎 9787302545651 定价：69.80 元
Python 数据科学零基础一本通 洪锦魁 9787302545392 定价：129 元
Python 语言程序设计 陈振 9787302547860 定价：49 元
Python 爬虫大数据采集与挖掘 -... 曾剑平 9787302540540 定价：59.80 元
中学生 Python 与 micro：bit 机器 ... 高旸、尚凯 9787302537625 定价：49 元
Python 人工智能 刘伟善 9787302547792 定价：59.80 元
Python 数据分析与可视化 - 微课 ... 魏伟一、李晓红 9787302546665 定价：49 元
Python 程序设计 翟萍、王军锋、9787302544388 定价：44.50 元
```

注意：因为 ×× 大学出版社官网中的数据是实时变化的，所以每次执行上述程序后的输出结果会有所不同。

2.2.3　使用 XPath 爬取网络数据

在 Python 程序中，可以使用 XPath 来解析爬虫数据。XPath 不但提供了非常简洁明了的路径选择表达式，而且还提供了超过上百个用于处理字符串、数值、时间的匹配以及节点、序列的内置函数；和 Beautiful Soup 相比，Xpath 的效率更高。

1. 安装 XPath

XPath 是一门在 XML 文档中查找信息的语言，可以遍历并提取 XML 文档中的元素和属性。我们可以使用如下命令来安装 XPath：

```
pip install lxml
```

在 Python 程序中使用 XPath 时，需要通过如下命令导入 XPath。

```
from lxml import etree
```

2. 使用 XPath

在下面的实例中，演示了使用 XPath 解析一段 HTML 代码的过程。实例文件 xp01.py 的具体实现代码如下所示：

```
from lxml import etree

wb_data = """
        <div>
            <ul>
                 <li class="item-0"><a href="link1.html">first item</a></li>
                 <li class="item-1"><a href="link2.html">second item</a></li>
                    <li class="item-inactive"><a href="link3.html">third item</
a></li>
                 <li class="item-1"><a href="link4.html">fourth item</a></li>
                 <li class="item-0"><a href="link5.html">fifth item</a>
             </ul>
         </div>
        """
html = etree.HTML(wb_data)
print(html)
result = etree.tostring(html)
print(result.decode("utf-8"))
```

执行后会输出：

```
<Element html at 0x1acbac0b708>
<html><body><div>
```

```
                <ul>
                    <li class="item-0"><a href="link1.html">first item</a></li>
                    <li class="item-1"><a href="link2.html">second item</a></li>
                       <li class="item-inactive"><a href="link3.html">third item</
a></li>
                    <li class="item-1"><a href="link4.html">fourth item</a></li>
                    <li class="item-0"><a href="link5.html">fifth item</a>
                 </li></ul>

             </div>
            </body></html>
```

由此可见，XPath 会在解析结果中自动补全缺少的 HTML 标记元素。在下面的实例中，演示了使用 XPath 解析 HTML 代码中的指定标签的过程。

实例文件 xp02.py 的具体实现代码如下所示：

```
    from lxml import etree

    wb_data = """
            <div>
                <ul>
                    <li class="item-0"><a href="link1.html">first item</a></li>
                    <li class="item-1"><a href="link2.html">second item</a></li>
                       <li class="item-inactive"><a href="link3.html">third item</
a></li>
                    <li class="item-1"><a href="link4.html">fourth item</a></li>
                    <li class="item-0"><a href="link5.html">fifth item</a>
                 </ul>
             </div>
            """
    html = etree.HTML(wb_data)
    html_data = html.xpath('/html/body/div/ul/li/a')
    print(html)
    for i in html_data:
        print(i.text)
```

执行后会解析出 html/body/div/ul/li/a 标签下的内容：

```
    first item
    second item
    third item
    fourth item
    fifth item
```

在下面的实例中，演示了使用 XPath 解析并获取网页数据的过程。

实例文件 xp03.py 的功能是，解析并获取 ×× 大学出版社官方网站中“重点推荐”的作者信息，并打印输出“重点推荐”的作者信息。文件 xp03.py 的具体实现代码如下所示。

```
    from lxml import etree

    html = etree.parse('http://www.tup.×××××.edu.cn/booksCenter/books_index.html',
etree.HTMLParser())
    result = html.xpath('//div[@class="m_b_right"]/p/text()')
    print(result)
```

执行上述代码会打印输出 ×× 大学出版社官方网站中“重点推荐”的作者信息：

```
    ['菠萝，本名李治中，清华大学生物系本科，美国杜克大学癌症生物学博士，现担任美国诺华制药癌症新药
开发部资深研究员，实验室负责人。爱好科普和公益事业。"健康不是闹着玩儿"公众号运营者之一，"向日葵"
儿童癌症公益平台发起者。\r\n                              ', '\r\n                    ']
```

2.2.4　实践案例：爬取体育新闻信息并保存到 XML 文件

在本节的内容中，将通过一个具体实例，详细讲解使用 Python 爬取新闻信息并保存到 XML 文件中的方法，然后讲解使用 Stanford CoreNLP 提取 XML 数据特征关系的过程。

1. 爬虫抓取数据

在本项目的“Scrap”目录中提供了多个爬虫文件，每一个文件都可以爬取指定网页的新闻信息，并且都可以将爬取的信息保存到 XML 文件中。例如通过文件 scrap1.py 可以抓取新浪体育某个页面中的新闻信息，并将抓取的信息保存到 XML 文件 news1.xml 中。文件 scrap1.py 的主要实现代码如下所示。

```
    doc=xml.dom.minidom.Document()
    root=doc.createElement('AllNews')
    doc.appendChild(root)
    #用于爬取新浪体育的网页
    urls2=['http://sports.example..com.cn/nba/25.shtml']

    def scrap():
        for url in urls2:
            count = 0  # 用于统计总共爬取新闻数量
            html = urlopen(url).read().decode('utf-8')
            #print(html)
                res=re.findall(r'<a href="(.*?)" target="_blank">(.+?)</
a><br><span>',html)#用于爬取超链接和新闻标题

            for i in res:
                try:
                    urli=i[0]
                    htmli=urlopen(urli).read().decode('utf-8')
                    time=re.findall(r'<span class="date">(.*?)</span>',htmli)
                    resp=re.findall(r'<p>(.*?)</p>',htmli)
                    #subHtml=re.findall('',htmli)
                    nodeNews=doc.createElement('News')
                    nodeTopic=doc.createElement('Topic')
                    nodeTopic.appendChild(doc.createTextNode('sports'))
                    nodeLink=doc.createElement('Link')
                    nodeLink.appendChild(doc.createTextNode(str(i[0])))
                    nodeTitle=doc.createElement('Title')
                    nodeTitle.appendChild(doc.createTextNode(str(i[1])))
                    nodeTime=doc.createElement('Time')
                    nodeTime.appendChild(doc.createTextNode(str(time)))
                    nodeText=doc.createElement('Text')
                    nodeText.appendChild(doc.createTextNode(str(resp)))
                    nodeNews.appendChild(nodeTopic)
                    nodeNews.appendChild(nodeLink)
                    nodeNews.appendChild(nodeTitle)
                    nodeNews.appendChild(nodeTime)
                    nodeNews.appendChild(nodeText)
                    root.appendChild(nodeNews)
                    print(i)
                    print(time)
                    print(resp)
                    count+=1
                except:
                    print(count)
                    break
    scrap()
    fp=open('news1.xml','w', encoding="utf-8")
    doc.writexml(fp, indent='', addindent='\t', newl='\n', encoding="utf-8")
```

执行后会将抓取的新浪体育的新闻信息保存到 XML 文件 news1.xml 中，如图 2-6 所示。


```xml
<?xml version="1.0" encoding="utf-8"?>
<AllNews>
    <News>
        <Topic>sports</Topic>
        <Link>https://sports.sina.com.cn/basketball/nba/2020-05-04/doc-iircuyvi1217318.shtml</Link>
        <Title>勇士拿维金斯+3首轮换字母哥？交易有个大前提</Title>
        <Time>['2020年05月04日 08:33']</Time>
        <Text>['\u3000\u3000北京时间5月4日，露天看台的记者普雷斯顿-埃利斯报道，如果雄鹿再次冲击总冠军失败，扬尼斯-阿德托孔博拒绝
    </News>
    <News>
        <Topic>sports</Topic>
        <Link>https://sports.sina.com.cn/basketball/nba/2020-05-02/doc-iirczymi9384207.shtml</Link>
        <Time>['2020年05月02日 07:43']</Time>
        <Text>['\u3000\u3000在人们印象中，克雷-汤普森是神射手的代名词，也是一个和各类负面消息绝缘的乖乖仔，这种印象大抵没错。但人
    </News>
    <News>
        <Topic>sports</Topic>
        <Link>https://sports.sina.com.cn/basketball/nba/2020-04-25/doc-iircuyvh9694055.shtml</Link>
        <Title>追梦曾跟主帅差点动手！科尔的处理情商太高了</Title>
        <Time>['2020年04月25日 08:44']</Time>
        <Text>['\u3000\u3000北京时间4月25日，德拉蒙德-格林在参加史蒂芬-杰克逊的节目时进一步爆料，他曾经跟主帅史蒂夫-科尔发生了
    </News>
    <News>
        <Topic>sports</Topic>
        <Link>https://sports.sina.com.cn/basketball/nba/2020-04-25/doc-iirczymi8216362.shtml</Link>
```


图 2-6　文件 news1.xml

2. 使用 Stanford CoreNLP 提取 XML 数据的特征关系

Stanford CoreNLP 是由斯坦福大学开源的一套 Java NLP 工具，提供了词性标注（part-of-speech（POS）tagger）、命名实体识别（named entity recognizer（NER））、情感分析（sentiment analysis）等功能。Stanford CoreNLP 为 Python 提供了对应的模块，可以通过如下命令进行安装：

```
pip install stanfordcorenlp
```

因为本项目抓取到的是中文信息，所以还需要在 Stanford CoreNLP 官网下载专门处理中文的软件包，例如 stanford-chinese-corenlp-2018-10-05-models.jar。

编写文件 nlpTest.py，功能是调用 Stanford CoreNLP 分析处理上面抓取到的数据文件 news1.xml，提取出数据中的人名、城市和组织等信息，主要实现代码如下所示。

```python
import os
from stanfordcorenlp import StanfordCoreNLP

nlp = StanfordCoreNLP(r'H:\stanford-corenlp-full-2018-10-05', lang='zh')

for line in open(r'H:\pythonshuju\2\2-5\pythonCrawler-master\venv\Scrap\news1.
xml','r',encoding='utf-8'):
    res=nlp.ner(line)
    person = ["PERSON:"]
    location=['LOCATION:']
    organization=['ORGNIZATION']
    gpe=['GRE']
    for i in range(len(res)):
        if res[i][1]=='PERSON':
            person.append(res[i][0])
    for i in range(len(res)):
        if res[i][1]=='LOCATION':
            location.append(res[i][0])
    for i in range(len(res)):
```

```
            if res[i][1]=='ORGANIZATION':
                organization.append(res[i][0])
        for i in range(len(res)):
            if res[i][1]=='GPE':
                gpe.append(res[i][0])
        print(person)
        print(location)
        print(organization)
        print(gpe)


    nlp.close()
```

执行后会输出提取分析后的数据：

```
    ['PERSON:', '凯文', '杜兰特', '杜兰特', '金州', '杜兰特', '曼尼-迪亚兹',
'Manny', 'Diaz', '迪亚兹', '杜兰特', '杜兰特', '威廉姆森', '巴雷特和卡', '雷蒂什']
    ['LOCATION:']
    ['ORGNIZATION', 'NCAA', '球队', '迈阿密', '大学', '橄榄', '球队', '迈阿密', '大
学', '新任', '橄榄球队', '迈阿密', '先驱者', '报', '杜克大学']
    ####### 后面省略好多信息
```

2.2.5　实践案例：抓取 ×× 百科中符合要求的信息

本实例文件 baike.py 能够抓取 ×× 百科网站中的热门信息，具体功能如下所示：

- 抓取 ×× 百科热门段子；
- 过滤带有图片的段子；
- 每当按一次回车键，显示一个段子的发布时间、发布人、段子内容和点赞数。

接下来，我们通过 3 个步骤来实现本案例。

（1）确定 URL 并抓取页面代码

首先我们确定好页面的 URL 是 http://www. 域名主页 .com/hot/page/1，其中最后一个数字 1 代表当前的页数，可以输入不同的值来获得某一页的段子内容。首先编写代码设置要抓取的目标首页和 user_agent 值，具体实现代码如下所示：

```
import urllib
import urllib.request
page = 1
url = 'http://www. 域名主页 .com/hot/page/' + str(page)
user_agent = 'Mozilla/4.0 (compatible; MSIE 5.5; Windows NT)'
headers = { 'User-Agent' : user_agent }
try:
    request = urllib.request.Request(url,headers = headers)
    response = urllib.request.urlopen(request)
    print (response.read())
except (urllib.request.URLError, e):
    if hasattr(e,"code"):
        print (e.code)
    if hasattr(e,"reason"):
        print e.reason
```

执行后会打印输出第一页的 HTML 代码。

（2）提取某一页的所有段子

在获取了 HTML 代码后，接下来开始获取某一页的所有段子。首先我们审查元素看一下，按浏览器的 F12，截图如图 2-7 所示。

```
<div class="articleGender manIcon">28</div>
</div>


<a href="/article/118348358" target="_blank" class='contentHerf' >
<div class="content">


<span>历史课上，女老师的鞋跟突然断了。她感慨道："我这鞋都穿五年了。"这时，角落里有人低声说道："不愧是历史老师！"</span>


</div>
</a>
```

图 2-7　http://www.域名主页.com/hot/ 的源码

由此可见，每一个段子都是被<div class="articleGender manIcon">…</div>包含的内容。如果想获取页面中的发布人、发布日期、段子内容以及点赞的个数，需要注意有些段子是带有图片的。因为不能在控制台中显示图片信息，所以需要删除带有图片的段子，确保保存只含文本的段子。为了实现这一功能，使用正则表达式方法 re.findall() 寻找所有匹配的内容。编写的正则表达式匹配语句如下所示：

```
        pattern = re.compile(
            '<div.*?author clearfix">.*?<h2>(.*?)</h2>.*?<div.*?content".*?<span>(.*?)</span>.*?</a>(.*?)<div class= "stats".*?number">(.*?)</i>', re.S)
```

（3）完整程序展现

上面的正则表达式是整个程序的核心，本实例的实现文件是 baike.py，具体实现代码如下所示：

```
import urllib.request
import re
class Qiubai:

    # 初始化，定义一些变量
    def __init__(self):
        # 初始页面为 1
        self.pageIndex = 1
        # 定义 UA
          self.user_agent = 'Mozilla/5.0 (Windows NT 6.1; WOW64) AppleWebKit/537.36 (KHTML, like Gecko) Chrome/55.0.2883.75 Safari/537.36'
        # 定义 headers
        self.headers = {'User-Agent': self.user_agent}
        # 存放段子的变量，每一个元素是每一页的段子
        self.stories = []
        # 程序是否继续运行
        self.enable = False
    def getPage(self, pageIndex):
        """
        传入某一页面的索引后的页面代码
        """
        try:
            url = 'http://www.域名主页.com/hot/page/' + str(pageIndex)
            # 构建 request
            request = urllib.request.Request(url, headers=self.headers)
            # 利用 urlopen 获取页面代码
            response = urllib.request.urlopen(request)
            # 页面转为 utf-8 编码
            pageCode = response.read().decode("utf8")
            return pageCode
            # 捕获错误原因
        except (urllib.request.URLError, e):
            if hasattr(e, "reason"):
```

```
                print (u"连接 ×× 百科失败，错误原因 ", e.reason)
                return None
    def getPageItems(self, pageIndex):
        """
        传入某一页代码，返回本页不带图的段子列表
        """
        # 获取页面代码
        pageCode = self.getPage(pageIndex)
        # 如果获取失败，返回 None
        if not pageCode:
            print ("页面加载失败 ...")
            return None
        # 匹配模式
        pattern = re.compile(
                '<div.*?author clearfix">.*?<h2>(.*?)</h2>.*?<div.*?content".*?<sp
an>(.*?)</span>.*?</a>(.*?)<div class= "stats".*?number">(.*?)</i>', re.S)
        # findall 匹配整个页面内容 ,items 匹配结果
        items = re.findall(pattern, pageCode)
        # 存储整页的段子
        pageStories = []
        # 遍历正则表达式匹配的结果 , 0 name, 1 content, 2 img, 3 votes
        for item in items:
            # 是否含有图片
            haveImg = re.search("img", item[2])
            # 不含，加入 pageStories
            if not haveImg:
                # 替换 content 中的 <br/> 标签为 \n
                replaceBR = re.compile('<br/>')
                text = re.sub(replaceBR, "\n", item[1])
                # 在 pageStories 中存储：名字、内容、点赞数
                pageStories.append(
                    [item[0].strip(), text.strip(), item[3].strip()])
        return pageStories
    def loadPage(self):
        """
        加载并提取页面的内容，加入列表中
        """
        # 如未看页数少于 2，则加载并抓取新一页补充
        if self.enable is True:
            if len(self.stories) < 2:
                pageStories = self.getPageItems(self.pageIndex)
                if pageStories:
                    # 添加到 self.stories 列表中
                    self.stories.append(pageStories)
                    # 实际访问的页码 +1
                    self.pageIndex += 1
    def getOneStory(self, pageStories, page):
        """
        调用该方法，回车输出段子，按下 q 结束程序的运行
        """
        # 循环访问一页的段子
        for story in pageStories:
            # 等待用户输入，回车输出段子，q 退出
            shuru = input()
            self.loadPage()
            # 如果用户输入 q 退出
            if shuru == "q":
                # 停止程序运行，start() 中 while 判定
                self.enable = False
                return
            # 打印 story:0 name, 1 content, 2 votes
            print (u"第 %d 页 \t 发布人 :%s\t\3:%s\n%s" % (page, story[0], story[2], story[1]))
    def start(self):
```

```
        """
        开始方法
        """
        print (u" 正在读取 ×× 百科，回车查看新段子，q 退出 ")
        # 程序运行变量 True
        self.enable = True
        # 加载一页内容
        self.loadPage()
        # 局部变量，控制当前读到了第几页
        nowPage = 0
        # 直到用户输入 q，self.enable 为 False
        while self.enable:
            if len(self.stories) > 0:
                # 输出一页段子
                pageStories = self.stories.pop(0)
                # 用于打印当前页面，当前页数 +1
                nowPage += 1
                # 输出这一页段子
                self.getOneStory(pageStories, nowPage)
if __name__ == '__main__':
    qiubaiSpider = Qiubai()
    qiubaiSpider.start()
```

执行效果如图 2-8 所示，每次按下回车键就会显示下一条热门 ×× 百科信息。

```
=========
正在读取××百科，回车查看新段子，q退出

第1页    发布人:GoodbyeMyLover    □:1682
带着5岁的小侄去ATM机取钱，他手上扎了个刺，正好借前面有人取钱的空档给他往出拔，那小子扭来扭去的怕疼，我万分恼火的吼道:“不许动”。只见前面取钱人头也不回地举起了双手

第1页    发布人:爷蹲街    □:2981
历史课上，女老师的鞋跟突然断了。她感慨道：“我这鞋都穿五年了。”这时，角落里有人低声说道：“不愧是历史老师！”

第1页    发布人:得得哒哒 □:2282
今天中午迷迷糊糊的时候手机响了，没看号码就接起来，结果手机那头高冷地说了句“老地方见”就直接挂了电话！一脸蒙圈地看了下电话号码……

快递小哥，你这样子我很惆怅啊！

第1页    发布人:&amp;小九    □:3999
读小学时，有一次手工课老师布置我们带一件衣服来学校洗，同学们都带干净的衣服做做样子，我带了件全是油渍的外套，关键还很厚，那节课，我一个人在讲台上搓了一节课的衣服...

第1页    发布人:些许我的南柯一梦 □:7002
老公出差要4天，就这么过了几天我家3岁的闺女像是忽然想起她还有个爹似的睁着大眼问我:你家那个老公嘞，你老公去哪了?当时给我乐的。。。。

第1页    发布人:丁一QQ    □:4792
发个热乎的就刚刚，放假了接女儿过来，她五岁。在长途站下车，怕她乱跑我一直抱着，她非要下来我不放，她急了大叫救命啊，要不是我大包小包的都是她衣服我就被警察叔叔带走了!
```

图 2-8　执行效果

2.3　使用专业爬虫库 Scrapy

因为爬虫应用程序的需求日益增长，所以在市面上诞生了很多第三方开源的爬虫框架，其中 Scrapy 是一个为了爬取网站数据、提取结构性数据而编写的专业框架。Scrapy 框架的用途十分广泛，可以用于数据挖掘、数据监测和自动化测试等工作。在本节的内容中，将简要讲解爬虫框架 Scrapy 的基本用法。

↑扫码看视频（本节视频课程时间：16 分 34 秒）

2.3.1　Scrapy 框架基础

框架 Scrapy 使用了 Twisted 异步网络库来处理网络通信，其整体架构如图 2-9 所示。

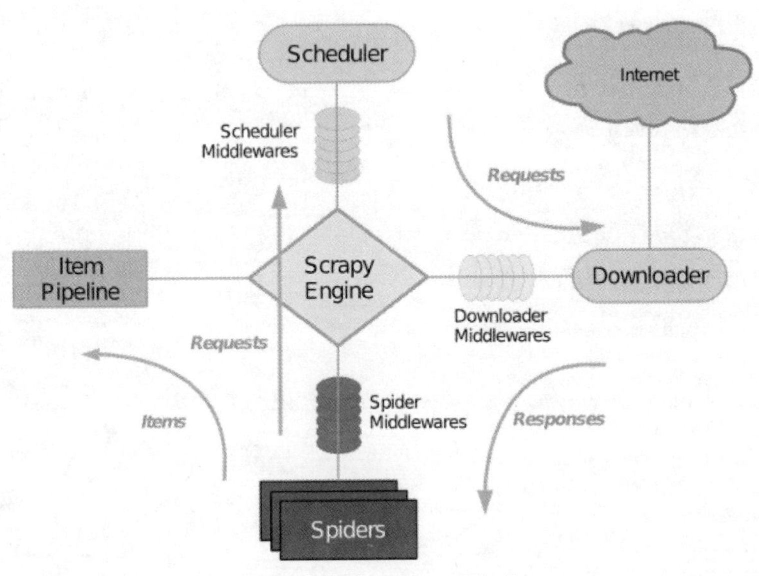

图 2-9　框架 Scrapy 的架构

1. Scrapy 框架组件

在 Scrapy 框架中，主要包括如表 2-4 所示的组件。

表 2-4

组件名称	说　明
Scrapy Engine（引擎）	用来处理整个系统的数据流，会触发到框架核心事务
调度器（Scheduler）	用来获取 Scrapy 发送过来的请求，然后将请求传入队列中，并在引擎再次请求的时候返回。调度器的功能是设置下一个要抓取的网址，并且还可以删除重复的网站
下载器（Downloader）	建立在高效的异步模型 Twisted 之上，功能是下载目标网址中的网页内容，并将网页内容返回给 Scrapy
爬虫（Spiders）	功能是从特定的网页中提取自己指定的信息，这些信息在爬虫领域中被称为实体（Item）
项目管道（Pipeline）	功能是处理从网页中提取的爬虫实体。当使用爬虫解析一个网页后，都会将实体发送到项目管道中进行处理，然后验证实体的有效性，并将不需要的信息删除
下载器中间件（Downloader Middlewares）	此模块位于 Scrapy 引擎和 Downloader 之间，为 Scrapy 引擎与下载器之间的请求及响应建立桥梁
爬虫中间件（Spider Middlewares）	此模块在 Scrapy 引擎和 Spiders 之间，功能是处理爬虫的响应输入和请求输出
调度中间件（Scheduler Middewares）	在 Scrapy 引擎和 Scheduler 之间，表示从 Scrapy 引擎发送到调度的请求和响应

2. Scrapy 框架运行流程

在使用 Scrapy 框架后，下面是大多数爬虫程序的运行流程。

（1）Scrapy Engine 从调度器中取出一个 URL 链接，这个链接将会作为接下来要抓取的目标。

（2）Scrapy Engine 将目标 URL 封装成一个 Request 请求并传递给下载器，下载器在下载 URL 资源后，将资源封装成 Response 应答包。

（3）使用爬虫解析 Response 应答包，如果解析出 Item 实体，则将结果交给实体管道进行进一步的处理。如果解析出的是 URL 链接，则把 URL 交给 Scheduler 等待下一步的抓取操作。

2.3.2 搭建 Scrapy 环境

在本地计算机安装 Python 后，可以使用“pip”命令或“easy_install”命令来安装 Scrapy，具体命令格式如下所示。

```
pip scrapy
easy_install scrapy
```

另外还需要确保已经安装了“win32api”模块，在安装此模块时必须安装和本地 Python 版本相对应的版本和位数（32 位或 64 位）。读者可以登录：http://www.lfd.uci.edu/~gohlke/pythonlibs/ 找到需要的版本，如图 2-10 所示。

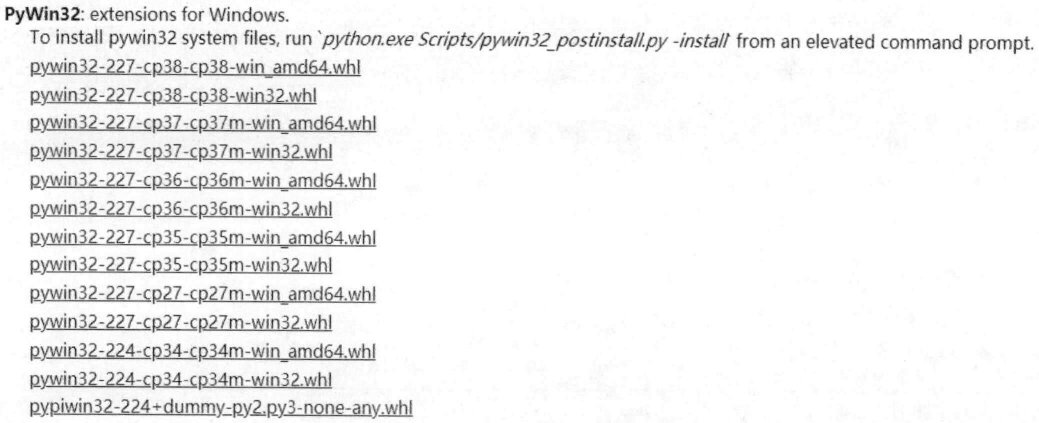

PyWin32: extensions for Windows.
To install pywin32 system files, run `*python.exe Scripts/pywin32_postinstall.py -install*` from an elevated command prompt.
pywin32-227-cp38-cp38-win_amd64.whl
pywin32-227-cp38-cp38-win32.whl
pywin32-227-cp37-cp37m-win_amd64.whl
pywin32-227-cp37-cp37m-win32.whl
pywin32-227-cp36-cp36m-win_amd64.whl
pywin32-227-cp36-cp36m-win32.whl
pywin32-227-cp35-cp35m-win_amd64.whl
pywin32-227-cp35-cp35m-win32.whl
pywin32-227-cp27-cp27m-win_amd64.whl
pywin32-227-cp27-cp27m-win32.whl
pywin32-224-cp34-cp34m-win_amd64.whl
pywin32-224-cp34-cp34m-win32.whl
pypiwin32-224+dummy-py2.py3-none-any.whl

图 2-10　下载“win32api”模块

下载后将得到一个“.whl”格式的文件，定位到此文件的目录，然后通过如下命令即可安装“win32api”模块。

```
python -m pip install --user ".whl"格式文件的全名
```

注意：如果遇到“ImportError: DLL load failed: 找不到指定的模块。”错误，需要将“Python\Python37\Lib\site-packages\win32”目录中的如下文件保存到本地系统盘中的“Windows\System32”目录下：

- pythoncom37.dll
- pywintypes37.dll

2.3.3 创建第一个 Scrapy 项目

请看下面的实例代码，演示了创建第一个 Scrapy 项目的过程。本实例用命令行来创建项目，然后将自动生成的文件稍加修改，即可完成一个颇具功能的爬虫程序。通过这个实例的实现过程，大家可以发现 Scrapy 的便捷性。

1. 创建项目

在开始爬取数据之前，必须先创建一个新的 Scrapy 项目。进入准备存储代码的目录中，然后运行如下所示的命令：

```
scrapy startproject tutorial
```

上述命令的功能是创建一个包含下列内容的“tutorial”目录。

```
tutorial/
    scrapy.cfg
    tutorial/
        __init__.py
        items.py
        pipelines.py
        settings.py
        spiders/
            __init__.py
            ...
```

对上述文件的具体说明如表 2-5 所示。

表 2-5

文件名称	说明
scrapy.cfg	项目的配置文件
tutorial/	该项目的 python 模块。之后您将在此加入代码
tutorial/items.py	项目中的 item 文件
tutorial/pipelines.py	项目中的 pipelines 文件
tutorial/settings.py	项目的设置文件
tutorial/spiders/	放置 spider 代码的目录

2. 定义 Item

Item 是保存爬取到的数据的容器，其使用方法和 Python 中的字典类似，并且提供了额外保护机制来避免拼写错误导致的未定义字段错误。我们可以通过创建一个 scrapy.Item 类，并且定义类型为 scrapy.Field 的类属性来定义一个 Item。

首先根据需要从 dmoz.org 获取到的数据对 item 进行建模。我们需要从 dmoz 中获取名字 url 以及网站的描述。对此，在 item 中定义相应的字段。编辑“tutorial”目录中的文件 items.py，具体实现代码如下所示。

```
import scrapy
class DmozItem(scrapy.Item):
    title = scrapy.Field()
    link = scrapy.Field()
    desc = scrapy.Field()
```

通过定义 item，可以很方便地使用 Scrapy 中的其他方法。而这些方法需要知道我们的 item 的定义。

3. 编写第一个爬虫（Spider）

Spider 是用户编写用于从单个网站（或者一些网站）爬取数据的类，其中包含了一个用于下载的初始 URL，如何跟进网页中的链接以及如何分析页面中的内容，提取生成 item 的方法。为创建一个 Spider，必须继承类 scrapy.Spider，且定义如下所示的三个属性。

（1）name：用于区别 Spider。该名字必须是唯一的，不可以为不同的 Spider 设定相同的名字。

（2）start_urls：包含了 Spider 在启动时进行爬取的 url 列表。因此，第一个被获取到的页面将是其中之一。后续的 URL 则从初始的 URL 获取到的数据中提取。

（3）parse()：是 spider 的一个方法。被调用时，每个初始 URL 完成下载后生成的 Response 对象将会作为唯一的参数传递给该函数。该方法负责解析返回的数据（response data），提取数据（生成 item）以及生成需要进一步处理的 URL 的 Request 对象。

下面是我们编写的第一个 Spider 代码，保存在“tutorial/spiders”目录下的文件 dmoz_spider.py 中，具体实现代码如下所示：

```
import scrapy
class DmozSpider(scrapy.Spider):
    name = "dmoz"
    allowed_domains = ["dmoz.org"]
    start_urls = [
        "http://www.dmoz.org/Computers/Programming/Languages/Python/Books/",
            "http://www.dmoz.org/Computers/Programming/Languages/Python/Resources/"
    ]
    def parse(self, response):
        filename = response.url.split("/")[-2]
        with open(filename, 'wb') as f:
            f.write(response.body)
```

4. 爬取

进入项目的根目录，执行下列命令启动 spider：

```
scrapy crawl dmoz
```

“crawl dmoz”是负责启动用于爬取“dmoz.org”网站的 spider，之后会得到如下所示的输出。

```
    2019-02-23 18:13:07-0400 [scrapy] INFO: Scrapy started (bot: tutorial)
    2019-02-23 18:13:07-0400 [scrapy] INFO: Optional features available: ...
    2019-02-23 18:13:07-0400 [scrapy] INFO: Overridden settings: {}
    2019-02-23 18:13:07-0400 [scrapy] INFO: Enabled extensions: ...
    2019-02-23 18:13:07-0400 [scrapy] INFO: Enabled downloader middlewares: ...
    2019-02-23 18:13:07-0400 [scrapy] INFO: Enabled spider middlewares: ...
    2019-02-23 18:13:07-0400 [scrapy] INFO: Enabled item pipelines: ...
    2019-02-23 18:13:07-0400 [dmoz] INFO: Spider opened
    2019-02-23 18:13:08-0400 [dmoz] DEBUG: Crawled (200) <GET http://www.dmoz.org/
Computers/Programming/Languages/Python/Resources/> (referer: None)
    2019-02-23 18:13:09-0400 [dmoz] DEBUG: Crawled (200) <GET http://www.dmoz.org/
Computers/Programming/Languages/Python/Books/> (referer: None)
    2019-02-23 18:13:09-0400 [dmoz] INFO: Closing spider (finished)
```

查看包含“dmoz”的输出，可以看到在输出的 log 中包含定义在“start_urls”的初始 URL，并且与 spider 是一一对应的。在 log 中可以看到其没有指向其他页面（(referer:None)）。除此之外，创建了两个包含 url 所对应的内容的文件：Book 和 Resources。

由此可见，Scrapy 为 Spider 的 start_urls 属性中的每个 URL 创建了 scrapy.Request 对象，并将 parse 方法作为回调函数（callback）赋值给了 Request。Request 对象经过调度，执行生成 scrapy.http.Response 对象并送回给 spider parse() 方法。

5. 提取 Item

有很多种从网页中提取数据的方法，Scrapy 使用了一种基于 XPath 和 CSS 的表达式机制：Scrapy Selectors。关于 selector 和其他提取机制的信息，建议读者参考 Selector 的官方文档。下面给出 XPath 表达式的例子及对应的含义：

- /html/head/title：选择 HTML 文档中 <head> 标签内的 <title> 元素；
- /html/head/title/text()：选择上面提到的 <title> 元素的文字；
- //td：选择所有的 <td> 元素；
- //div[@class=” mine”]：选择所有具有 class="mine" 属性的 div 元素。

上面仅仅是列出了几个简单的 XPath 例子，XPath 实际上要比这强大得多。为了配合 XPath，Scrapy 除提供了 Selector 之外，还提供了方法来避免每次从 response 中提取数据时生成 selector 的麻烦。

在 Selector 中有如下 5 个最基本的方法。

（1）xpath()：用于选取指定的标签内容，例如下面的代码表示选取所有的 book 标签：

```
selector.xpath('//book')
```

（2）css()：传入 CSS 表达式，用于选取指定的 CSS 标签内容。

（3）extract()：返回选中内容的 Unicode 字符串，返回结果是列表。

（4）re()：根据传入的正则表达式提取数据，返回 Unicode 字符串格式的列表。

（5）re_first()：返回 SelectorList 对象中的第一个 Selector 对象调用的 re 方法。

接下来使用内置的 Scrapy shell，首先进入本实例项目的根目录，然后执行如下命令来启动 shell。

```
scrapy shell "http://www.dmoz.org/Computers/Programming/Languages/Python/Books/"
```

此时 shell 将会输出类似如下所示的内容。

```
 [ ... Scrapy log here ... ]
2019-02-23 17:11:42-0400 [default] DEBUG: Crawled (200) <GET http://www.dmoz.
org/Computers/Programming/Languages/Python/Books/> (referer: None)
[s] Available Scrapy objects:
[s]   crawler    <scrapy.crawler.Crawler object at 0x3636b50>
[s]   item       {}
[s]    request     <GET http://www.dmoz.org/Computers/Programming/Languages/
Python/Books/>
[s]    response    <200 http://www.dmoz.org/Computers/Programming/Languages/
Python/Books/>
[s]   settings   <scrapy.settings.Settings object at 0x3fadc50>
[s]   spider     <Spider 'default' at 0x3cebf50>
[s] Useful shortcuts:
[s]   shelp()           Shell help (print this help)
[s]   fetch(req_or_url) Fetch request (or URL) and update local objects
[s]   view(response)    View response in a browser
In [1]:
```

当载入 shell 后会得到一个包含 response 数据的本地 response 变量。输入“response.body”命令后会输出 response 的包体，输入“response.headers”后可以看到 response 的报头。更为重要的是，当输入 response.selector 时，将获取到一个可以用于查询返回数据的 selector（选择器），以及映射到 response.selector.xpath()、response.selector.css() 的快捷方法（shortcut）:response.xpath() 和 response.css()。同时，shell 根据 response 提前初始化了变量 sel。该 selector 根据 response 的类型自动选择最合适的分析规则（XML vs HTML）。

6. 提取数据

接下来尝试从这些页面中提取出一些有用的数据，可以在终端中输入 response.body 来观察 HTML 源码并确定合适的 XPath 表达式。但是这个任务非常无聊且不容易，可以考虑使用 Firefox 的 Firebug 扩展来简化工作。

在查看了网页的源码后，会发现网站的信息是被包含在第二个 <ul> 元素中。可以通过下面的代码选择该页面中网站列表里的所有 <li> 元素。

```
response.xpath('//ul/li')
```

通过如下命令获取对网站的描述：

```
response.xpath('//ul/li/text()').extract()
```

通过如下命令获取网站的标题：

```
response.xpath('//ul/li/a/text()').extract()
```

2.3.4 实践案例：抓取某电影网的热门电影信息

本实例是使用 Scrapy 爬虫抓取某电影网中热门电影信息的过程。首先使用命令行创建一个 Scrapy 项目，然后设置要爬取的 URL 范围和过滤规则，最后编写功能函数获取热门电影的详细信息。

（1）在具体爬取数据之前，必须先创建一个新的 Scrapy 项目。首先进入准备保存项目代码的目录中，然后运行如下所示的命令：

```
scrapy startproject scrapydouban
```

（2）编写文件 moviedouban.py，设置要爬取的 URL 范围和过滤规则，主要实现代码如下所示：

```
class MoviedoubanSpider(CrawlSpider):
    name = "moviedouban"
    allowed_domains = ["movie.域名主页 .com"]
    start_urls = ["https://movie.域名主页 .com/"]

    rules = (
         Rule(LinkExtractor(allow=r"/subject/\d+/($|\?\w+)"),
             callback="parse_movie", follow=True),
     )

    def __init__(self):
            self.page_number = 1


    def parse_movie(self, response):
            print("RESPONSE: {}".format(response))
```

（3）编写执行脚本文件 pyrequests_douban.py，功能是编写功能函数获取热门电影的详细信息，包括电影名、URL 链接地址、导演信息、主演信息等。文件 pyrequests_douban.py 的主要实现代码如下所示：

```
def get_celebrity(url):
    print("Sending request to {}".format(url))
    celebrity = {}
    res = get_request(url)
    if res.status_code == 200:
            html = BeautifulSoup(res.content, "html.parser")
            celebrity['avatars'] = {}
            if html.find("div", {'class': "pic"}):
                    for s in ["small", "medium", "large"]:
                            celebrity['avatars'][s] = [dt['src'].strip() for dt in
html.find("div", {'class': "pic"}).findAll("img", src=re.compile(r'{}'.format(s)))]
    return celebrity


def get_initial_release(html):
    return html.find("div", {'id': "info"}).find("span", {'property':
"v:initialReleaseDate"}).string
```

```
    def get_photos(html, url):
        photos = []
        if html.find("a", {"href": url}):
                print("Sending request for {} to get all photos".format(url))
                res = get_request(url)
                if res.status_code == 200:
                        html = BeautifulSoup(res.content, 'html.parser')
                        if html.find("div", {'class': "article"}):
                                all_photos_html = html.find("div", {'class':
"article"}).findAll("div", {'class': "mod"})[0]
                                for li in all_photos_html.findAll("li")[:10]:
                                        photo = {}
                                        photo['url'] = li.find("a")['href']
                                        photo['photo'] = li.find("img")['src']
                                        photos.append(photo)
        return photos


    def get_request(url, s=0):
        time.sleep(s)
        return requests.get(url)


    def get_runtime(html):
        return html.find("div", {'id': "info"}).find("span", {'property':
"v:runtime"}).string


    def get_screenwriters(html):
        directors = []
        if html.find("div", {'class': "info"}):
                for d in html.find("div", {'class': "info"}).findChildren()[2].
findAll("a"):
                        director = {}
                        director['id'] = d['href'].split('/')[2]
                        director['name'] = d.string
                        director['alt'] = "{}{}".format(domain, d['href'])
                        director['avatars'] = get_celebrity(director['alt'])
                        directors.append(director)
        return directors

    def get_starring(html):
        starrings = []
        if html.find("span", {'class': "actor"}):
                for star in html.find("span", {'class': "actor"}).find("span",
{'class': "attrs"}).findAll("a"):
                        starring = {}
                        starring['id'] = star['href'].split('/')[2]
                        starring['name'] = star.string.strip()
                        starring['alt'] = "{}{}".format(domain, star['href'])
                        starring['avatars'] = get_celebrity(starring['alt'])
                        starrings.append(starring)
        return starrings


    def get_trailer(html, url):
        trailers = []
        if html.find("a", {"href": url}):
                print("Sending request for {} to get trailers".format(url))
                res = get_request(url)
                if res.status_code == 200:
                        html = BeautifulSoup(res.content, 'html.parser')
                        if html.find("div", {'class': "article"}):
```

```
                    for d in html.find("div", {'class': "article"}).
findAll("div", {'class': "mod"}):
                        trailer = {}
                        if "预告片" in d.find("h2").string:
                            trailer['url'] = d.find("a", {'class':
"pr-video"})['href']
                            trailer['view_img'] = d.find("a",
{'class': "pr-video"}).find("img")['src']
                            trailer['duration'] = d.find("a",
{'class': "pr-video"}).find("em").string
                            trailer['title'] = d.find("p").
find("a").string.strip()
                            trailer['date'] = d.find("p", {'class':
"trail-meta"}).find("span").string.strip()
                            trailer['responses_url'] = d.find("p",
{'class': "trail-meta"}).find("a")['href']
                        trailers.append(trailer)
        return trailers


    def get_types(html):
        return ", ".join([sp.string for sp in html.find("div", {"id": "info"}).
findAll("span", {'property': "v:genre"})])


    def manual_scrape():
        print("Sending request to {}".format(url.format(page_limit, page_start)))
        response = get_request(url.format(page_limit, page_start), 5)
        if response.status_code == 200:
            subjects = json.loads(response.text)['subjects']
            for subject in subjects:
                subject['api_return'] = {}
                print("Sending request to {}".format(public_api_url.
format(subject['id'])))
                api_response = get_request(public_api_url.
format(subject['id']))
                if api_response.status_code == 200:
                    subject['api_return'] = api_response.json()
                subject['url_content'] = {}
                ## Call Happy API to post data
                print("Sending request to {}".format(subject['url']))
                content_response = get_request(subject['url'], 5)
                if content_response.status_code == 200:
                    content = BeautifulSoup(content_response.content,
'html.parser')
                    subject['url_content']['screenwriters'] = get_
screenwriters(content)
                    subject['url_content']['starring'] = get_
starring(content)
                    subject['url_content']['types'] = get_types(content)
                    subject['url_content']['initial_release'] = get_
initial_release(content)
                    subject['url_content']['runtime'] = get_
runtime(content)
                    subject['url_content']['trailer'] = get_
trailer(content, "{}trailer".format(subject['url']))
                    subject['url_content']['photos'] = get_photos(content,
"{}all_photos".format(subject['url']))
                pprint.pprint(subject)


    if __name__ == "__main__":
```

```
    global url, page_limit, page_start, domain, public_api_url

    domain = "https://movie. 域名主页 .com"
    public_api_url = "https://api. 域名主页 .com/v2/movie/subject/{}"
    url = "https://movie. 域名主页 .com/j/search_subjects?type=movie&tag=%E7%83%AD
%E9%97%A8&sort=recommend&page_limit={}&page_start={}"
    page_limit = 20
    page_start = 0

    manual_scrape()
```

执行后会输出显示爬取到的热门电影信息，如图 2-11 所示。

```
Sending request to https://movie.douban.com/celebrity/1373451/
Sending request for https://movie.douban.com/subject/30377703/all_photos to get all photos
{'api_return': {'aka': ['中国版《完美陌生人》', '手机狂响', 'Kill Mobile'],
                'alt': 'https://movie.douban.com/subject/30377703/',
                'casts': [{'alt': 'https://movie.douban.com/celebrity/1009179/',
                           'avatars': {'large': 'https://img3.doubanio.com/view/celebrity/s_ratio_celebrity/public/p1542346320.44.jpg',
                                       'medium': 'https://img3.doubanio.com/view/celebrity/s_ratio_celebrity/public/p1542346320.44.jpg',
                                       'small': 'https://img3.doubanio.com/view/celebrity/s_ratio_celebrity/public/p1542346320.44.jpg'},
                           'id': '1009179',
                           'name': '佟大为'},
                          {'alt': 'https://movie.douban.com/celebrity/1319032/',
                           'avatars': {'large': 'https://img3.doubanio.com/view/celebrity/s_ratio_celebrity/public/p1444800807.11.jpg',
                                       'medium': 'https://img3.doubanio.com/view/celebrity/s_ratio_celebrity/public/p1444800807.11.jpg',
                                       'small': 'https://img3.doubanio.com/view/celebrity/s_ratio_celebrity/public/p1444800807.11.jpg'},
                           'id': '1319032',
                           'name': '马丽'},
                          {'alt': 'https://movie.douban.com/celebrity/1000145/',
                           'avatars': {'large': 'https://img3.doubanio.com/view/celebrity/s_ratio_celebrity/public/p2520.jpg',
                                       'medium': 'https://img3.doubanio.com/view/celebrity/s_ratio_celebrity/public/p2520.jpg',
                                       'small': 'https://img3.doubanio.com/view/celebrity/s_ratio_celebrity/public/p2520.jpg'},
                           'id': '1000145',
                           'name': '霍思燕'},
                          {'alt': 'https://movie.douban.com/celebrity/1316368/',
                           'avatars': {'large': 'https://img3.doubanio.com/view/celebrity/s_ratio_celebrity/public/p1473410979.5.jpg',
                                       'medium': 'https://img3.doubanio.com/view/celebrity/s_ratio_celebrity/public/p1473410979.5.jpg',
                                       'small': 'https://img3.doubanio.com/view/celebrity/s_ratio_celebrity/public/p1473410979.5.jpg'},
```

图 2-11　爬取到的热门电影信息

2.3.5　实践案例：抓取某网站中的照片并保存到本地

本实例的功能是使用 Scrapy 爬虫抓取某网站中的照片信息，并将抓取到的照片保存到本地硬盘中。编写文件 art.py 设置要爬取的 URL 范围和内容元素，在 start_urls 中设置要抓取的起始 URL 地址，然后使用函数 json.loads() 解析 JSON 数据，在 JSON 数据中保存了每个照片的地址，然后下载对应地址的照片。主要实现代码如下所示：

```
import scrapy
from scrapy import Request
import json

class ImagesSpider(scrapy.Spider):
    BASE_URL='http:// 域名主页 .so.com/zj?ch=beauty&sn=%s&listtype=new&temp=1'# 注意
这里 %S 改过了
    #BASE_URL='http:// 域名主页 .so.com/j?q=%E9%BB%91%E4%B8%9D&src=srp&correct=%E9
%BB%91%E4%B8%9D&pn=60&ch=&sn=%s&sid=fc52f43bfb771f78907396c3167f10ad&ran=0&ras=0&cn
=0&gn=10&kn=50'
    start_index=0

    MAX_DOWNLOAD_NUM=1000

    name="images"
    start_urls=[BASE_URL %0]

    def parse(self,response):
```

```
        infos=json.loads(response.body.decode('utf-8'))
        for info in infos['list']:
            yield {'image_urls':[info['qhimg_url']]}

        self.start_index+=30
        if infos['count']>0 and self.start_index<self.MAX_DOWNLOAD_NUM:
            yield Request(self.BASE_URL % self.start_index)
```

执行后会显示抓取目标网站图片的过程，如图 2-12 所示。

```
2019-02-11 15:31:23 [scrapy.core.engine] DEBUG: Crawled (200) <GET http://p3.so.qhmsg.com/t0154687f04bb6e87c8.jpg> (referer: None)
2019-02-11 15:31:23 [scrapy.pipelines.files] DEBUG: File (downloaded): Downloaded file from <GET http://p3.so.qhmsg.com/t0154687f04bb6e87c8.jpg> referred in <None>
2019-02-11 15:31:23 [scrapy.core.engine] DEBUG: Crawled (200) <GET http://p4.so.qhmsg.com/t013862476b8fbf7969.jpg> (referer: None)
2019-02-11 15:31:23 [scrapy.pipelines.files] DEBUG: File (downloaded): Downloaded file from <GET http://p4.so.qhmsg.com/t013862476b8fbf7969.jpg> referred in <None>
2019-02-11 15:31:23 [scrapy.core.engine] DEBUG: Crawled (200) <GET http://p4.so.qhmsg.com/t01c382465efd9da959.jpg> (referer: None)
2019-02-11 15:31:23 [scrapy.pipelines.files] DEBUG: File (downloaded): Downloaded file from <GET http://p4.so.qhmsg.com/t01c382465efd9da959.jpg> referred in <None>
2019-02-11 15:31:23 [scrapy.core.scraper] DEBUG: Scraped from <200 http://image.so.com/zj?ch=beauty&sn=0&listtype=new&temp=1>
{'image_urls': ['http://p3.so.qhmsg.com/t0154687f04bb6e87c8.jpg'], 'images': [{'url': 'http://p3.so.qhmsg.com/t0154687f04bb6e87c8.jpg', 'path':
 'full/bf820e4ed8c044f94431988051e3402f721f431f.jpg', 'checksum': 'b10e0a58b31b053eda738597b0c46722'}]}
2019-02-11 15:31:23 [scrapy.core.scraper] DEBUG: Scraped from <200 http://image.so.com/zj?ch=beauty&sn=0&listtype=new&temp=1>
{'image_urls': ['http://p4.so.qhmsg.com/t013862476b8fbf7969.jpg'], 'images': [{'url': 'http://p4.so.qhmsg.com/t013862476b8fbf7969.jpg', 'path':
 'full/209d862baa417e36c248c38e39fd503d7d550eef.jpg', 'checksum': 'df568ec4c2ec907a40851fc6de25768f'}]}
2019-02-11 15:31:23 [scrapy.core.scraper] DEBUG: Scraped from <200 http://image.so.com/zj?ch=beauty&sn=0&listtype=new&temp=1>
{'image_urls': ['http://p4.so.qhmsg.com/t01c382465efd9da959.jpg'], 'images': [{'url': 'http://p4.so.qhmsg.com/t01c382465efd9da959.jpg', 'path':
 'full/fe3ab046bd19a05aa8df53d7d53fe15674187000.jpg', 'checksum': 'b92548a0ed535a424124041590f712c9'}]}
2019-02-11 15:31:23 [scrapy.core.engine] DEBUG: Crawled (200) <GET http://p0.so.qhmsg.com/t017abc1de09c20ca55.jpg> (referer: None)
2019-02-11 15:31:23 [scrapy.pipelines.files] DEBUG: File (downloaded): Downloaded file from <GET http://p0.so.qhmsg.com/t017abc1de09c20ca55.jpg> referred in <None>
2019-02-11 15:31:24 [scrapy.core.scraper] DEBUG: Scraped from <200 http://image.so.com/zj?ch=beauty&sn=0&listtype=new&temp=1>
{'image_urls': ['http://p0.so.qhmsg.com/t017abc1de09c20ca55.jpg'], 'images': [{'url': 'http://p0.so.qhmsg.com/t017abc1de09c20ca55.jpg', 'path':
 'full/a99c7328f98d78764752ead804edf2160bbe97ab.jpg', 'checksum': '27e9e9a362df067d33e7340571a737ee'}]}
2019-02-11 15:31:24 [scrapy.core.engine] DEBUG: Crawled (200) <GET http://p0.so.qhimgs1.com/t010165ed1120ab3834.jpg> (referer: None)
2019-02-11 15:31:24 [scrapy.pipelines.files] DEBUG: File (downloaded): Downloaded file from <GET http://p0.so.qhimgs1.com/t010165ed1120ab3834.jpg> referred in
```

图 2-12　抓取过程

并将抓取到的照片保存到本地文件夹“download_images”中，如图 2-13 所示。

00a035f9d5da2a7123c651cf05e68b1873c179f7.jpg
0c08846a93e387c05588eb66090b812d779b1172.jpg
0d88fe04085299b5e8fc70abb98e599763111cd8.jpg
0e60aaee09d0ad84fcd608c0ca2bfa9b156de4bd.jpg
0f29c9d67deafaa154d0f2989918aa61a1a1ccf0.jpg
2cedfa6058832b78d282354e8f5137d9dc939339.jpg
2e44c30c155d6a69bc12adb7015badfd5c2a3602.jpg
3a5540b113f21ef56c3e34698a8a5a92e26c4b65.jpg
3d021a37a9acc8959f85ff403af73e386f75d2b0.jpg
3e87cbefc4300baa8733f66a32108d978ec7a9f1.jpg
04c99f0e59464082de4076f3a6248c00f290c24d.jpg
4abb880acf03d877032d26c866330e9bb270eca7.jpg
4d2e58b300f3c37a2f821e53d1ef7e85e0710089.jpg
4d16bf98e1b59004b9df079aff3d06568998f0b5.jpg
4f64e2dc11f1dc43af2c8edf0c8fbb8386f0ff66.jpg
5b6a0722322c2b30018b3635899694fe96d3500e.jpg
5e8496e12f956ed3a7474665028dc96670ae91d9.jpg
5f11a4deecaf30e40631f59e6df61ba652aefbd8.jpg
06a1020d5703d7478f648c25844cb029e3d3b36c.jpg
6b8f89541d72eadc7d6e37f6567e8713f802fbee.jpg
6d37adf796f7bc9eb95d54834ef194f577899548.jpg
6e791b75cb5ec639c74c9823ffddbeb2810228f5.jpg
6f1a76263c759bea724c7d4742ce8c15c31d3877.jpg
07c2a4f3ca0047be8cf32e2cda8b9cf3714385ff.jpg
7a3517769b6b7bc46ebbe1991b784906bcdafb15.jpg
7cad9a501e42e079e5b2d54f4b43633a22a6e075.jpg
7da24afa6ebc249cec84249af3237b347e75a8d6.jpg
8a8b74d995152860488adbc15ae1cfb13dd03b84.jpg
8a22f9b7b4421837a2cd99d7c2adb90fb79a4594.jpg
8b95ac2458c81eec5dc0c3cabf6e8d1bb114a99f.jpg
8d7dce7a348f3c959c2a5e1a84b05767afd7a1d2.jpg
8f21b8d4c93d3da8d6af193c4b24963eb9d29b01.jpg
8fee047886ff301d8b1662321abb479645c2578d.jpg
09e67f3092a5d3ee3175577b0685e0bc511035ce.jpg
9e47008091c7fd460440989c402a58c96b4e2eb8.jpg
9f9350e53bb0b646d43e7c144abf823ea3baef49.jpg
11da4d830c25593262010463c758c7437d47c9a6.jpg
14fa0f532708535d6b11b8c016446cabc2e5172c.jpg
18a0a40bb3ed732da0a6de861af67886b2df28f4.jpg
22ef83c60610be24d7a3fb0055ab9d235e29c9fa.jpg
27a66c39ecafcc2794df8bfaf3d2e47715499503.jpg
41cad27e4361582f7e160bbdf723d6cf1c4b106b.jpg
56a85a291c3ad7561a73cfc2dadde23783af4fb5.jpg
59ce90fe50c98bba8d3314de7df968e0fb832fa1.jpg
061f6644bfd38b8bcfed42848782bc2506bc3403.jpg
66c0c3edb6aac5e919cd72057724ebb3519cccff.jpg
79d8fe082112fceab713c44f63b2162ea1dd5abb.jpg
79f13fbb181b6920babefe118faa886eac7e0fe2.jpg
81c96f1f4e6f1e32e0d9f663461480d17ddfc2b2.jpg
83bcb74d5cf446f9bf3276d4cc25bc66b382751e.jpg
086d8ed6a55b4fd6f49c9e100b96a2a658ec3bf9.jpg
88b0bacfa2c864e79b27de13fcbbb534d5601c68.jpg
89d666a99122419529e9bba506848c7f798274e3.jpg
93eebdaa0d3d2a89e1cfc911a58198fea04e7bfe.jpg
98fa8b4f0b8967c7aa346794e6a19bae11ac150a.jpg
99bf5bfe07b2bb93dbd5ce9a1997f0e0d3aa29b4.jpg
130d8d3319ba50900003041bfdee80b91f88f00a.jpg
209d862baa417e36c248c38e39fd503d7d550eef.jpg
345e15addf9042ad8e237f6afce76edd12a357cd.jpg
375ee3b9005c57dc9d8de2eceda3df639003d93a.jpg
409bd8b2e365b36031b802d0d9a169111a4b98b4.j
539da0891b37220f19c8e1cc5eac0c71fbc90dbe.jpg
0601a84234c0db8a1650bb1cfa376e68ef422880.jpg
633ebe7db80cd754574808ea900d5946c00da039.jp
711b06b2ac6f53d207992425d6b33adea3dd5963.jp
758ae34a48d27bcae0711e828f30d138d6ae7c0b.jp
881cfb63ed65910df86ee6559a29e259a338dff6.jpg
899ea6ffce13e6b5e47cfe4a4bf313320ded60f1.jpg
961c16528449e775b27e6cb118a043963a7d1314.jp
2484f5a55e0277bb616f9d0bd7ebc737ee9f9d7f.jpg
7896bf169fbb3b9f69578309db9e8c3e32394436.jpg
13193b5b562db5cb5e1ea9e34042fa2da27bcb9d.jp
67869aa1e173e3bf1d5090aca31eab8ea2b133c4.jp
82796a5c38b16b426014b718a2557b22c7871fc2.jp
416671d8b252248b9809e52f1154f01c6c2713f3.jpg
564115c19bcda2bf70618bdc827ecb0a0c5f28a8.jpg
0578567af335a9b12e03c57abe34dfa48b650d7b.jp
620933b5266f33e974ceb79e56683375c0142221.jpg
876442f49661bf0be789f7bed9d2554e4351d1df.jpg
902317ba0f6e43bd7bca0604116fc31d0118375e.jpg
2314739cf5d2b789ad860e7dfa7a37f996a80b7a.jpg
6161459ff6876bfbd98293d579721e4fe083d724.jpg
8418515ffb39fa9aecc1bebf03c217a0b350a8c0.jpg
9959079dda2dc92b1cc1a0cc4e9e4b6a042c51d6.jp
62795447e6468ab30d533a2d29853e75fecbcda5.jp
99650045aa7d77953e267793972cff6b722cc3e5.jpg
762054158fc875a5f2d207490e3f890c9f2c3840.jpg
939917816b2fbbc9da3de772350d18e9f75e2a06.jp
628570064584908fd6b36f15ed6a995e8743fa6f.jpg
a89c9338b761cba5d2dac1d43f9640a44fa0f3ea.jpg

图 2-13　在本地硬盘保存抓取到的图片

第 3 章

数据可视化技术：matplotlib 基础

matplotlib 是 Python 语言中著名的数据可视化工具包，通过使用 matplotlib，可以非常方便地实现和数据统计相关的图形，例如折线图、散点图、直方图等。正因为 matplotlib 在绘图领域的强大功能，所以在 Python 数据挖掘方面得到了重视。在本章的内容中，将详细讲解在 Python 语言中使用 matplotlib 实现数据分析的知识，为读者步入本书后面知识的学习打下基础。

3.1 搭建 matplotlib 环境

在 Python 程序中使用库 matplotlib 之前，需要先确保安装了 matplotlib 库。在 Windows 系统中安装 matplotlib 之前，首先需要确保已经安装了 Visual Studio.NET。在安装 Visual Studio.NET 后，可以安装 matplotlib 了。

↑扫码看视频（本节视频课程时间：2 分 03 秒）

安装 matplotlib 时，最简单的安装方式是使用如下“pip”命令或“easy_install”命令。

```
easy_install matplotlib
pip install matplotlib
```

虽然上述两种安装方式比较简单、省心，但是并不能保证安装的 matplotlib 适合我们安装的 Python。例如笔者在写作本书时使用的是 Python 3.9，当时最新的 matplotlib 版本是 3.3.0。建议读者登录 https://pypi.org/project/matplotlib/#files，如图 3-1 所示。在这个页面中查找与你使用的 Python 版本匹配的 wheel 文件（扩展名为“.whl”的文件）。例如，如果使用的是 64 位的 Python 3.8，则需要下载 matplotlib-3.3.0rc1-cp38-cp38-win_amd64.whl。

matplotlib-3.2.2-cp36-cp36m-win32.whl (9.0 MB)	Wheel	cp36	Jun 18, 2020	View
matplotlib-3.2.2-cp36-cp36m-win_amd64.whl (9.2 MB)	Wheel	cp36	Jun 18, 2020	View
matplotlib-3.2.2-cp37-cp37m-macosx_10_9_x86_64.whl (12.5 MB)	Wheel	cp37	Jun 18, 2020	View
matplotlib-3.2.2-cp37-cp37m-manylinux1_x86_64.whl (12.4 MB)	Wheel	cp37	Jun 18, 2020	View
matplotlib-3.2.2-cp37-cp37m-win32.whl (9.0 MB)	Wheel	cp37	Jun 18, 2020	View
matplotlib-3.2.2-cp37-cp37m-win_amd64.whl (9.2 MB)	Wheel	cp37	Jun 18, 2020	View
matplotlib-3.2.2-cp38-cp38-macosx_10_9_x86_64.whl (12.5 MB)	Wheel	cp38	Jun 18, 2020	View
matplotlib-3.2.2-cp38-cp38-manylinux1_x86_64.whl (12.4 MB)	Wheel	cp38	Jun 18, 2020	View
matplotlib-3.2.2-cp38-cp38-win32.whl (9.0 MB)	Wheel	cp38	Jun 18, 2020	View

图 3-1　登录 https://pypi.python.org/pypi/matplotlib/

注意：如果登录 https://pypi.python.org/pypi/matplotlib/ 找不到适合自己的 matplotlib，还可以尝试登录 https://www.lfd.uci.edu/~gohlke/pythonlibs/，如图 3-2 所示。这个网站发布安装程序的时间通常比 matplotlib 官网要早一段时间。

matplotlib-3.3.0rc1-cp38-cp38-win_amd64.whl
matplotlib-3.3.0rc1-cp38-cp38-win32.whl
matplotlib-3.3.0rc1-cp37-cp37m-win_amd64.whl
matplotlib-3.3.0rc1-cp37-cp37m-win32.whl
matplotlib-3.3.0rc1-cp36-cp36m-win_amd64.whl
matplotlib-3.3.0rc1-cp36-cp36m-win32.whl
matplotlib-3.2.2-pp36-pypy36_pp73-win32.whl
matplotlib-3.2.2-cp39-cp39-win_amd64.whl
matplotlib-3.2.2-cp39-cp39-win32.whl
matplotlib-3.2.2-cp38-cp38-win_amd64.whl
matplotlib-3.2.2-cp38-cp38-win32.whl
matplotlib-3.2.2-cp37-cp37m-win_amd64.whl
matplotlib-3.2.2-cp37-cp37m-win32.whl
matplotlib-3.2.2-cp36-cp36m-win_amd64.whl
matplotlib-3.2.2-cp36-cp36m-win32.whl
matplotlib-2.2.5-pp273-pypy_73-win32.whl
matplotlib-2.2.5-cp38-cp38-win_amd64.whl
matplotlib-2.2.5-cp38-cp38-win32.whl
matplotlib-2.2.5-cp37-cp37m-win_amd64.whl
matplotlib-2.2.5-cp37-cp37m-win32.whl
matplotlib-2.2.5-cp36-cp36m-win_amd64.whl

图 3-2　登录 http://www.lfd.uci.edu

例如笔者当时下载得到的文件是 matplotlib-3.3.0rc1-cp38-cp38-win_amd64.whl，将这个文件保存在“H:\matp”目录下，接下来需要打开一个命令窗口，并切换到该项目文件夹“H:\matp”，然后使用如下所示的“pip”命令即可安装 matplotlib：

```
python -m pip install --user matplotlib-3.3.0rc1-cp38-cp38-win_amd64.whl
```

3.2　绘制散点图

在数据可视化分析工作中，经常需要绘制和数据统计相关的图形，例如折线图、散点图、直方图等；其中散点图通常用两组数据构成多个坐标点，考察坐标点的分布，判断两个变量之间是否存在某种关联或总结坐标点的分布模式。

↑扫码看视频（本节视频课程时间：10 分 45 秒）

3.2.1　绘制一个简单的点

假设你有一堆的数据样本，想要找出其中的异常值，那么最直观的方法就是将它们画成散点图。在下面的实例文件 dian01.py 中，演示了使用 matplotlib 绘制含有两个点的散点图的过程。

```
import matplotlib.pyplot as plt                     #导入 pyplot 包，并缩写为 plt
#定义 2 个点的 x 集合和 y 集合
x=[1,2]
y=[2,4]
```

```
plt.scatter(x,y)                                    # 绘制散点图
plt.show()                                          # 展示绘画框
```

在上述实例代码中绘制了拥有两个点的散点图，向函数 scatter() 传递了两个分别包含 *x* 值和 *y* 值的列表。执行效果如图 3-3 所示。

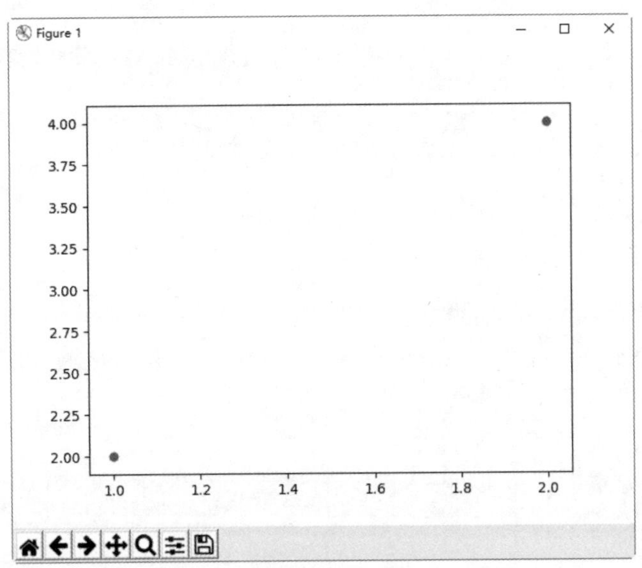

图 3-3　执行效果

在上述实例中，可以进一步调整一下坐标轴的样式，例如可以加上如下所示的代码：

```
#[] 里的 4 个参数分别表示 x 轴起始点，x 轴结束点，y 轴起始点，y 轴结束点
plt.axis([0,10,0,10])
```

3.2.2　添加标题和标签

我们可以设置散点图的输出样式，例如添加标题，给坐标轴添加标签，并确保所有文本都大到能够看清。请看下面的实例文件 dian02.py，功能是使用 matplotlib 函数 scatter() 绘制一系列点，然后设置显示标题和标签。

```
""" 使用 scatter() 绘制散点图 """
import matplotlib.pyplot as plt

x_values = range(1, 6)
y_values = [x*x for x in x_values]
'''
scatter()
x: 横坐标 y: 纵坐标 s: 点的尺寸
'''
plt.scatter(x_values, y_values, s=50)

# 设置图表标题并给坐标轴加上标签
plt.title('Square Numbers', fontsize=24)
plt.xlabel('Value', fontsize=14)
plt.ylabel('Square of Value', fontsize=14)

# 设置刻度标记的大小
plt.tick_params(axis='both', which='major', labelsize=14)
plt.show()
```

执行效果如图 3-4 所示。

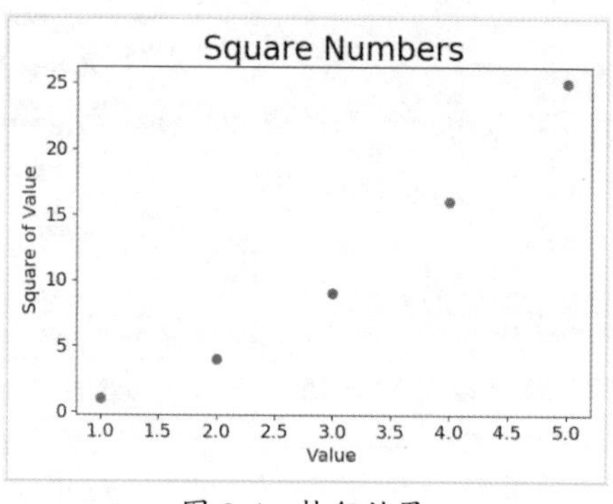

图 3-4　执行效果

3.2.3　绘制 10 个点

请看下面的实例文件 dian03.py，功能是使用 matplotlib 函数 scatter() 绘制 10 个点的散点图。

```
import matplotlib.pyplot as plt
import numpy as np
# 保证图片在浏览器内正常显示
N = 10                                                  # 10 个点
x = np.random.rand(N)
y = np.random.rand(N)
plt.scatter(x, y)
plt.show()
```

在上述代码中，首先使用函数 random.rand() 分别生成了 10 个点的横坐标和纵坐标，然后使用函数 scatter() 绘制这 10 个点。执行效果如图 3-5 所示。

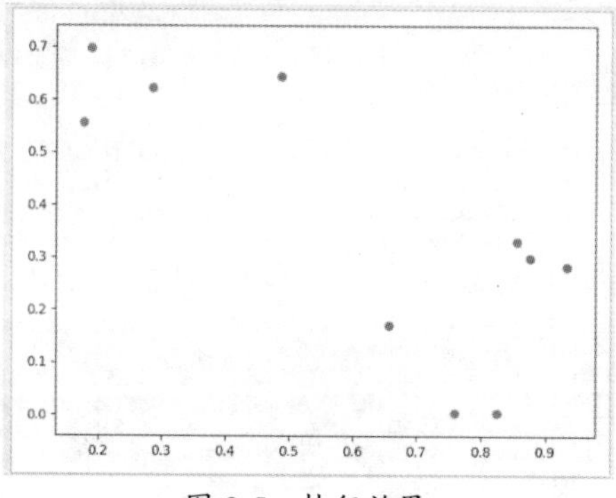

图 3-5　执行效果

3.2.4　修改散点的大小

请看下面的实例文件 dian04.py，功能是使用 matplotlib 函数 scatter() 绘制 10 个点并设置点的大小。

```
import matplotlib.pyplot as plt
import numpy as np

N = 10                                  # 10 个点
x = np.random.rand(N)
y = np.random.rand(N)

s = (30*np.random.rand(N))**2           # 每个点随机大小
```

```
plt.scatter(x, y, s=s)
plt.show()
```

在上述代码中，首先使用函数 random.rand() 分别生成了 10 个点的横坐标和纵坐标，然后通过 s 重新随机修改了点的大小，最后使用函数 scatter() 绘制这 10 个点。执行效果如图 3-6 所示。

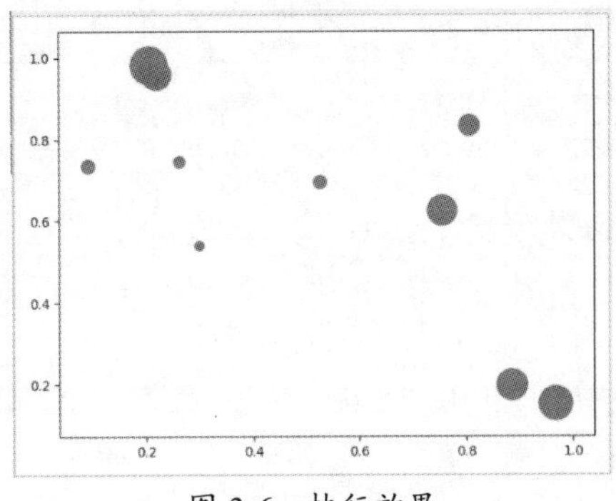

图 3-6　执行效果

3.2.5　设置散点的颜色和透明度

请看下面的实例文件 dian05.py，功能是使用 matplotlib 函数 scatter() 绘制散点图，分别设置散点的颜色和透明度。

```
import matplotlib.pyplot as plt
import numpy as np

# 10 个点
N = 10
x = np.random.rand(N)
y = np.random.rand(N)
s = (30*np.random.rand(N))**2                    # 每个点随机大小

c = np.random.rand(N)                            # 随机颜色
plt.scatter(x, y, s=s, c=c, alpha=0.5)
plt.show()
```

在上述代码中，随机设置了 10 个点的颜色，然后使用函数 scatter() 绘制这 10 个点，并通过 alpha 设置每个点的透明度。执行效果如图 3-7 所示。

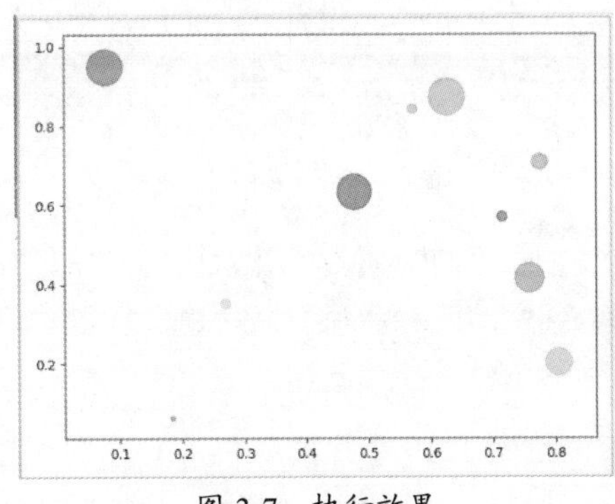

图 3-7　执行效果

3.2.6 修改散点的形状

请看下面的实例文件 dian06.py，功能是使用 matplotlib 函数 scatter() 绘制散点图，然后使用函数 scatter() 的属性 marker 设置散点的形状。

```
import matplotlib.pyplot as plt
import numpy as np

N = 10                                                    # 10 个点
x = np.random.rand(N)
y = np.random.rand(N)
s = (30*np.random.rand(N))**2
c = np.random.rand(N)
plt.scatter(x, y, s=s, c=c, marker='^', alpha=0.5)
plt.show()
```

在上述代码中，随机设置了 10 个点的颜色，然后使用函数 scatter() 绘制这 10 个点，并通过 alpha 设置每个点的透明度，使用 marker='^' 设置这 10 个点的形状为三角形。执行效果如图 3-8 所示。

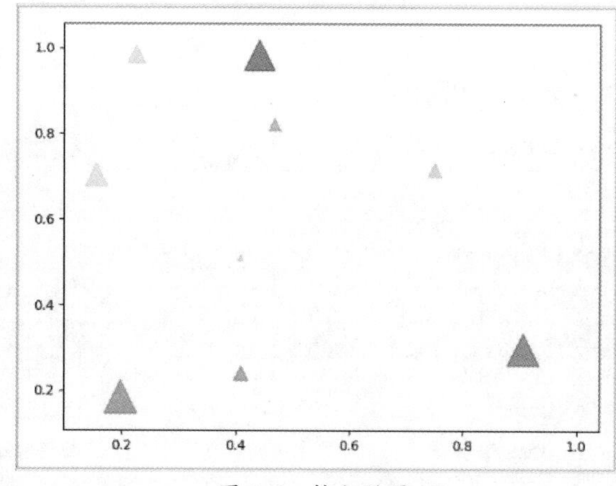

图 3-8 执行效果

3.2.7 绘制两组数据的散点图

请看下面的实例文件 dian07.py，功能是使用 matplotlib 函数 scatter() 绘制两组数据的散点图，然后使用函数 scatter() 的属性 marker 设置这两组散点图的形状。

```
import matplotlib.pyplot as plt
import numpy as np
N = 10                                    # 10 个点
x1 = np.random.rand(N)
y1 = np.random.rand(N)
x2 = np.random.rand(N)
y2 = np.random.rand(N)
plt.scatter(x1, y1, marker='o')
plt.scatter(x2, y2, marker='^')
plt.show()
```

在上述代码中，首先随机设置了 10 个点的颜色，并分别创建两组数据 x1、y1 和 x2、y2，然后基于这两组数据绘制散点图并分别将 marker 设置为“o”和“^”，表示将两组散点图的形状设置为三角形分别设置为圆形和三角形。执行效果如图 3-9 所示。

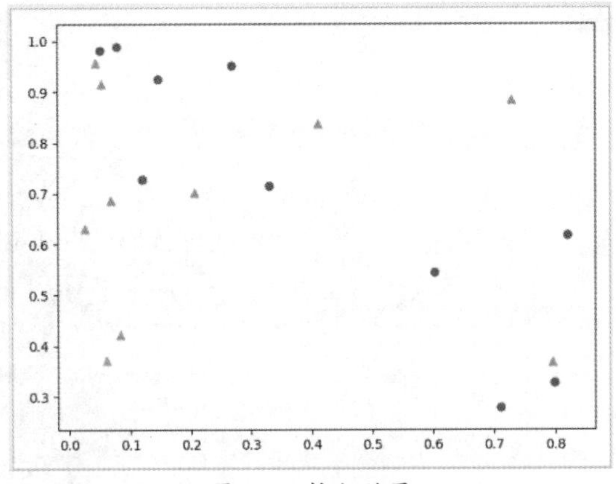

图 3-9　执行效果

3.2.8　为散点图设置图例

图例是集中于地图一角或一侧的地图上各种符号和颜色所代表的内容与指标的说明，有助于更好地认识地图。请看下面的实例文件 dian03.py，首先使用函数 scatter() 绘制两组数据的散点图，然后使用 matplotlib 函数 scatter() 的属性 marker 设置这两组散点图的形状，最后用函数 scatter() 的属性 label 设置这两组散点图的图例。

```
import matplotlib.pyplot as plt
import numpy as np

N = 10                              # 10 个点
x1 = np.random.rand(N)
y1 = np.random.rand(N)
x2 = np.random.rand(N)
y2 = np.random.rand(N)
plt.scatter(x1, y1, marker='o', label="circle")
plt.scatter(x2, y2, marker='^', label="triangle")
plt.legend(loc='best')
plt.show()
```

执行效果如图 3-10 所示。

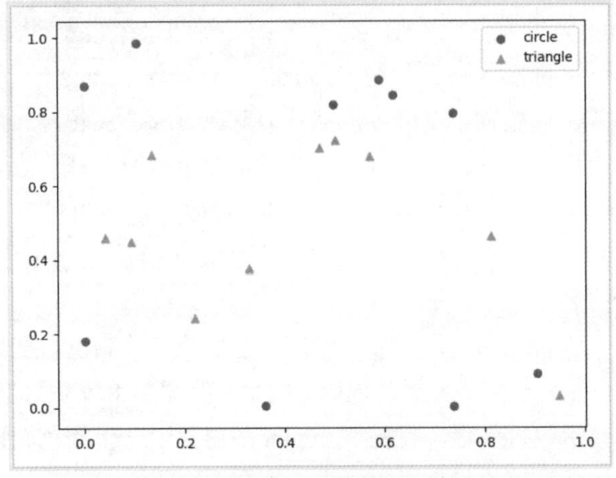

图 3-10　执行效果

3.2.9 自定义散点图样式

在现实应用中，经常需要绘制散点图并设置各个数据点的样式。例如，可能先以一种颜色显示较小的值，用另一种颜色显示较大的值。当绘制大型数据集时，还需要对每个点都设置同样的样式，再使用不同的样式选项重新绘制某些点，这样可以突出显示它们的效果。在matplotlib 库中，可以使用函数 scatter() 绘制单个点，通过传递 *x* 点和 *y* 点坐标的方式在指定的位置绘制一个点。例如在下面的实例文件 dian09.py 中，演示了使用 matplotlib 绘制指定样式散点图的过程。

```
①import matplotlib.pyplot as plt
②from pylab import *
③mpl.rcParams['font.sans-serif'] = ['SimHei'] #指定默认字体

④mpl.rcParams['axes.unicode_minus'] = False #解决保存图像是负号'-'显示为方块的问题
⑤x_values = list(range(1, 1001))
⑥y_values = [x**2 for x in x_values]

⑦plt.scatter(x_values, y_values, c=(0, 0, 0.8), edgecolor='red', s=40)

⑧# 设置图标标题，并设置坐标轴标签 .
⑨plt.title(" 大中华区销售统计表 ", fontsize=24)
⑩plt.xlabel(" 节点 ", fontsize=14)
⑪plt.ylabel(" 销售数据 ", fontsize=14)

⑫# 设置刻度大小 .
⑬plt.tick_params(axis='both', which='major', labelsize=14)

⑭# 设置每个坐标轴的取值范围
⑮plt.axis([0, 110, 0, 1100])

⑯plt.show()
```

下面我们对上述代码中的主要代码作一下解析。

（1）第 ②、③、④ 行代码：导入字体库，设置中文字体，并解决负号“-”显示为方块的问题。

（2）第 ⑤ 和第 ⑥ 行代码：使用 Python 循环实现自动计算数据功能，首先创建一个包含 *x* 值的列表，其中包含数字 1 ～ 1000。接下来创建一个生成 *y* 值的列表解析，它能够遍历 *x* 值（for x in x_values），计算其平方值（*x***2），并将结果存储到列表 y_values 中。

（3）第 ⑦ 行代码：将输入列表和输出列表传递给函数 scatter()。另外，因为 matplotlib 允许给散列点图中的各个点设置一个颜色，默认为蓝色点和黑色轮廓。所以当在散列点图中包含的数据点不多时效果会很好。但是当需要绘制很多个点时，这些黑色的轮廓可能会粘连在一起，此时需要删除数据点的轮廓。所以在本行代码中，在调用函数 scatter() 时传递了实参：edgecolor='none'。为了修改数据点的颜色，在此向函数 scatter() 传递参数 c，并将其设置为要使用的颜色的名称“red”。

注意：颜色映射（Colormap）是一系列颜色，它们从起始颜色渐变到结束颜色。在可视化视图模型中，颜色映射用于突出数据的规律，例如可能需要用较浅的颜色来显示较小的值，并使用较深的颜色来显示较大的值。在模块 pyplot 中内置了一组颜色映射，要想使用这些颜色映射，需要告诉 pyplot 应该如何设置数据集中每个点的颜色。

（4）第 ⑮ 行代码：因为这个数据集较大，所以将点设置得较小，在本行代码中使用函数 axis() 指定了每个坐标轴的取值范围。函数 axis() 要求提供 4 个值：x 和 y 坐标轴的最小值和最大值。此处将 x 坐标轴的取值范围设置为 0 ～ 110，并将 y 坐标轴的取值范围设置为 0 ～ 1100。

（5）第 ⑯ 行（最后一行）代码：使用函数 plt.show() 显示绘制的图形。当然也可以让程序自动将图表保存到一个文件中，此时只需将对 plt.show() 函数的调用替换为对 plt.savefig() 函数的调用即可。

```
plt.savefig (' plot.png' , bbox_inches='tight' )
```

在上述代码中，第 1 个实参用于指定要以什么样的文件名保存图表，这个文件将存储到当前实例文件 dianyang.py 所在的目录中。第 2 个实参用于指定将图表多余的空白区域裁剪掉。如果要保留图表周围多余的空白区域，可以省略这个实参。

执行效果如图 3-11 所示。

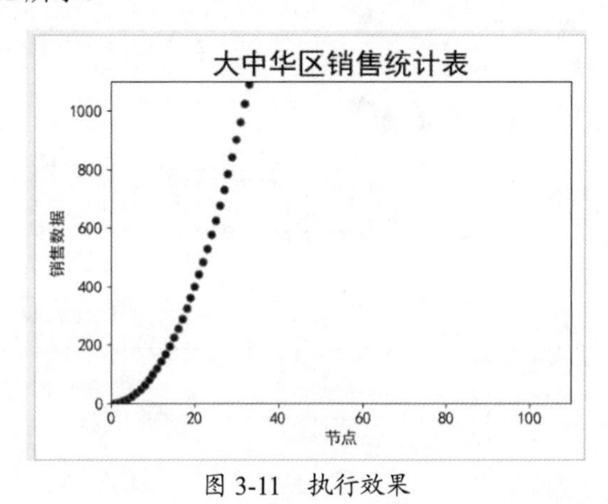

图 3-11　执行效果

3.3　绘制折线图

除了前面介绍的散点图外，在数据可视化工作中还经常需要绘制折线图。在数据分析中，折线图可以显示随时间（根据常用比例设置）而变化的连续数据，因此非常适用于显示在相等时间间隔下数据的趋势变化。在本节的内容中，将详细讲解在 Python 程序中使用各种库绘制折线图的知识。

↑ 扫码看视频（本节视频课程时间：9 分 08 秒）

3.3.1　绘制最简单的折线

我们可以使用库 matplotlib 绘制折线图，例如在下面的实例文件 zhe01.py 中，使用 matplotlib 绘制了一个简单的折线图，现实样式是默认的效果。

```
import matplotlib.pyplot as plt
squares = [1, 4, 9, 16, 25]
plt.plot(squares)
plt.show()
```

在上述实例代码中，使用平方数序列 1、4、9、16 和 25 来绘制一个折线图，在具体实现时，只需向 matplotlib 提供这些平方数序列数字就能完成绘制工作，工作流程如下：

（1）导入模块 pyplot，并给它指定了别名 plt，以免反复输入 pyplot，在模块 pyplot 中包含很多用于生成图表的函数；

（2）创建了一个列表，在其中存储了前述平方数；

（3）将创建的列表传递给函数 plot()，这个函数会根据这些数字绘制出有意义的图形；

（4）通过函数 plt.show() 打开 matplotlib 查看器，并显示绘制的图形。

执行效果如图 3-12 所示。

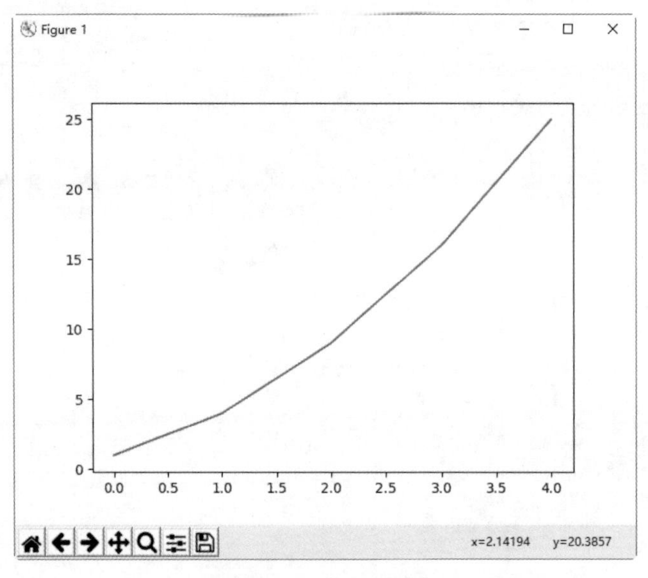

图 3-12　执行效果

3.3.2　设置标签文字和线条粗细

本章前面实例文件 zhe01.py 的执行效果不够完美，开发者可以对绘制的线条样式进行灵活设置。例如可以设置线条的粗细、实现数据准确性校正等操作。例如在下面的实例文件 zhe02.py 中，演示了使用 matplotlib 绘制指定样式折线图效果的过程。

```
①import matplotlib.pyplot as plt                          # 导入模块
②input_values = [1, 2, 3, 4, 5]
③squares = [1, 4, 9, 16, 25]
④plt.plot(input_values, squares, linewidth=5)
⑤# 设置图表标题，并在坐标轴上添加标签
⑥plt.title("Numbers", fontsize=24)
⑦plt.xlabel("Value", fontsize=14)
⑧plt.ylabel("ARG Value", fontsize=14)
⑨# 设置单位刻度的大小
⑩plt.tick_params(axis='both', labelsize=14)
⑪plt.show()
```

上述实例中参数设置较为复杂，我们来解析一下主要代码行。

（1）第 ④ 行代码中的“linewidth=5”：设置线条的粗细。

（2）第 ④ 行代码中的函数 plot()：当向函数 plot() 提供一系列数字时，它会假设第一个数据点对应的 x 坐标值为 0，但是实际上我们的第一个点对应的 x 值为 1。为改变这种默认行

为，可以给函数 plot() 同时提供输入值和输出值，这样函数 plot() 可以正确地绘制数据，因为同时提供了输入值和输出值，所以无须对输出值的生成方式进行假设，所以最终绘制出的图形是正确的。

（3）第 ⑥ 行代码中的函数 title()：设置图表的标题。

（4）第 ⑥ 到 ⑧ 行中的参数 fontsize：设置图表中的文字大小。

（5）第 ⑦ 行中的函数 xlabel() 和第 ⑧ 行中的函数 ylabel()：分别设置 x 轴的标题和 y 轴的标题。

（6）第 ⑩ 行中的函数 tick_params()：设置刻度样式，其中指定的实参将影响 x 轴和 y 轴上的刻度（axis='both'），并将刻度标记的字体大小设置为 14（labelsize=14）。

执行效果如图 3-13 所示。

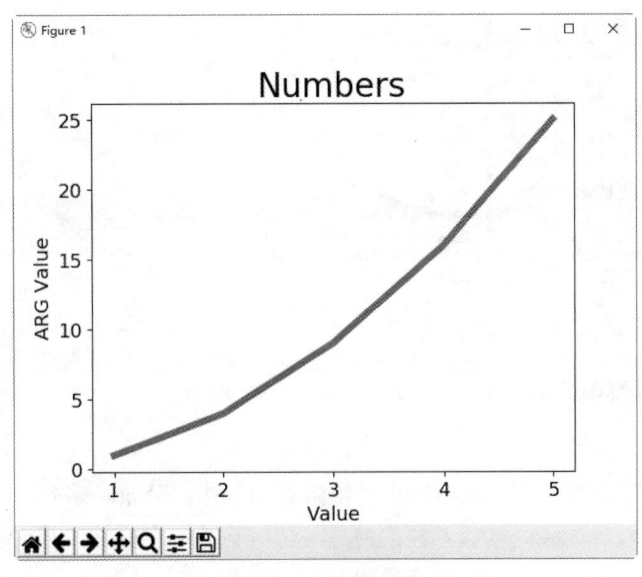

图 3-13　执行效果

3.3.3　绘制由 1 000 个点组成的折线图

请看下面的实例文件 zhe03.py，功能是使用 matplotlib 函数 scatter() 绘制 1 000 个散点组成折线图。

```
import matplotlib.pyplot as plt

x_values = range(1, 1001)
y_values = [x*x for x in x_values]
'''
scatter()
x:横坐标 y:纵坐标 s:点的尺寸
'''
plt.scatter(x_values, y_values, s=10)

# 设置图表标题并给坐标轴加上标签
plt.title('Square Numbers', fontsize=24)
plt.xlabel('Value', fontsize=14)
plt.ylabel('Square of Value', fontsize=14)
```

```
# 设置刻度标记的大小
plt.tick_params(axis='both', which='major', labelsize=14)

# 设置每个坐标轴的取值范围
plt.axis([0, 1100, 0, 1100000])
plt.show()
```

在上述代码中，函数 axis() 有 4 个参数值 [xmin, xmax, ymin, ymax]，分别表示 *x*、*y* 坐标轴的最小值和最大值。执行效果如图 3-14 所示。

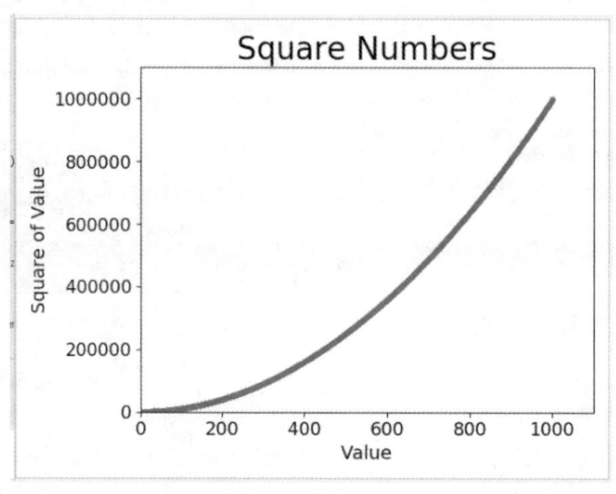

图 3-14　执行效果

3.3.4　绘制渐变色的折线图

使用颜色映射可以绘制渐变色的折线图，颜色映射是一系列颜色，它们从起始颜色渐变到结束颜色。在可视化中，颜色映射用于突出数据的规律。例如，你可以用较浅的颜色来显示较小的值，使用较深的颜色来显示较大的值。请看下面的实例文件 zhe04.py，功能是使用 matplotlib 函数 scatter() 绘制渐变色的折线图。

```
import matplotlib.pyplot as plt

x_values = range(1, 1001)
y_values = [x*x for x in x_values]
'''
scatter()
x:横坐标 y:纵坐标 s:点的尺寸
'''
plt.scatter(x_values, y_values, c=y_values, cmap=plt.cm.Blues, edgecolors='none', s=10)

# 设置图表标题并给坐标轴加上标签
plt.title('Square Numbers', fontsize=24)
plt.xlabel('Value', fontsize=14)
plt.ylabel('Square of Value', fontsize=14)

# 设置刻度标记的大小
plt.tick_params(axis='both', which='major', labelsize=14)

# 设置每个坐标轴的取值范围
plt.axis([0, 1100, 0, 1100000])
plt.show()
```

在上述代码中，将参数 c 设置成了一个 y 值列表，并使用参数 cmap 告诉 pyplot 使用哪个颜色映射。这些代码将 y 值较小的点显示为浅蓝色，并将 y 值较大的点显示为深蓝色。执行效果如图 3-15 所示。

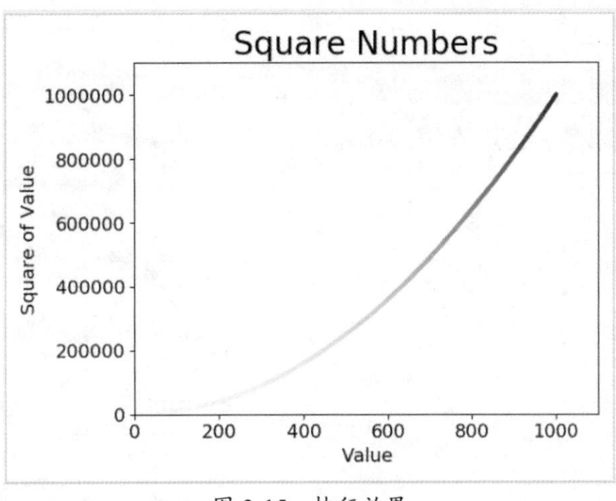

图 3-15　执行效果

3.3.5　绘制多幅子图

在 Matplotlib 绘图系统中，可以显式地控制图像、子图和坐标轴。matplotlib 库中的“图像”指的是用户界面看到的整个窗口内容。在图像里面有所谓“子图”，子图的位置是由坐标网格确定的，而“坐标轴”却不受此限制，可以放在图像的任意位置。当调用 plot() 函数的时候，matplotlib 调用 gca() 函数以及 gcf() 函数来获取当前的坐标轴和图像。如果无法获取图像，则会调用 figure() 函数来创建一个。从严格意义上来说，是使用 subplot（1，1，1）创建一个只有一个子图的图像。

在 matplotlib 绘图系统中，所谓“图像”就是 GUI 里以“Figure #”为标题的那些窗口。图像编号从 1 开始，与 MATLAB 的风格一致，而与 Python 从 0 开始编号的风格不同。表 3-1 中的参数是图像的属性。

表 3-1　图像的属性

参数	默认值	描述
num	1	图像的数量
figsize	figure.figsize	图像的长和宽（英寸）
dpi	figure.dpi	分辨率（点 / 英寸）
facecolor	figure.facecolor	绘图区域的背景颜色
edgecolor	figure.edgecolor	绘图区域边缘的颜色
frameon	True	是否绘制图像边缘

例如在下面的实例文件 zhe05.py 中，演示了让一个折线图和一个散点图同时出现在同一个绘画框中的过程。

```
    import matplotlib.pyplot as plt #将绘画框进行对象化
    fig=plt.figure() #将p1定义为绘画框的子图，211表示将绘画框划分为2行1列，最后的1表示第一
幅图
```

```
p1=fig.add_subplot(211)
x=[1,2,3,4,5,6,7,8]
y=[2,1,3,5,2,6,12,7]
p1.plot(x,y) #将p2定义为绘画框的子图，212表示将绘画框划分为2行1列，最后的2表示第二幅图
p2=fig.add_subplot(212)
a=[1,2]
b=[2,4]
p2.scatter(a,b)
plt.show()
```

上述代码执行后的效果如图 3-16 所示。

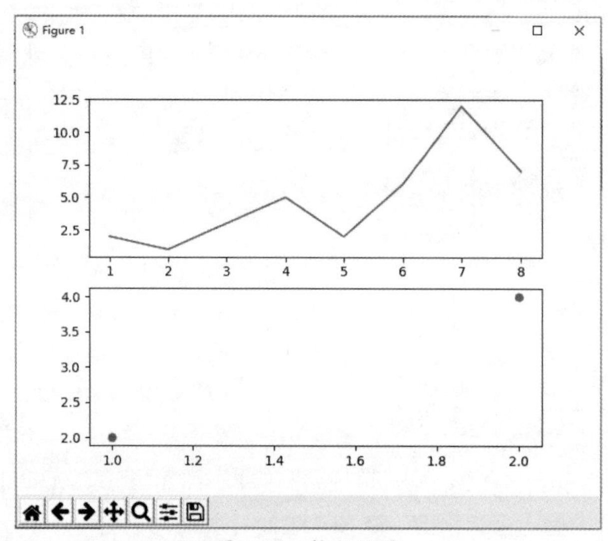

图 3-16　执行效果

3.3.6　绘制正弦函数和余弦函数曲线

在数据分析领域，经常需要正 / 余弦曲线表示周期性的数据。在 Python 程序中，最简单绘制曲线的方式是使用数学中的正弦函数或余弦函数。例如在下面的实例文件 zhe06.py 中，演示了使用正弦函数和余弦函数绘制曲线的过程。

```
from pylab import *
X = np.linspace(-np.pi, np.pi, 256,endpoint=True)
C,S = np.cos(X), np.sin(X)
plot(X,C)
plot(X,S)
show()
```

执行后的效果如图 3-17 所示。

在上述实例中，展示的是使用的 matplotlib 默认配置的效果。其实开发者可以调整大多数的默认配置，例如图片大小和分辨率（dpi）、线宽、颜色、风格、坐标轴、坐标轴以及网格的属性、文字与字体属性等。但是，matplotlib 的默认配置在大多数情况下已经做得足够好，开发人员可能只在很少的情况下才会想更改这些默认配置。例如在下面的实例文件 zhe07.py 中，展示了使用 matplotlib 的默认配置和自定义绘图样式的过程。在代码中设置 point 的分辨率为 80，设置使用宽度为 1 （像素）的蓝色连续线条绘制余弦曲线，使用宽度为 1 （像素）的绿色连续线条绘制正弦曲线。

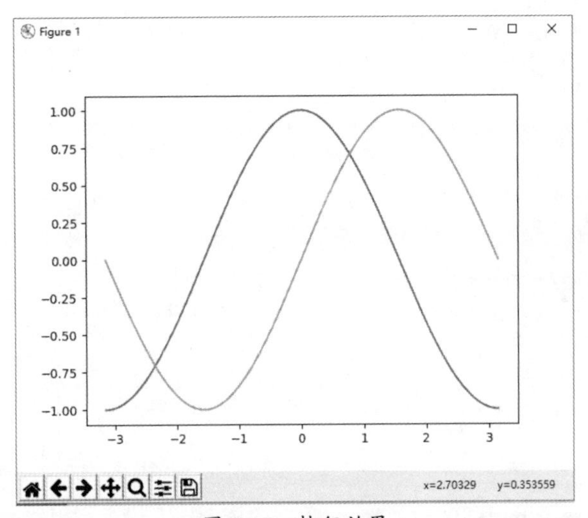

图 3-17　执行效果

```
# 导入 matplotlib 的所有内容（nympy 可以使用 np 这个名字）
from pylab import *
# 创建一个 8 * 6 点（point）的图，并设置分辨率为 80
figure(figsize=(8,6), dpi=80)
# 创建一个新的 1 * 1 的子图，接下来的图样绘制在其中的第 1 块（也是唯一的一块）
subplot(1,1,1)
X = np.linspace(-np.pi, np.pi, 256,endpoint=True)
C,S = np.cos(X), np.sin(X)
# 绘制余弦曲线，使用蓝色的、连续的、宽度为 1 （像素）的线条
plot(X, C, color="blue", linewidth=1.0, linestyle="-")
# 绘制正弦曲线，使用绿色的、连续的、宽度为 1 （像素）的线条
plot(X, S, color="green", linewidth=1.0, linestyle="-")
# 设置横轴的上下限
xlim(-4.0,4.0)
# 设置横轴记号
xticks(np.linspace(-4,4,9,endpoint=True))
# 设置纵轴的上下限
ylim(-1.0,1.0)
# 设置纵轴记号
yticks(np.linspace(-1,1,5,endpoint=True))
# 以分辨率 72 来保存图片
# savefig("exercice_2.png",dpi=72)
# 在屏幕上显示
show()
```

上述实例代码中的配置与默认配置完全相同，我们可以在交互模式中修改其中的值来观察效果。执行后的效果如图 3-18 所示。

在绘制曲线时可以改变线条的颜色和粗细，例如以蓝色和红色分别表示余弦和正弦函数，然后将线条变粗一点，接着在水平方向拉伸一下整个图。

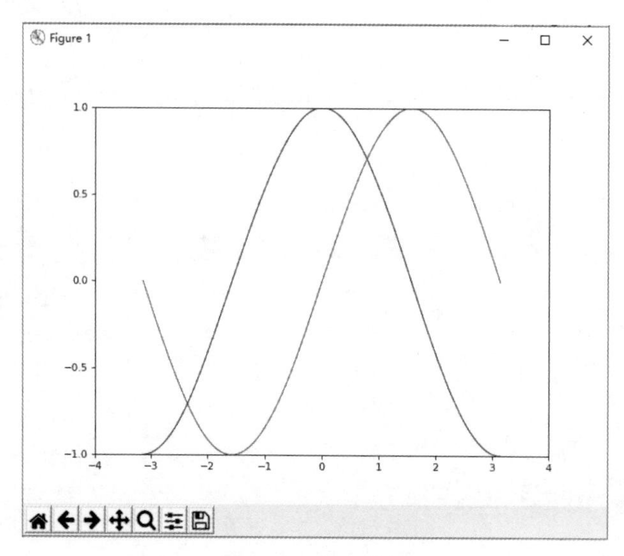

图 3-18　执行效果

```
...
figure(figsize=(10,6), dpi=80)
plot(X, C, color="blue", linewidth=2.5, linestyle="-")
plot(X, S, color="red",  linewidth=2.5, linestyle="-")
...
```

此时的执行效果如图 3-19 所示。

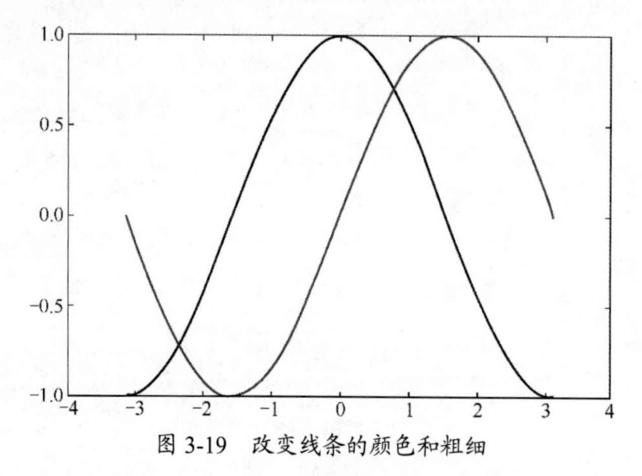

图 3-19　改变线条的颜色和粗细

请看下面的实例文件 zhe07-1.py，功能是使用 Matplotlib 分别绘制正弦曲线和余弦曲线，并分别在绘制的曲线上标注正弦和余弦。

```
import matplotlib.pyplot as plt   # 导入绘图模块
import numpy as np                #  导入需要生成数据的 numpy 模块

'''
第一种方式 text()
 text(x,y,s,fontdict=None, withdash=False)
    参数说明：（1）x,y 坐标位置
             (2) 显示的文本
'''
x = np.arange(0, 2 * np.pi, 0.01)
plt.plot(np.sin(x))
'''x,y 代表着坐标系中数值'''
```

```
plt.text(20, 0, 'sin(0) = 0')
'''
第二种方式　figtext()
    使用 figtext 时候，x,y 代表相对值，图片的宽度

'''
x2 = np.arange(0, 2 * np.pi, 0.01)
plt.plot(np.cos(x2))
''''''
plt.figtext(0.5, 0.5, 'cos(0)=0')
plt.show()
```

执行效果如图 3-20 所示。

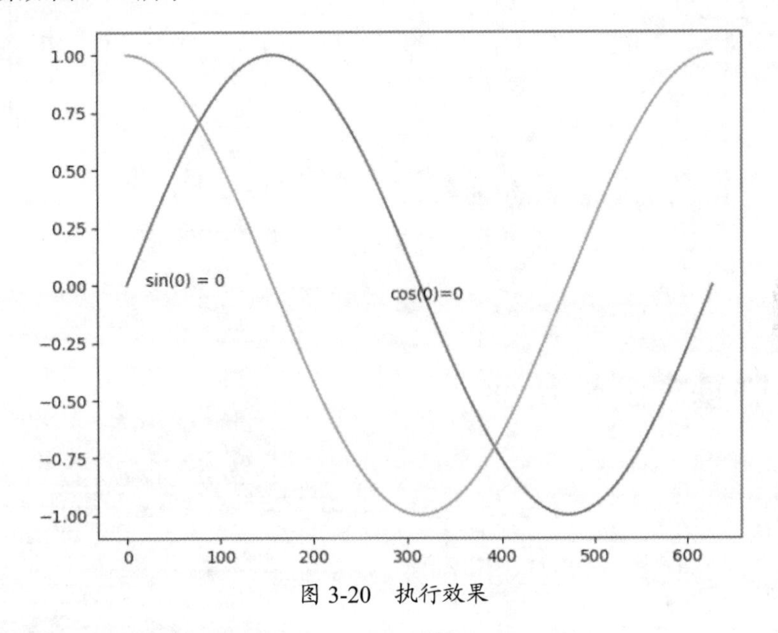

图 3-20　执行效果

3.3.7　绘制 3 条不同的折线

在使用 matplotlib 绘制折线时，可以使用函数 plot() 设置绘制不同样式的折线，例如红色虚线、蓝色正方形和绿色三角形等。请看下面的实例文件 zhe03.py，功能是使用 matplotlib 绘制 3 条不同的折线。

```
import numpy as np
import matplotlib.pyplot as plt

# 间隔 200ms 均匀采样
t = np.arange(0., 5., 0.2)

#红色虚线、蓝色正方形和绿色三角形
plt.plot(t, t, 'r--', t, t ** 2, 'bs', t, t ** 3, 'g^')
plt.show()
```

执行效果如图 3-21 所示。

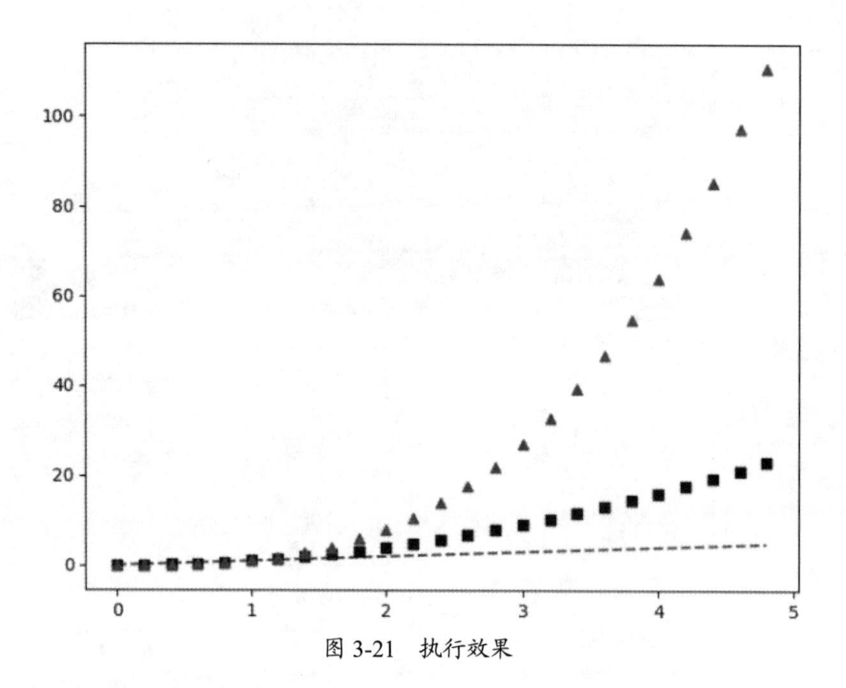

图 3-21 执行效果

3.4 绘制基本的柱状图

柱状图是一种长方形的，以长度为变量的统计图表，用途较为广泛。在数据可视化分析工作中，经常需要绘制和数据统计相关的柱状图。在本节的内容中，将详细讲解在 Python 程序中使用各种库绘基本柱状图的知识。

↑扫码看视频（本节视频课程时间：6 分 15 秒）

3.4.1 绘制只有一个柱子的柱状图

在 Python 程序中，可以使用 matplotlib 很容易地绘制出一个柱状图。例如在线的实例文件 bar01.py 中，只需使用 3 行代码就可以绘制出一个简单的柱状图。

```
import matplotlib.pyplot as plt
plt.bar(x = 0,height = 1)
plt.show()
```

在上述代码中，首先使用 import 导入了 matplotlib.pyplot ，然后直接调用其 bar() 函数绘制柱状图，最后使用 show() 函数显示图像。其中在函数 bar() 中存在如下两个参数：

- left：柱形的左边缘的位置，如果指定为 1，那么当前柱形的左边缘的 x 值就是 1.0；
- height：这是柱形的高度，也就是 y 轴的值。

执行上述代码后会绘制出一个柱状图，如图 3-22 所示。

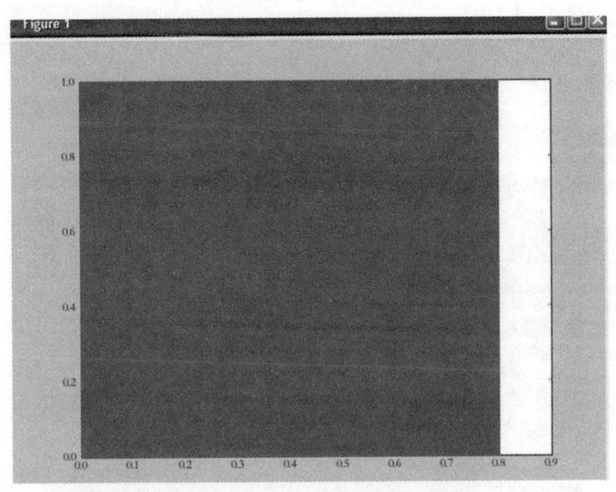

图 3-22　执行效果

3.4.2　绘制有两个柱形的柱状图

在具体实践中，有时需要绘制多个柱形来表示不同的可视化结果。在绘制函数 bar() 中，参数 *x* 和 height 除了可以使用单独的值（此时是一个柱形）外，还可以使用元组来替换（此时代表多个柱形）。例如在下面的实例文件 bar02.py 中，演示了使用 matplotlib 绘制两个柱形的柱状图的过程。

```
import matplotlib.pyplot as plt                    # 导入模块
plt.bar(x = (0,1),height = (1,0.5))                # 绘制两个柱形图
plt.show()                                         # 显示绘制的图
```

执行效果如图 3-23 所示。

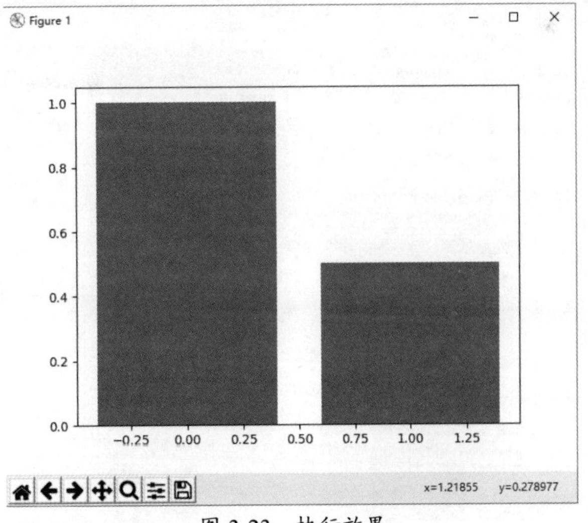

图 3-23　执行效果

在上述实例代码中，*x*＝（0,1）的意思是总共有两个矩形，其中第一个的左边缘为 0，第二个的左边缘为 1。参数 height 的含义也是同理。当然，此时有的读者可能觉得这两个矩形“太宽”了，不够美观。可以通过指定函数 bar() 中的 width 参数来设置它们的宽度。请看下面的实例文件 bar03.py，功能是设置柱子的宽度 width 为 0.35。

```
import matplotlib.pyplot as plt
plt.bar(x = (0,1),height = (1,0.5),width = 0.35)
plt.show()
```

执行效果如图 3-24 所示

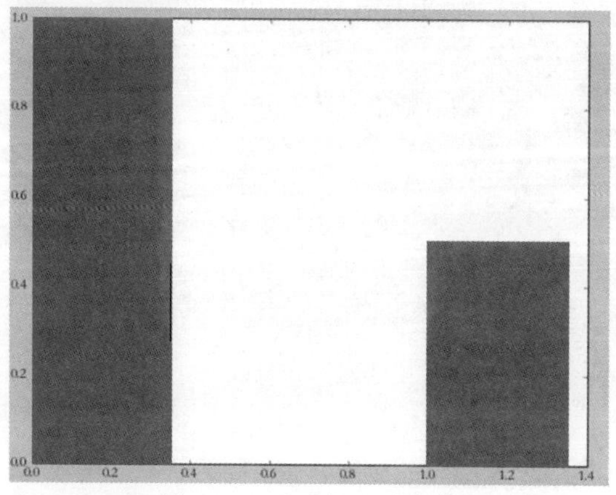

图 3-24　设置柱状图宽度

3.4.3　设置柱状图的标签

柱状图绘制完毕后，有的读者会问：如何标明 x 和 y 轴的说明信息呢，比如使用 x 轴表示性别，使用 y 轴表示人数。在下面的实例文件 bar04.py 中，演示了使用库 matplotlib 绘制出含有说明信息柱状图的过程。

```
import matplotlib.pyplot as plt
from pylab import *
mpl.rcParams['font.sans-serif'] = ['SimHei']        # 指定默认字体
mpl.rcParams['axes.unicode_minus'] = False          # 解决保存图像时负号 '-' 显示为方块的
问题
plt.xlabel(u'性别')                                 #x 轴的说明信息
plt.ylabel(u'人数')                                 #y 轴的说明信息
plt.bar(x = (0,1),height = (1,0.5),width = 0.35)
plt.show()
```

上述代码执行后的效果如图 3-25 所示。

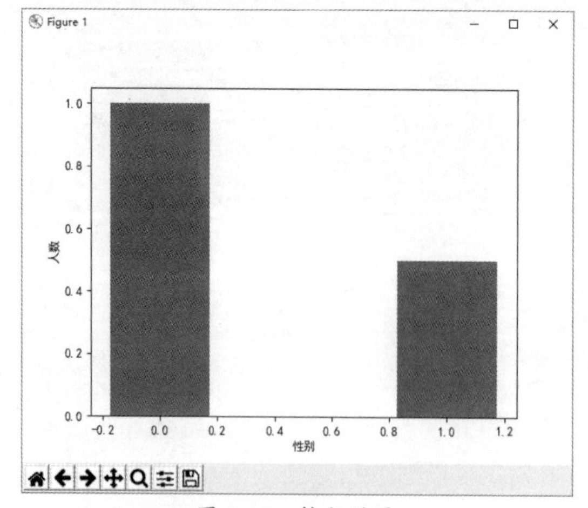

图 3-25　执行效果

接下来可以对 x 轴上的每个 bar 进行说明，例如设置第一个柱状图是“男”，第二个柱状图是“女”。此时可以通过如下实例文件 bar05.py 实现：

```
plt.xlabel(u'性别')
plt.ylabel(u'人数')
plt.xticks((0,1),(u'男',u'女'))
plt.bar(x = (0,1),height = (1,0.5),width = 0.35)
plt.show()
```

在上述代码中，函数 plt.xticks() 的用法和前面使用的 x 和 height 的用法差不多。如果你有几个 bar，那么就对应几维的元组，其中第一个参数表示文字的位置，第二个参数表示具体的文字说明。不过这里有个问题，例如有时我们指定的位置有些“偏移”，最理想的状态应该在每个矩形的中间。我们可以通过直接指定函数 bar() 里面的 align="center" 就可以让文字居中了。

```
plt.xlabel(u'性别')
plt.ylabel(u'人数')
plt.xticks((0,1),(u'男',u'女'))
plt.bar(x = (0,1),height = (1,0.5),width = 0.35,align="center")
plt.show()
```

此时的执行效果如图 3-26 所示。

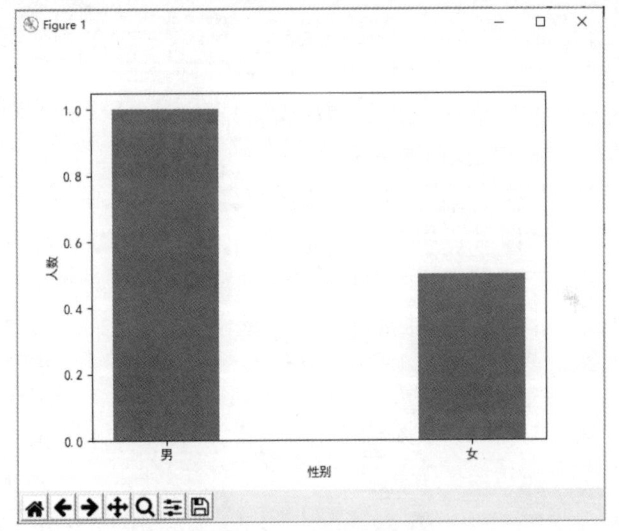

图 3-26　执行效果

接下来可以通过如下代码给柱状图表加入一个标题：

```
plt.title(u"性别比例分析")
```

为了使整个程序显得更加科学合理，接下来我们可以通过如下代码设置一个图例：

```
plt.xlabel(u'性别')
plt.ylabel(u'人数')
plt.title(u"性别比例分析")
plt.xticks((0,1),(u'男',u'女'))
rect = plt.bar(left = (0,1),height = (1,0.5),width = 0.35,align="center")
plt.legend((rect,),(u"图例",))
plt.show()
```

在上述代码中用到了函数 legend()，里面的参数必须是元组。即使只有一个图例也必须是元组，否则显示不正确。此时的执行效果如图 3-27 所示。

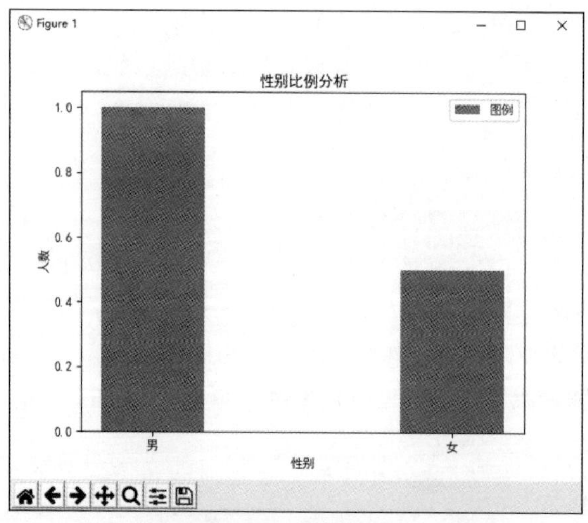

图 3-27　执行效果

接下来还可以在每个矩形的上面标注对应的 y 值，此时需要使用如下所示的方法实现：

```
    def autolabel(rects):
        for rect in rects:
            height = rect.get_height()
                plt.text(rect.get_x()+rect.get_width()/2., 1.03*height, '%s' %
float(height))
```

在上述实例代码中，其中 plt.text 有三个参数，分别是：x 坐标、y 坐标以及要显示的文字。调用函数 autolabel() 的具体实现代码如下所示：

```
    autolabel(rect)
```

为了避免绘制矩形柱状图紧靠着顶部，最好能够空出一段距离，此时可以通过函数 bar() 的属性参数 yerr 来设置。一旦设置了这个参数，那么对应的矩形上面就会有一个竖着的线。当把 yerr 这个值设置得很小的时候，上面的空白就自动出来了。

```
    rect = plt.bar(left = (0,1),height = (1,0.5),width = 0.35,align="center",ye
rr=0.0001)
```

到此为止，一个比较美观的柱状图绘制完毕，将代码整理并保存在如下实例文件 bar06.py 中，具体实现代码如下所示：

```
    import matplotlib.pyplot as plt
    from pylab import *
    mpl.rcParams['font.sans-serif'] = ['SimHei']    # 指定默认字体
    mpl.rcParams['axes.unicode_minus'] = False  # 解决保存图像时负号 '-' 显示为方块的问题
    def autolabel(rects):
        for rect in rects:
            height = rect.get_height()
                plt.text(rect.get_x()+rect.get_width()/2., 1.03*height, '%s' %
float(height))
    plt.xlabel(u' 性别 ')
    plt.ylabel(u' 人数 ')
    plt.title(u" 性别比例分析 ")
    plt.xticks((0,1),(u' 男 ',u' 女 '))
    # 绘制柱形图
    rect = plt.bar(x = (0,1),height = (1,0.5),width = 0.35,align="center",ye
rr=0.0001)
    plt.legend((rect,),(u" 图例 ",))
    autolabel(rect)
    plt.show()
```

上述代码执行后的效果如图 3-28 所示。

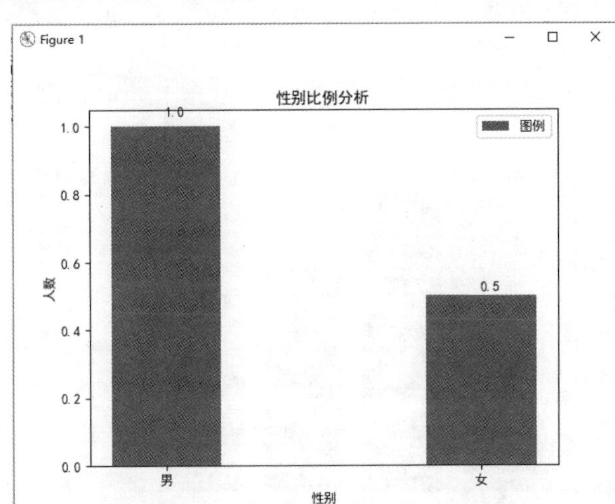

图 3-28　执行效果

3.4.4　设置柱状图的颜色

在柱状图的可视化显示中，颜色的加入是比较常用的。请看下面的实例文件 bar06-1.py，功能是使用库 matplotlib 绘制绿颜色的柱状图。

```
import matplotlib.pyplot as plt

data = [5, 20, 15, 25, 10]
plt.bar(range(len(data)), data, fc='g')
plt.show()
```

执行效果如图 3-29 所示。

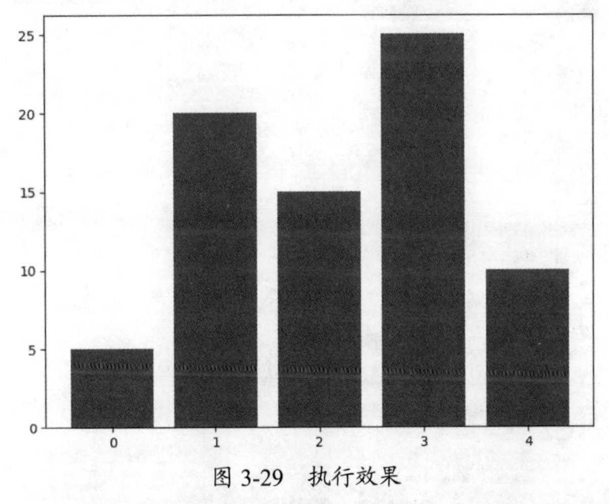

图 3-29　执行效果

我们再看下面的实例文件 bar06-2.py，功能是使用库 matplotlib 绘制不同颜色的柱状图。

```
import matplotlib.pyplot as plt

data = [5, 20, 15, 25, 10]
```

```
plt.bar(range(len(data)), data, color='rgb')  # or `color=['r', 'g', 'b']`
plt.show()
```

执行效果如图 3-30 所示。

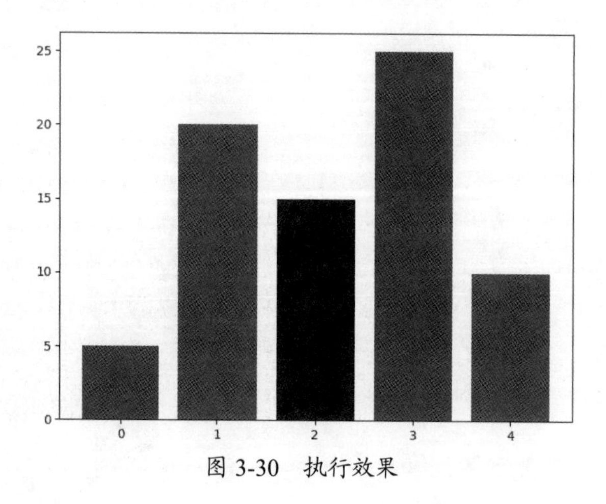

图 3-30　执行效果

3.4.5　绘制堆叠柱状图

堆叠柱状图是柱状图的变形，它可以清晰地比较某一个维度数据中不同类型数据之间的差异，亦可比较总数之间的差异。请看下面的实例文件 bar06-3.py，功能是使用库 matplotlib 绘制堆叠柱状图。

```
import numpy as np
import matplotlib.pyplot as plt

size = 5
x = np.arange(size)
a = np.random.random(size)
b = np.random.random(size)
plt.bar(x, a, label='a')
plt.bar(x, b, bottom=a, label='b')
plt.legend()
plt.show()
```

执行效果如图 3-31 所示。

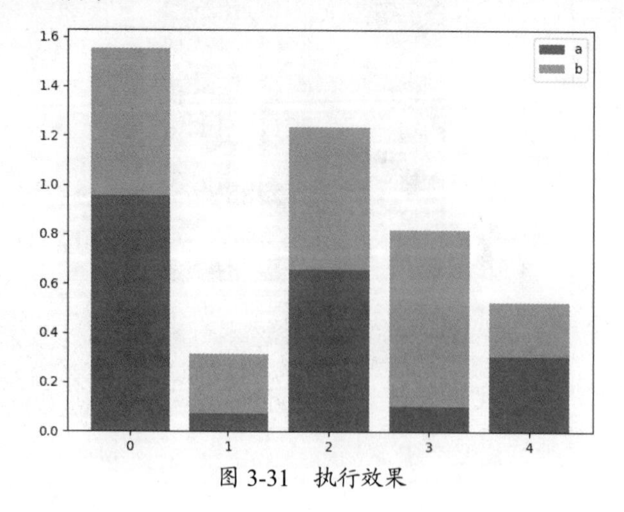

图 3-31　执行效果

3.4.6　绘制并列柱状图

请看下面的实例文件 bar06-4.py，功能是使用库 matplotlib 绘制并列的柱状图。

```
import numpy as np
import matplotlib.pyplot as plt

size = 5
x = np.arange(size)
a = np.random.random(size)
b = np.random.random(size)
c = np.random.random(size)

total_width, n = 0.8, 3
width = total_width / n
x = x - (total_width - width) / 2

plt.bar(x, a, width=width, label='a')
plt.bar(x + width, b, width=width, label='b')
plt.bar(x + 2 * width, c, width=width, label='c')
plt.legend()
plt.show()
```

执行效果如图 3-32 所示。

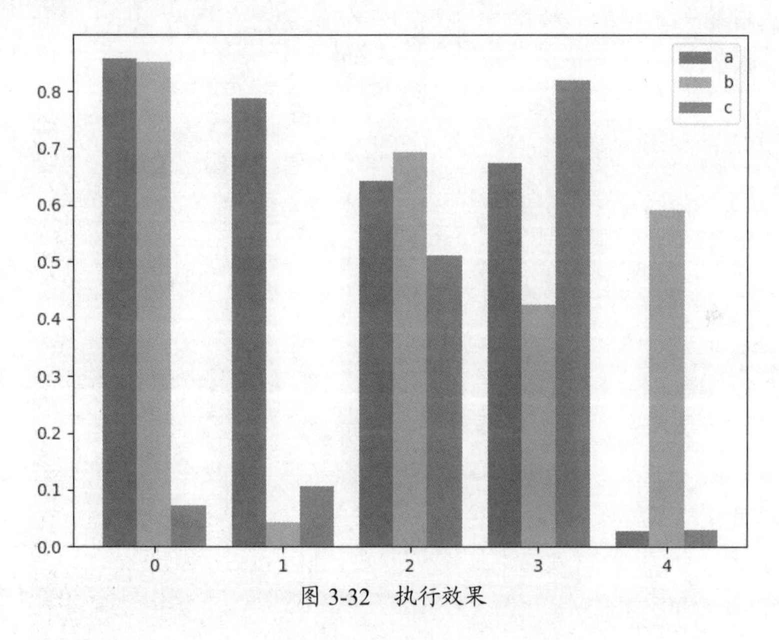

图 3-32　执行效果

3.5　绘制其他类型的散点图和折线图

除了在本章前面介绍的散点图和折线图外，还可以使用 Python 绘制功能更加强大的散点图和折线图。例如带有动画效果的统计图，基于爬虫数据绘制的统计图等。在本节的内容中，将详细讲解绘制其他类型的散点图和折线图的知识。

↑扫码看视频（本节视频课程时间：11 分 31 秒）

3.5.1 绘制随机漫步图

随机漫步（random walk）是一种数学统计模型，它由一连串轨迹组成。其中每一次都是随机的，它能用来表示不规则的变动形式。气体或液体中分子活动的轨迹等可作为随机漫步的模型。如同一个人酒后乱步所形成的随机记录。在 1903 年由卡尔·皮尔逊首次提出随机漫步这一概念，目前已经被广泛应用于生态学、经济学、心理学、计算科学、物理学、化学和生物学等领域，用来说明这些领域内观察到的行为和过程，是记录随机活动的基本模型。

1. 在 Python 程序中生成随机漫步数据

在 Python 程序中生成随机漫步数据后，可以使用 Matplotlib 以灵活方便的方式将这些数据展现出来。随机漫步的行走路径很有自己的特色，每次行走动作完全都是随机的，没有任何明确的方向，漫步结果是由一系列随机决策决定的。例如，漂浮在水滴上的花粉因不断受到水分子的挤压而在水面上移动。水滴中的分子运动是随机的，因此花粉在水面上的运动路径犹如随机漫步。

为了在 Python 程序中模拟随机漫步的过程，在下面的实例文件 random_walk.py 中创建一个名为 RandomA 的类，此类可以随机地选择前进方向。类 RandomA 需要用到三个属性，其中一个是存储随机漫步次数的变量，其他两个是列表，分别用于存储随机漫步经过的每个点的 x 坐标和 y 坐标。

```
①from random import choice
②class RandomA():
③    """ 能够随机生成漫步数据的类 """
④    def __init__(self, num_points=5100):
⑤        """ 初始化随机漫步属性 """
⑥        self.num_points = num_points
⑦        # 所有的随机漫步开始于 (0, 0).
⑧        self.x_values = [0]
⑨        self.y_values = [0]
⑩    def shibai(self):
⑪        """ 计算在随机漫步中包含的所有的点 """
⑫        # 继续漫步，直到达到所需长度为止 .
⑬        while len(self.x_values) < self.num_points:
⑭            # 决定前进的方向，沿着这个方向前进的距离 .
⑮            x_direction = choice([1, -1])
⑯            x_distance = choice([0, 1, 2, 3, 4])
⑰            x_step = x_direction * x_distance
⑱            y_direction = choice([1, -1])
⑲            y_distance = choice([0, 1, 2, 3, 4])
⑳            y_step = y_direction * y_distance
㉑            # 不能原地踏步
㉒            if x_step == 0 and y_step == 0:
㉓                continue
㉔            # 计算下一个点的坐标，即 x 值和 y 值
㉕            next_x = self.x_values[-1] + x_step
㉖            next_y = self.y_values[-1] + y_step
㉗            self.x_values.append(next_x)
㉘            self.y_values.append(next_y)
```

在上述代码中，类 RandomA 包含两个函数：__init__ () 和 shibai()，其中前者实现初始化处理；后者用于计算随机漫步经过的所有点。

接下来，我们详细分析一下上述的代码段。

（1）为了能够做出随机决策，首先将所有可能的选择都存储在一个列表中。在每次做

出具体决策时，通过“from random import choice”代码使用函数 choice() 来决定使用哪种选择。

（2）接下来将随机漫步包含的默认点数设置为 5 100，这个数值能够确保足以生成有趣的模式，同时也能够确保快速地模拟随机漫步。

（3）在第 8 行和第 9 行代码中创建了两个用于存储 x 和 y 值的列表，并设置每次漫步都从点（0，0）开始出发。

（4）第 13 行：使用 while 语句建立了一个循环，这个循环可以不断运行，直到漫步包含所需数量的点为止。这个函数的主要功能是告知 Python 应该如何模拟四种漫步决定：向右走还是向左走？沿指定的方向走多远？向上走还是向下走？沿选定的方向走多远？

（5）第 15 行：使用 choice([1, −1]) 给 x_direction 设置一个值，在漫步时要么表示向右走的 1，要么表示向左走的 −1。

（6）第 16 行：使用 choice([0, 1, 2, 3, 4]) 随机地选择一个 0 ～ 4 之间的整数，告诉 Python 沿指定的方向走的距离（x_distance）。通过包含 0，不但可以沿两个轴进行移动，而且还可以沿着 y 轴进行移动。

（7）第 17 行到第 20 行，将移动方向乘以移动距离，以确定沿 x 轴移动的距离。如果 x_step 为正则向右移动，如果为负则向左移动，如果为 0 则垂直移动；如果 y_step 为正则向上移动，如果为负则向下移动，如果为零则水平移动。

（8）第 22 行和第 23 行：开始执行下一次循环。如果 x_step 和 y_step 都为零则原地踏步，在我们的程序中必须杜绝这种原地踏步的情况发生。

（9）第 25 到第 28 行：为了获取漫步中下一个点的 x 值，将 x_step 与 x_values 中的最后一个值相加，对 y 值进行相同的处理。获得下一个点的 x 值和 y 值之后，将它们分别附加到列表 x_values 和 y_values 的末尾。

2. 在 Python 程序中绘制随机漫步图

在前面的实例文件 random_walk.py 中，已经创建了一个名为 RandomA 的类。在下面的实例文件 yun.py 中，可以借助于 matplotlib 将类 RandomA 中生成的漫步数据绘制出来，最终生成一个随机漫步图。

```
①import matplotlib.pyplot as plt
②from random_walk import RandomA
③# 只要当前程序是活动的，就要不断模拟随机漫步过程
④while True:
⑤    # 创建一个随机漫步实例，将包含的点都绘制出来
⑥    rw = RandomA(51000)
⑦    rw.shibai()
⑧    # 设置绘图窗口的尺寸大小
⑨    plt.figure(dpi=128, figsize=(10, 6))
⑩    point_numbers = list(range(rw.num_points))
⑪    plt.scatter(rw.x_values, rw.y_values, c=point_numbers, cmap=plt.cm.Blues,
⑫        edgecolors='none', s=1)
⑬    # 用特别的样式（红色、绿色和粗点）突出起点和终点 .
⑭    plt.scatter(0, 0, c='green', edgecolors='none', s=100)
⑮    plt.scatter(rw.x_values[-1], rw.y_values[-1], c='red', edgecolors='none',
⑯        s=100)
⑰    # 隐藏坐标轴 .
⑱    plt.axes().get_xaxis().set_visible(False)
⑲    plt.axes().get_yaxis().set_visible(False)
```

```
⑳        plt.show()
㉑        keep_running = input("哥，还继续漫步吗？ (y/n): ")
㉒        if keep_running == 'n':
㉓            break
```

（1）第 1 行和第 2 行：分别导入模块 pyplot 和前面编写的类 RandomA。

（2）第 5 行：创建一个 RandomA 实例，并将其存储到 rw 中。

（3）第 6 行：设置点数的数目。

（4）第 7 行：调用函数 shibai()。

（5）第 9 行：使用函数 figure() 设置图表的宽度、高度、分辨率。

（6）第 10 行：使用颜色映射来指出漫步中各点的先后顺序，并删除每个点的黑色轮廓，这样可以让它们的颜色显示得更加明显。为了根据漫步中各点的先后顺序进行着色，需要传递参数 c，并将其设置为一个列表，其中包含各点的先后顺序。由于这些点是按顺序绘制的，因此给参数 c 指定的列表只需包含数字 1 ～ 51 000 即可。使用函数 range() 生成一个数字列表，其中包含的数字个数与漫步包含的点数相同。接下来，我们将这个列表存储在 point_numbers 中，以便后面使用它来设置每个漫步点的颜色。

（7）第 11 行到第 12 行：将随机漫步包含的 *x* 和 *y* 值传递给函数 scatter()，并选择合适的点尺寸。将参数 c 设置为在第 10 行中创建的 point_numbers，用于设置使用颜色映射 Blues，并传递实参 edgecolors='none' 删除每个点周围的轮廓。

（8）第 14 到第 16 行：在绘制随机漫步图后重新绘制起点和终点，目的是突出显示随机漫步过程中的起点和终点。在程序中让起点和终点变得更大，并用不同的颜色显示。为了实现突出显示的功能，使用绿色绘制点（0，0），设置这个点比其他的点都粗大（设置为 s=100）。在突出显示终点时，在漫步包含的最后一个坐标的 *x* 值和 *y* 值的位置绘制一个点，并设置它的颜色是红色，并将其粗大值 s 设置为 100。

（9）第 18 行和第 19 行：隐藏图表中的坐标轴，使用函数 plt.axes() 将每条坐标轴的可见性都设置为 False。

（10）第 21 行到第 23 行：实现模拟多次随机漫步功能，因为每次随机漫步都不同，要想在不多次运行程序的情况下使用前面的代码实现模拟多次随机漫步的功能，最简单的办法是将这些代码放在一个 while 循环中。这样通过本实例模拟一次随机漫步后，在 matplotlib 查看器中可以浏览漫步结果，接下来可以在不关闭查看器的情况下暂停程序的执行，并询问你是否要再模拟一次随机漫步。如果输入 y 则可以模拟多次随机漫步。这些随机漫步都在起点附近进行，大多数是沿着特定方向偏离起点，漫步点分布不均匀等。要结束程序的运行，只需输入 n 即可。

本实例最终执行后的效果如图 3-33 所示。

图 3-33　执行效果

3.5.2　数据可视化分析某地的天气情况

假设存在一个 CSV 文件“death_valley_2014.csv”，在里面保存了 2014 年某地全年每一天各个时段的温度，下面开始可视化分析这个 CSV 文件。

（1）可视化展示 2014 年 4 月的温度

编写 Python 文件 4month.py，使用库 matplotlib 绘制统计折线图，可视化显示 2014 年 4 月的温度。文件 4month.py 的具体实现代码如下所示：

```
import csv
from datetime import datetime
from matplotlib import pyplot as plt
plt.rcParams['font.sans-serif'] = ['SimHei'] # 指定默认字体
plt.rcParams['axes.unicode_minus'] = False # 解决保存图像是负号 '-' 显示为方块的问题
filename='./csv/death_valley_2014.csv'
with open(filename,'r')as file:
    #1.创建阅读器对象
    reader=csv.reader(file)
    #2.读取文件头信息
    header_row=next(reader)

    #3.保存最高气温数据
    dates,hights=[],[]
    for row in reader:
    current_date=datetime.strptime(row[0],"%Y-%m-%d")
    dates.append(current_date)
    #4.将字符串转换为整型数据
    hights.append(row[1])
    #5.根据数据绘制图形
    fig=plt.figure(dpi=128,figsize=(10,6))
    #6.将列表 hights 传给 plot() 方法
    plt.plot(dates,hights,c='red')
    #7.设置图形的格式
    plt.title('2014 年 4 月份的温度 ',fontsize=24)
    plt.xlabel('',fontsize=26)
    # 3.绘制斜线日期标签
    fig.autofmt_xdate()
    plt.ylabel(' 华氏摄氏度 F',fontsize=16)
    plt.tick_params(axis='both',which='major',labelsize=16)
    plt.show()
```

本实例的执行效果如图 3-34 所示。

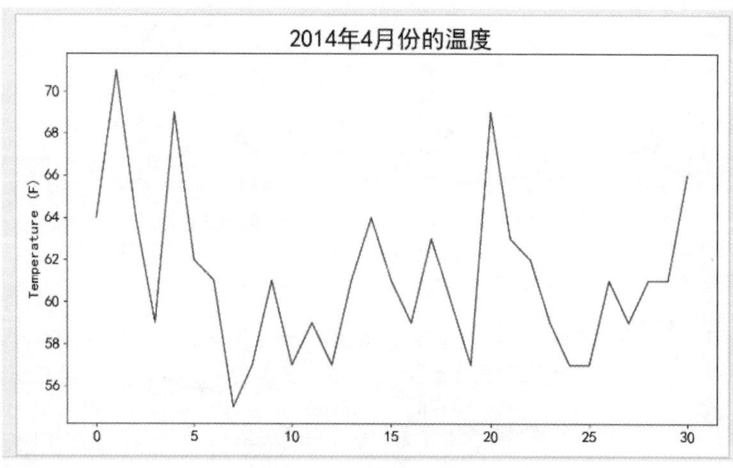

图 3-34　执行效果

（2）可视化展示全年天气数据

编写 Python 文件 year.py，使用库 matplotlib 绘制统计折线图，可视化显示 2014 年全年的温度。文件 year.py 的具体实现代码如下所示：

```
import csv
from datetime import datetime
from matplotlib import pyplot as plt
plt.rcParams['font.sans-serif'] = ['SimHei'] # 指定默认字体
plt.rcParams['axes.unicode_minus'] = False # 解决保存图像是负号 '-' 显示为方块的问题
filename='./csv/death_valley_2014.csv'
with open(filename,'r')as file:
    #1. 创建阅读器对象
    reader=csv.reader(file)
    #2. 读取文件头信息
    header_row=next(reader)

    #3. 保存最高气温数据
    dates,hights=[],[]
    for row in reader:
    current_date=datetime.strptime(row[0],"%Y-%m-%d")
    dates.append(current_date)
    #4. 将字符串转换为整型数据
    hights.append(row[1])
    #5. 根据数据绘制图形
    fig=plt.figure(dpi=128,figsize=(10,6))
    #6. 将列表 hights 传给 plot() 方法
    plt.plot(dates,hights,c='red')
    #7. 设置图形的格式
    plt.title('2014 全年的温度 ',fontsize=24)
    plt.xlabel('',fontsize=26)
    # 3. 绘制斜线日期标签
    fig.autofmt_xdate()
    plt.ylabel(' 华氏摄氏度 F',fontsize=16)
    plt.tick_params(axis='both',which='major',labelsize=16)
    plt.show()
```

本实例的执行效果如图 3-35 所示。

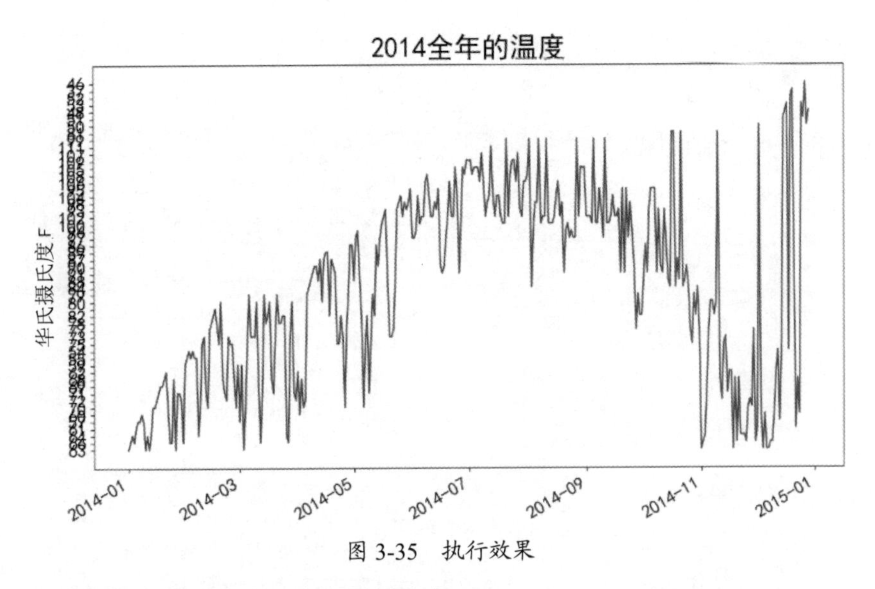

图 3-35　执行效果

（3）可视化展示某年最高温度和最低温度

编写 Python 文件 high_lows.py，使用库 matplotlib 绘制统计折线图，统计出 2014 年的最高温度和最低温度。文件 high_lows.py 的具体实现代码如下所示：

```
import csv
from matplotlib import pyplot as plt
from datetime import datetime

file = './csv/death_valley_2014.csv'
with open(file) as f:
    reader = csv.reader(f)
    header_row = next(reader)
    # 从文件中获取最高气温
    highs,dates,lows = [], [], []
    for row in reader:
        try:
            date = datetime.strptime(row[0],"%Y-%m-%d")
            high = int(row[1])
            low = int(row[3])
        except ValueError:
            print(date,'missing data')
        else:
            highs.append(high)
            dates.append(date)
            lows.append(low)

# 根据数据绘制图形
fig = plt.figure(figsize=(10,6))
plt.plot(dates,highs,c='r',alpha=0.5)
plt.plot(dates,lows,c='b',alpha=0.5)
plt.fill_between(dates,highs,lows,facecolor='b',alpha=0.2)
# # 设置图形的格式
plt.title('Daily high and low temperatures-2014',fontsize=16)
plt.xlabel('',fontsize=12)
fig.autofmt_xdate()
plt.ylabel('Temperature(F)',fontsize=12)
plt.tick_params(axis='both',which='major',labelsize=20)
plt.show()
```

执行后的效果如图 3-36 所示。

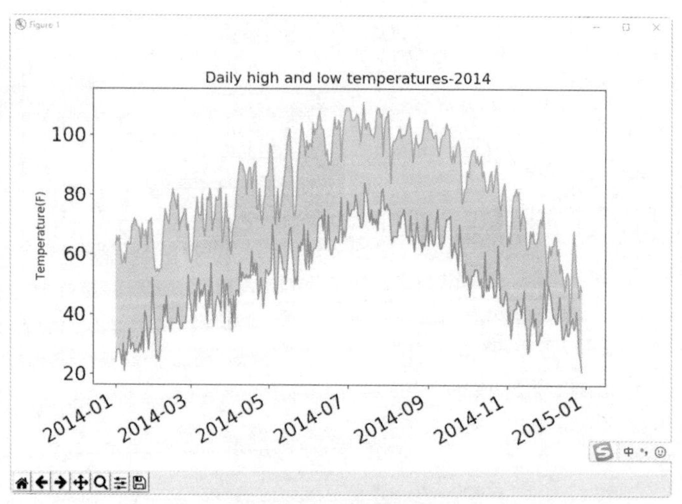

图 3-36　执行效果

3.5.3　在 Tkinter 中使用 matplotlib 绘制图表

在下面的实例文件 123.py 中，演示了在标准 GUI 程序 Tkinter 中使用 matplotlib 绘制图表的过程。首先通过函数 initialize(self) 设置了绘图的基本属性信息，绘制了默认大小刻度的图表；然后监听用户拖动滑动条，并根据滑动条的值绘制图表。

```
class App(tk.Tk):
    def __init__(self, parent=None):
        tk.Tk.__init__(self, parent)
        self.parent = parent
        self.initialize()

    def initialize(self):
        self.title("在 Tkinter 中使用 Matplotlib！")
        button = tk.Button(self, text="退出", command=self.on_click)
        button.grid(row=1, column=0)
        self.mu = tk.DoubleVar()
        self.mu.set(5.0)  #参数的默认值是"mu"
        slider_mu = tk.Scale(self,
                            from_=7, to=0, resolution=0.1,
                            label='mu', variable=self.mu,
                            command=self.on_change
                            )
        slider_mu.grid(row=0, column=0)
        self.n = tk.IntVar()
        self.n.set(512)  #参数的默认值是"n"
        slider_n = tk.Scale(self,
                           from_=512, to=2,
                           label='n', variable=self.n, command=self.on_change
                           )
        slider_n.grid(row=0, column=1)

        fig = Figure(figsize=(6, 4), dpi=96)
        ax = fig.add_subplot(111)
        x, y = self.data(self.n.get(), self.mu.get())
        self.line1, = ax.plot(x, y)
        self.graph = FigureCanvasTkAgg(fig, master=self)
        canvas = self.graph.get_tk_widget()
        canvas.grid(row=0, column=2)
```

```
    def on_click(self):
        self.quit()

    def on_change(self, value):
        x, y = self.data(self.n.get(), self.mu.get())
        self.line1.set_data(x, y)   # 更新 data 数据
        # 更新 graph
        self.graph.draw()

    def data(self, n, mu):
        lst_y = []
        for i in range(n):
            lst_y.append(mu * random.random())
        return range(n), lst_y

if __name__ == "__main__":
    app = App()
    app.mainloop()
```

执行后可以拖动左侧的滑动条控制绘制的图表，执行效果如图 3-37 所示。

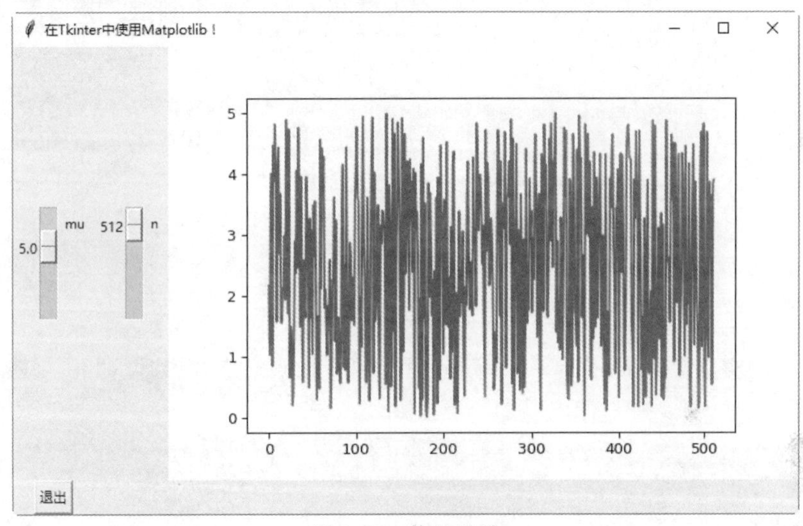

图 3-37　执行效果

3.5.4　绘制包含点、曲线、注释和箭头的统计图

请看下面的实例文件 jian.py，功能是使用 matplotlib 在绘制的曲线中添加注释和箭头。

```
import matplotlib.pyplot as plt   # 导入绘图模块
import numpy as np  # 导入需要生成数据的 numpy 模块

'''
添加注释  annotate()
    参数 : (1)x   : 注释文本
           (2)xy:
           (3) xytext:
           (4) 设置箭头, arrowprops
                 arrowprops : 是一个 dict (字典)
            第一种方式:{'width':宽度,'headwidth':箭头宽,'headlength':箭头长,
                     'shrink':两端收缩总长度分数}
                        例如:arrowprops={'width':5,'headwidth':10,'headlength':10,'
shrink':0.1}
            第二种方式:'arrowstyle':样式
```

```
                例如：
              有关arrowstyle的样式：'-' 、'->'、'<-'、'-['、'|-|'、'-|>'、'<|-'、'<->'
                                    'fancy','simple','wedge'
    '''
    x = np.random.randint(0,30,size=10)
    x[5] = 30  # 把索引为5的位置改为30
    plt.figure(figsize=(12,6))
    plt.plot(x)
    plt.ylim([-2,35]) # 设置y轴的刻度
    plt.annotate(s='this point is important',xy=(5,30),xytext=(7,31),
                arrowprops={'arrowstyle':'->'})
    plt.show()
```

执行效果如图3-38所示。

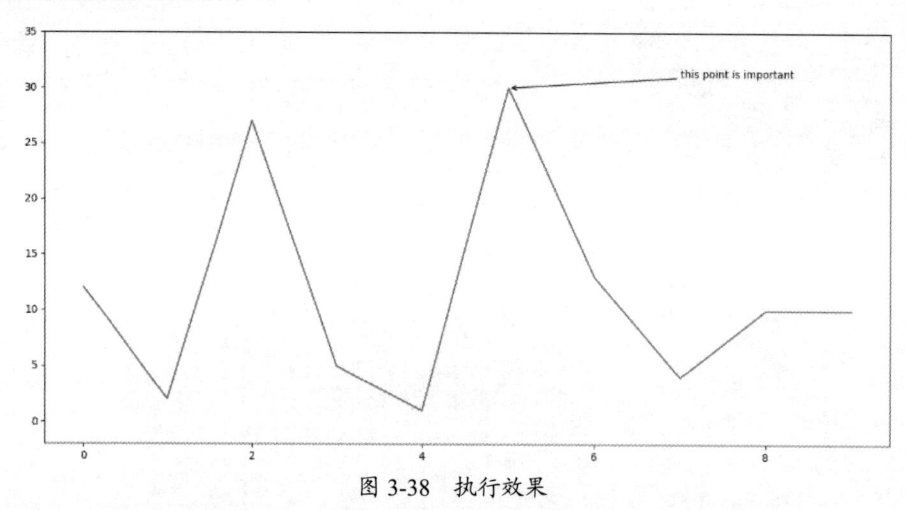

图3-38 执行效果

再看下面的实例文件zong.py，功能是使用matplotlib绘制包含点、曲线、注释和箭头的统计图。

```
    from matplotlib import pyplot as plt
    import numpy as np

    # 绘制曲线
    x = np.linspace(2, 21, 20)  # 取闭区间[2, 21]之间的等差数列，列表长度20
    y = np.log10(x) + 0.5
    plt.figure()  # 添加一个窗口。如果只显示一个窗口，可以省略该句。
    plt.plot(x, y)  # plot在一个figure窗口中添加一个图，绘制曲线，默认颜色

    # 绘制离散点
    plt.plot(x, y, '.y')  # 绘制黄色的点，为了和曲线颜色不一样
    x0, y0 = 15, np.log10(15) + 0.5
    plt.annotate('Interpolation point', xy=(x0, y0), xytext=(x0, y0 - 1), arrowprop
s=dict(arrowstyle='->'))  # 添加注释
    for x0, y0 in zip(x, y):
        plt.quiver(x0, y0 - 0.3, 0, 1, color='g', width=0.005)  # 绘制箭头

    x = range(2, 21, 5)
    y = np.log10(x) + 0.5
    plt.plot(x, y, 'om')  # 绘制紫红色的圆形的点
    x0, y0 = 7, np.log10(7) + 0.5
    plt.annotate('Original point', xy=(x0, y0), xytext=(x0, y0 - 1), arrowprops=dic
t(arrowstyle='->'))
    for x0, y0 in zip(x, y):
        plt.quiver(x0, y0 + 0.3, 0, -1, color='g', width=0.005)  # 绘制箭头
```

```
# 设置坐标范围
plt.xlim(2, 21)  # 设置 x 轴范围
plt.xticks(range(0, 23, 2))  # 设置 x 轴坐标点的值，为 [0, 22] 之间的以 2 为差值的等差数组
plt.ylim(0, 3)  # 设置 y 轴范围

# 显示图形
plt.show()  # 显示绘制出的图
```

对上述代码的具体说明如下：

（1）导入 matplotlib 模块的 pyplot 类，这里主要使用了 pyplot 里的一些方法。导入 numpy 用于生成一些数列，分别给 pyplot 和 numpy 记个简洁的别名 plt 和 n;

（2）通过方法 np.linspace(start, stop, num) 生成闭区间 [stop, stop] 里的数组长度为 num 的等差数列，在本例中想作为插值点显示出来；

（3）通过方法 plt.figure() 添加窗口。如果把所有图形绘制在一个窗口里则可以省略该行代码，因为 figure(1) 会被默认创建。如果想添加窗口，则需要再添加一行代码 plt.figure()，plt.figure(num) 的窗口序号 num 会自动自增加 1；

（4）通过方法 plt.plot() 在窗口中绘制曲线，传递 x, y 参数，分别表示横轴和纵轴。

本实例的执行效果如图 3-39 所示。

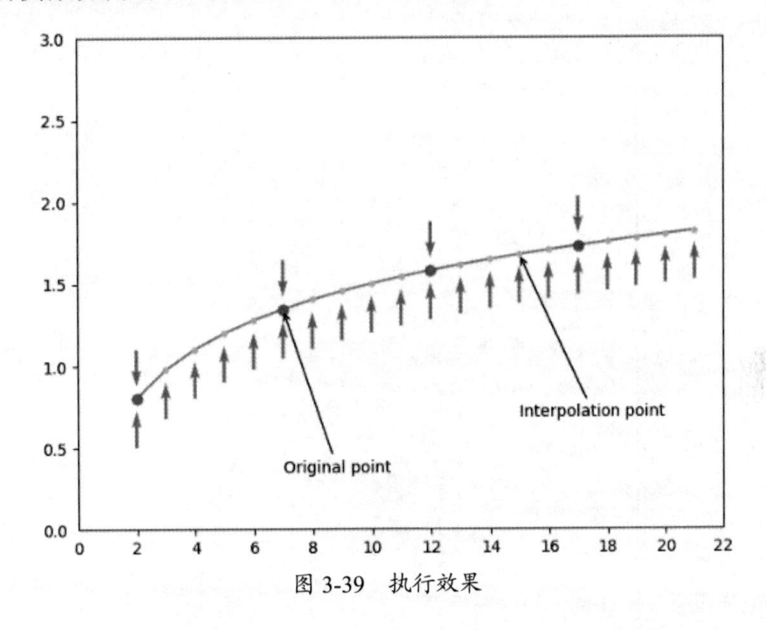

图 3-39　执行效果

3.5.5　在两栋房子之间绘制箭头指示符

请看下面的实例文件 zhishi.py，首先使用散点图表示房子的位置，然后在两栋房子（住宅 A 和住宅 B）之间绘制箭头指示符。

```
import numpy as np
import matplotlib.pyplot as plt

fig, ax = plt.subplots(figsize=(5, 5))
ax.set_aspect(1)

x1 = -1 + np.random.randn(100)
y1 = -1 + np.random.randn(100)
```

```
x2 = 1. + np.random.randn(100)
y2 = 1. + np.random.randn(100)

ax.scatter(x1, y1, color="r")
ax.scatter(x2, y2, color="g")

bbox_props = dict(boxstyle="round", fc="w", ec="0.5", alpha=0.9)
ax.text(-2, -2, "住宅 A", ha="center", va="center", size=20,
        bbox=bbox_props, fontproperties="SimSun")
ax.text(2, 2, "住宅 B", ha="center", va="center", size=20,
        bbox=bbox_props,fontproperties="SimSun")

bbox_props = dict(boxstyle="rarrow", fc=(0.8, 0.9, 0.9), ec="b", lw=2)
t = ax.text(0, 0, "Direction", ha="center", va="center", rotation=45,
            size=15,
            bbox=bbox_props)

bb = t.get_bbox_patch()
bb.set_boxstyle("rarrow", pad=0.6)

ax.set_xlim(-4, 4)
ax.set_ylim(-4, 4)

plt.show()
```

本实例的执行效果如图 3-40 所示。

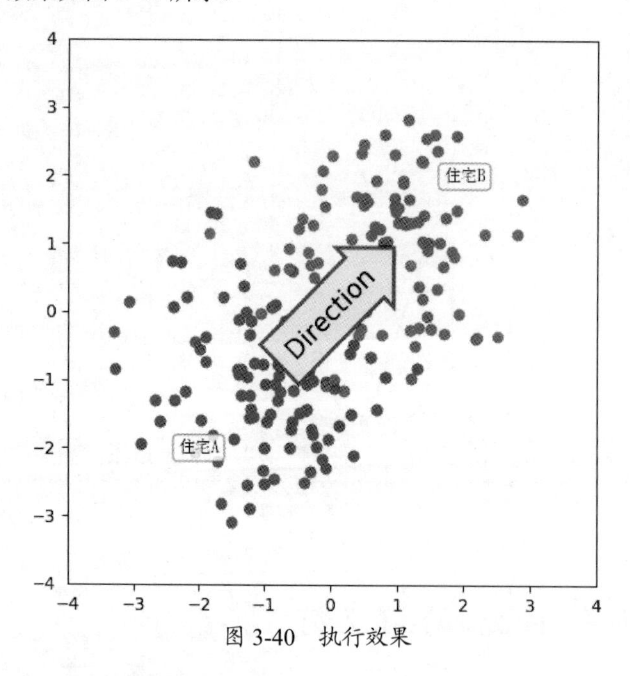

图 3-40　执行效果

3.5.6 根据坐标绘制行走路线图

请看下面的实例文件 road.py，功能是根据预先设置的位置坐标绘制行走路线图。

```
import matplotlib.pyplot as plt
import numpy
import matplotlib.colors as colors
import matplotlib.cm as cmx

_locations = [
```

```
        (4, 4),  # depot
        (4, 4),  # unload depot_prime
        (4, 4),  # unload depot_second
        (4, 4),  # unload depot_fourth
        (4, 4),  # unload depot_fourth
        (4, 4),  # unload depot_fifth
        (2, 0),
        (8, 0),  # locations to visit
        (0, 1),
        (1, 1),
        (5, 2),
        (7, 2),
        (3, 3),
        (6, 3),
        (5, 5),
        (8, 5),
        (1, 6),
        (2, 6),
        (3, 7),
        (6, 7),
        (0, 8),
        (7, 8)
    ]

    plt.figure(figsize=(10, 10))
    p1 = [l[0] for l in _locations]
    p2 = [l[1] for l in _locations]
    plt.plot(p1[:6], p2[:6], 'g*', ms=20, label='depot')
    plt.plot(p1[6:], p2[6:], 'ro', ms=15, label='customer')
    plt.grid(True)
    plt.legend(loc='lower left')

    way = [[0, 12, 18, 17, 16, 4, 14, 10, 11, 13, 5], [0, 6, 9, 8, 20, 3], [0, 19,
21, 15, 7, 2]]   #

    cmap = plt.cm.jet
    cNorm  = colors.Normalize(vmin=0, vmax=len(way))
    scalarMap = cmx.ScalarMappable(norm=cNorm,cmap=cmap)

    for k in range(0, len(way)):
        way0 = way[k]
        colorVal = scalarMap.to_rgba(k)
        for i in range(0, len(way0)-1):
            start = _locations[way0[i]]
            end = _locations[way0[i+1]]
    #           plt.arrow(start[0], start[1], end[0]-start[0], end[1]-start[1],
length_includes_head=True,
    #                   head_width=0.2, head_length=0.3, fc='k', ec='k', lw=2,
ls=lineStyle[k], color='red')
            plt.arrow(start[0], start[1], end[0]-start[0], end[1]-start[1],
                      length_includes_head=True, head_width=0.2, lw=2,
                      color=colorVal)
    plt.show()
```

在上述代码中，使用 _locations 保存了表示位置的坐标，使用 cmap 表示绘制的颜色库，通过 cNorm 设置颜色的范围，有几条线路就设置几种颜色，scalarMap 表示颜色生成完毕。在绘图时根据索引获得相应的颜色。本实例的执行效果如图 3-41 所示。

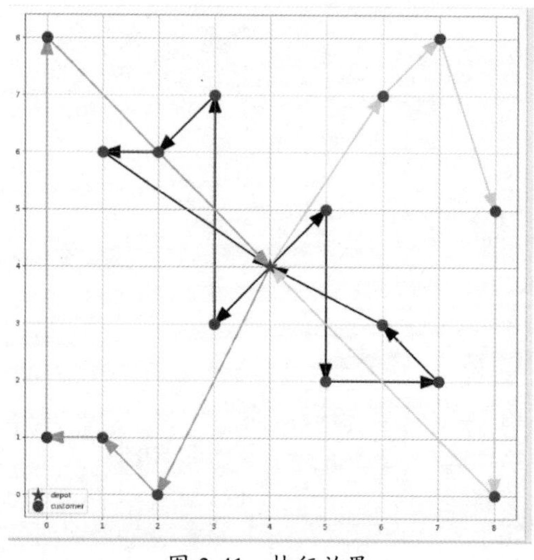

图 3-41　执行效果

3.5.7　绘制方程式曲线图

请看下面的实例文件 xian.py，功能是使用库 matplotlib 绘制如下两个方程式的曲线图。

$f(t)=e^{-t}cos(2t)$

$g(t)=cos(2t)sin(2t)\ cos(3t)$

具体实现代码如下：

```
import numpy as np
import matplotlib.pyplot as plt

def f1(t):
    return np.exp(-t) * np.cos(2 * np.pi * t)

def f2(t):
    return np.sin(2 * np.pi * t) * np.cos(3 * np.pi * t)

t = np.arange(0.0, 5.0, 0.02)

plt.figure(figsize=(8, 7), dpi=98)
p1 = plt.subplot(211)
p2 = plt.subplot(212)

label_f1 = "$f(t)=e^{-t} \cos (2 \pi t)$"
label_f2 = "$g(t)=\sin (2 \pi t) \cos (3 \pi t)$"

p1.plot(t, f1(t), "g-", label=label_f1)
p2.plot(t, f2(t), "r-.", label=label_f2, linewidth=2)

p1.axis([0.0, 5.01, -1.0, 1.5])

p1.set_ylabel("v", fontsize=14)
p1.set_title("A simple example", fontsize=18)
p1.grid(True)
# p1.legend()

tx = 2
ty = 0.9
```

```
    p1.text(tx, ty, label_f1, fontsize=15, verticalalignment="top",
horizontalalignment="right")

    p2.axis([0.0, 5.01, -1.0, 1.5])
    p2.set_ylabel("v", fontsize=14)
    p2.set_xlabel("t", fontsize=14)
    # p2.legend()
    tx = 2
    ty = 0.9
    p2.text(tx, ty, label_f2, fontsize=15, verticalalignment="bottom",
horizontalalignment="left")

    p2.annotate('', xy=(1.8, 0.5), xytext=(tx, ty), arrowprops=dict(arrowstyle="->",
connectionstyle="arc3"))

    plt.show()
```

在上述代码中，首先定义函数 f1(t) 返回 e-tcos(2πt) 的值，定义函数 f2(t) 返回 cos(2πt) sin(2πt) cos(3πt) 的值，然后分别调用函数 plt.subplot() 绘制两幅子图 p1 和 p2，子图 p1 和 p2 分别用于绘制两个函数的曲线图。本实例的执行效果如图 3-42 所示。

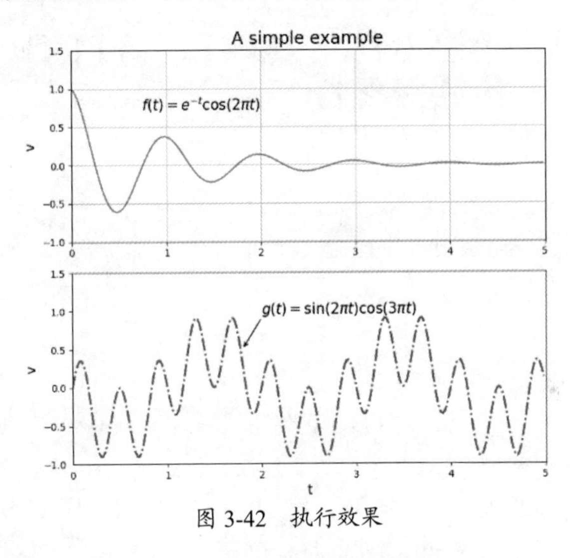

图 3-42　执行效果

3.5.8　绘制星空图

请看下面的实例文件 cloud.py，功能是使用库 matplotlib 绘制不同样式的小星星。

```
import matplotlib.pyplot as plt   # 导入绘图模块
import numpy as np  # 导入需要生成数据的 numpy 模块

x = np.random.randn(100)
y = np.random.randn(100)
''' 设置每一个点的颜色随机生成 '''
color = np.random.random(300).reshape((100, 3))  # 一千行三列
''' 设置每一个点的大小随机生成 '''
size = np.random.randint(0, 100, 100)
plt.scatter(x, y, color=color, s=size, marker='*')
plt.show()
```

执行效果如图 3-43 所示。

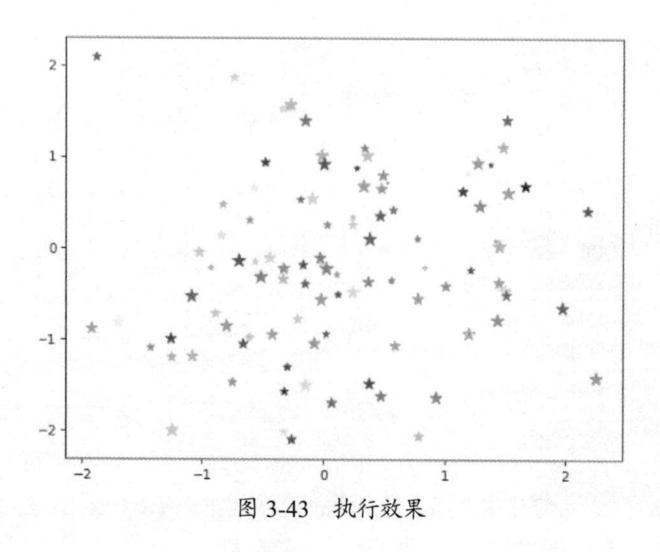

图 3-43　执行效果

3.6　实践案例：绘制 BTC（比特币）和 ETH（以太币）的价格走势图

在本节的内容中，将远程获取当前国际市场中 BTC（比特币）和 ETH（以太币）的实时价格，并绘制 BTC（比特币）和 ETH（以太币）的价格走势曲线图。

↑扫码看视频（本节视频课程时间：1 分 55 秒）

3.6.1　抓取数据

编写实例文件 Assignment_Step1.py，功能是抓取权威网站中 BTC 和 ETH 的报价数据，并打印输出显示 BTC 和 ETH 的当前价格。首先编写函数 price()，功能是获取参数 url 中的数据信息，使用 json() 解析 BTC 和 ETH 的报价数据。

```
import requests
def price(symbol, comparison_symbols=['USD'], exchange=''):
    url = 'https://min-api.cryptocompare.com/data/price?fsym={}&tsyms={}'\
            .format(symbol.upper(), ','.join(comparison_symbols).upper())
    if exchange:
        url += '&e={}'.format(exchange)
    page = requests.get(url)
    data = page.json()
    return data

print(" 当前 BTC 的美元价格为 : "+str(price('BTC')))
print(" 当前 ETH 的美元价格为 : "+str(price('ETH')))
```

2020 年 5 月 13 日执行本实例后会输出：

```
当前 BTC 的美元价格为 : {'USD': 8910.06}
当前 ETH 的美元价格为 : {'USD': 190.47}
```

3.6.2　绘制 BTC/ 美元价格曲线

编写实例文件 Assignment_Step2.py，功能是根据当前的 BTC 价格，使用库 matplotlib 绘制 BTC/ 美元价格曲线图。首先使用 from 语句导入文件 Assignment_Step1.py 中的函数 price()，然后通过函数 price() 获取 BTC 和 ETH 的报价数据，并根据报价绘制价格曲线图。

```
from Assignment_Step1 import price
import datetime
import matplotlib.pyplot as plt

x=[0]
y=[0]
fig = plt.gcf()
fig.show()
fig.canvas.draw()
plt.ylim([0, 20000])
i=0
while(True):
    data = price('BTC')
    i+=1
    x.append(i)
    y.append(data['USD'])
    plt.title("BTC vs USD, Last Update is: "+str(datetime.datetime.now()))
    plt.plot(x,y)
    fig.canvas.draw()
    plt.pause(1000)
```

执行后的效果如图 3-44 所示。

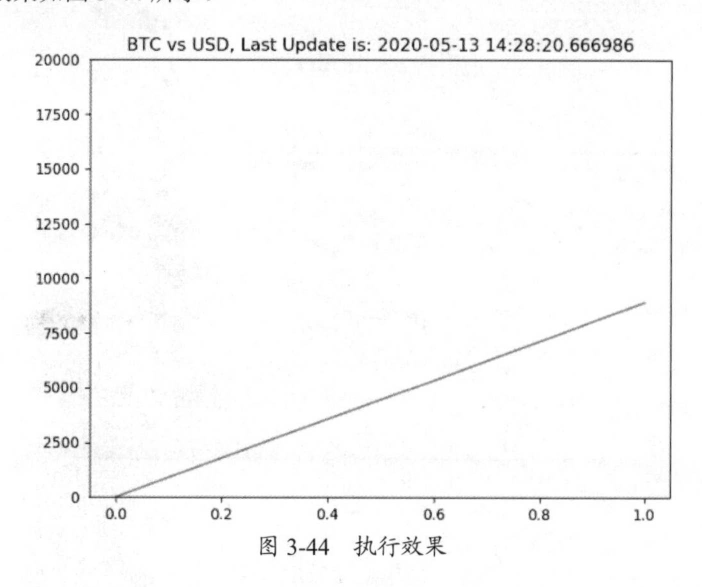

图 3-44　执行效果

3.6.3　绘制 BTC 和 ETH 的历史价格曲线图

编写实例文件 Assignment_Step3.py，功能是首先爬取权威网站中的 BTC 和 ETH 的历史价格数据，然后使用库 matplotlib 和 pandas 绘制 BTC 和 ETH 的历史价格曲线图。首先编写函数 hourly_price_historical()，功能是获取参数 url 中的数据信息，使用 json() 解析 BTC 和 ETH 的历史报价数据；然后编写函数 plotchart()，功能是绘制 BTC 和 ETH 的历史价格曲线图。

```
import requests
import datetime
import pandas as pd
import matplotlib.pyplot as plt

def hourly_price_historical(symbol, comparison_symbol, limit, aggregate, 
exchange=''):
      url = 'https://min-api.cryptocompare.com/data/histohour?fsym={}&tsym={}&lim
it={}&aggregate={}'\
              .format(symbol.upper(), comparison_symbol.upper(), limit, aggregate)
     if exchange:
         url += '&e={}'.format(exchange)
     print(url)
     page = requests.get(url)
     data = page.json()['Data']
     df = pd.DataFrame(data)
     df['timestamp'] = [datetime.datetime.fromtimestamp(d) for d in df.time]

     return df

def plotchart(axis, df, symbol, comparison_symbol):
     axis.plot(df.timestamp, df.close)

df1 = hourly_price_historical('BTC','USD', 2000, 1)
df2 = hourly_price_historical('ETH','USD', 2000, 1)
f, axarr = plt.subplots(2)


plotchart(axarr[0],df1,'BTC','USD')
plotchart(axarr[1],df2,'ETH','USD')

plt.show()
```

执行后的效果如图 3-45 所示。

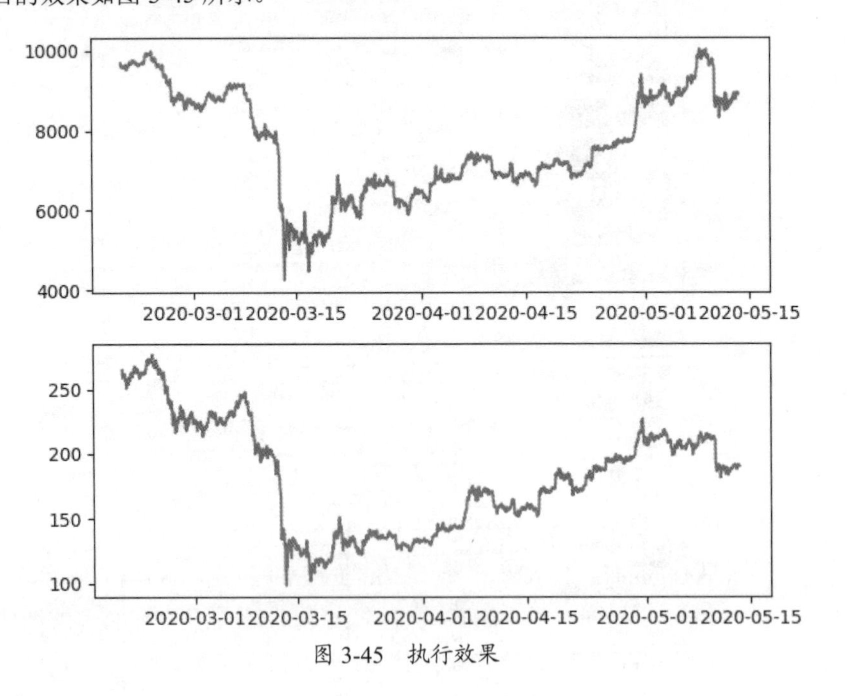

图 3-45　执行效果

第4章 使用 matplotlib 绘制其他类型的统计图

除了前面介绍的散点图、折线图和柱状图之外，在数据可视化分析应用中还经常绘制其他类型的统计图，例如饼形图、雷达图、热力图等。在本章的内容中，将详细讲解使用 Python 绘制其他类型统计图的知识，为读者步入本书后面知识的学习打下基础。

4.1 绘制基本的饼状图

饼状图常用于数据统计和分析领域，通常分为 2D 与 3D 饼状图。饼状图显示一个数据系列（数据系列：在图表中绘制的相关数据点，这些数据源自数据表的行或列。在现实应用中，经常使用饼状图来展示数据分析的结果，这样可以更加直观地展示数据分析结果。在本节的内容中，将详细讲解使用 Python 绘制饼状图的知识。

↑扫码看视频（本节视频课程时间：8 分 56 秒）

4.1.1 绘制简易的饼状图

在 Python 程序中，可以使用 matplotlib 中的函数 matplotlib.pyplot.pie() 绘制一个饼状图。请看下面的实例文件 pie01.py，功能是使用 matplotlib 绘制一个饼状图，可视化展示 ×× 网站会员用户教育水平的分布信息。

```
import matplotlib.pyplot as plt
plt.rcParams['font.sans-serif'] = ['SimHei'] # 指定默认字体
# 构造数据
edu = [0.2515,0.3724,0.3336,0.0368,0.0057]
labels = ['中专','大专','本科','硕士','其他']
"""
绘制饼图
explode：设置各部分突出
label：设置各部分标签
labeldistance：设置标签文本距圆心位置，1.1 表示 1.1 倍半径
autopct：设置圆里面文本
shadow：设置是否有阴影
startangle：起始角度，默认从 0 开始逆时针转
pctdistance：设置圆内文本距圆心距离
返回值
l_text：圆内部文本，matplotlib.text.Text object
p_text：圆外部文本
"""
```

```
# 绘制饼图
plt.pie(x = edu, # 绘图数据
    labels=labels, # 添加教育水平标签
    autopct='%.1f%%' # 设置百分比的格式，这里保留一位小数
)
# 添加图标题
plt.title('×× 网站会员用户的教育水平分布 ')
plt.show()                    # 显示图形
```

执行效果如图 4-1 所示。

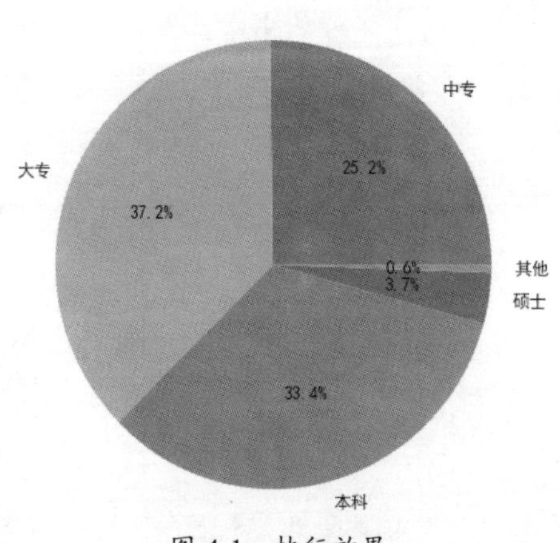

图 4-1　执行效果

4.1.2　修饰饼状图

请看下面的实例文件 pie02.py，功能是以前面的例文件 pie01.py 为基础进行升级，修饰了使用 matplotlib 绘制的饼状图，可视化展示 ×× 网站会员用户教育水平的分布信息。在本实例中，使用 wedgeprops 设置了饼图内外边界线条的宽度和颜色。

```
import matplotlib.pyplot as plt

plt.rcParams['font.sans-serif'] = ['SimHei']                    # 指定默认字体
edu = [0.2515,0.3724,0.3336,0.0368,0.0057]
labels = [' 中专 ',' 大专 ',' 本科 ',' 硕士 ',' 其他 ']
# 添加修饰的饼图
explode = [0,0.1,0,0,0]# 生成数据，用于突出显示大专学历人群
colors=['#9999ff','#ff9999','#7777aa','#2442aa','#dd5555']      # 自定义颜色

# 中文乱码和坐标轴负号的处理
plt.rcParams['font.sans-serif'] = ['Microsoft YaHei']
plt.rcParams['axes.unicode_minus'] = False

# 将横、纵坐标轴标准化处理，确保饼图是一个正圆，否则为椭圆
plt.axes(aspect='equal')
# 绘制饼图
plt.pie(x = edu,                                    # 绘图数据
    explode=explode,                                # 突出显示大专人群
    labels=labels,                                  # 添加教育水平标签
    colors=colors,                                  # 设置饼图的自定义填充色
    autopct='%.1f%%',                               # 设置百分比的格式，这里保留一位小数
    pctdistance=0.8,                                # 设置百分比标签与圆心的距离
    labeldistance = 1.1,                            # 设置教育水平标签与圆心的距离
    startangle = 180,                               # 设置饼图的初始角度
```

```
    radius = 1.2,                                   # 设置饼图的半径
    counterclock = False,                           # 是否逆时针，这里设置为顺时针方向
    wedgeprops = {'linewidth': 1.5, 'edgecolor':'green'}, # 设置饼图内外边界的属性值
    textprops = {'fontsize':10, 'color':'black'},         # 设置文本标签的属性值
)

# 添加图标题
plt.title('×× 网站会员用户的教育水平分布统计图 ')
# 显示图形
plt.show()
```

执行效果如图 4-2 所示。

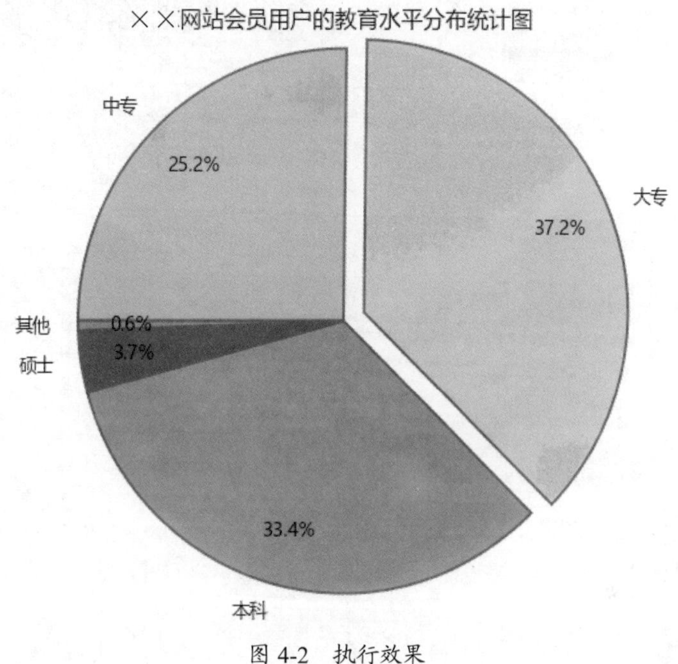

图 4-2　执行效果

4.1.3　突出显示某个饼状图的部分

在前面的实例文件 pie02.py 中，我们通过设置函数 matplotlib.pyplot.pie() 中的参数 explode 实现了某个饼状部分的突出显示。请看下面的实例文件 pie03.py，功能是使用 matplotlib 绘制饼状图，并设置了第二个饼状图部分突出显示。

```
import matplotlib.pyplot as plt
plt.rcParams['font.sans-serif']='SimHei'          #设置中文显示
plt.figure(figsize=(6,6))                          #将画布设定为正方形，则绘制的饼图是正圆
label=['第一','第二','第三']                       #定义饼图的标签，标签是列表
explode=[0.01,0.2,0.01]                            #设定各项距离圆心 n 个半径
#plt.pie(values[-1,3:6],explode=explode,labels=label,autopct='%1.1f%%')#绘制饼图
values=[4,7,9]
plt.pie(values,explode=explode,labels=label,autopct='%1.1f%%')#绘制饼图
plt.title('2018 年饼图 ')
plt.savefig('./2018 年饼图 ')
plt.show()
```

执行效果如图 4-3 所示。

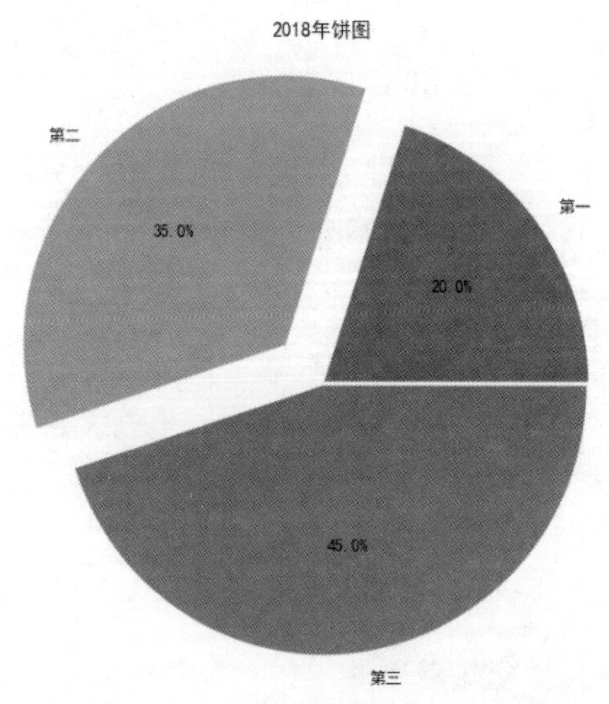

图 4-3　执行效果

4.1.4　为饼状图添加图例

在使用 matplotlib 绘制饼状图时，可以通过内置函数 legend() 为绘制的饼状图添加图例，以增加可视信息的完善程度。请看下面的实例文件 pie04.py，功能是使用 matplotlib 绘制的饼状图，并为绘制的饼状图添加图例说明。

```
    import matplotlib.pyplot as plt
    import matplotlib

    matplotlib.rcParams['font.sans-serif'] = ['SimHei']
    matplotlib.rcParams['axes.unicode_minus'] = False

    label_list = ["第一部分", "第二部分", "第三部分"]     # 各部分标签
    size = [55, 35, 10]     # 各部分大小
    color = ["red", "green", "blue"]     # 各部分颜色
    explode = [0.05, 0, 0]    # 各部分突出值

    patches, l_text, p_text = plt.pie(size, explode=explode, colors=color,
labels=label_list, labeldistance=1.1, autopct="%1.1f%%", shadow=False,
startangle=90, pctdistance=0.6)
    plt.axis("equal")     # 设置横轴和纵轴大小相等，这样饼才是圆的
    plt.legend()
    plt.show()
```

执行效果如图 4-4 所示。

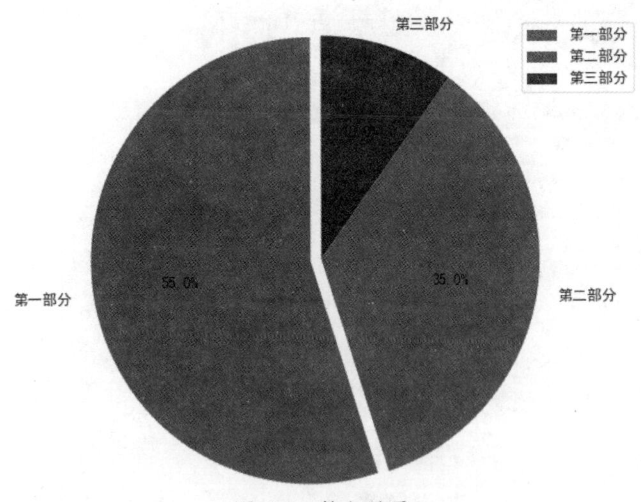

图 4-4　执行效果

4.1.5 使用饼状图可视化展示某地区程序员的工龄

请看下面的实例文件 pie05.py，功能是使用 matplotlib 绘制的饼状图，可视化展示某地区程序员的工龄分部信息。在 4 个列表中分别存储 4 个城市的数据，在列表 label 中保存 4 个范围的工龄，在列表 color 中保存 4 种颜色，最后根据上述城市数据、label 和 color 绘制饼状图。

```
import matplotlib.pyplot as plt
from matplotlib.font_manager import FontProperties
font = FontProperties(fname=r"C:\Windows\Fonts\simhei.ttf", size=14)
beijing = [17,17,23,43]
shanghai = ['19%','4%','23%','54%']
guangzhou = ['53%','25%','13%','9%']
shenzhen = ['41%','22%','20%','17%']
label = ['2-3 years','3-4 years','4-5 years','5+ years']
color = ['red','green','yellow','purple']
indic = []

# 我们将数据最大的突出显示
for value in beijing:
    if value == max(beijing):
        indic.append(0.1)
    else:
        indic.append(0)

plt.pie(
    beijing,
    labels=label,
    colors=color,
    startanglo=90,
    shadow=True,
    explode=tuple(indic),#tuple 方法用于将列表转化为元组
    autopct='%1.1f%%'# 是数字 1，不是 l
)

plt.title(u' 统计 ×× 地区程序员的工龄 ', FontProperties=font)

plt.show()
```

执行效果如图 4-5 所示。

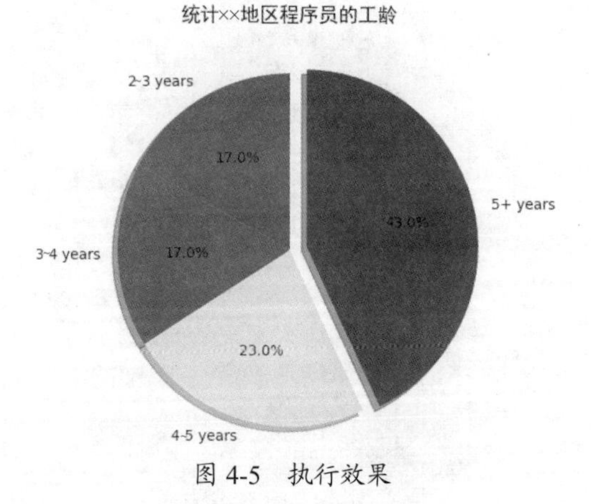

图 4-5　执行效果

4.1.6　绘制多个饼状图

（1）请看下面的实例文件 pie06.py，功能是使用 matplotlib 同时绘制 4 个饼状图。

```
    import matplotlib.pyplot as plt
    import matplotlib.font_manager as fm
    myfont = fm.FontProperties(fname=r'C:\Windows\Fonts\simhei.ttf')
    fig = plt.figure("tu")
    ax_list = []
    labels_list= [['娱乐','育儿','饮食','房贷','交通','其他',"坚持"],['Sur',
"Fea", "Dis", "Hap", "Sad", "Ang", "Nat"]]
    sizes_list= [[2,5,12,70,2,2,7]]
    explode_list=[ (0,0,0,0.1,0,0,0)]
    title_list=["头部左右","头部上下","头部垂面旋转","表情"]
    for ax_one in range(4):
        ax1 = fig.add_subplot(2,2, ax_one + 1)
         ax1.pie(sizes_list[0],explode=explode_list[0],labels=labels_list[1],autopct
='%1.1f%%',shadow=False,startangle=150)
        plt.title(title_list[ax_one],fontproperties=myfont)

    plt.show()
    if __name__ == '__main__':
        pass
```

执行效果如图 4-6 所示。

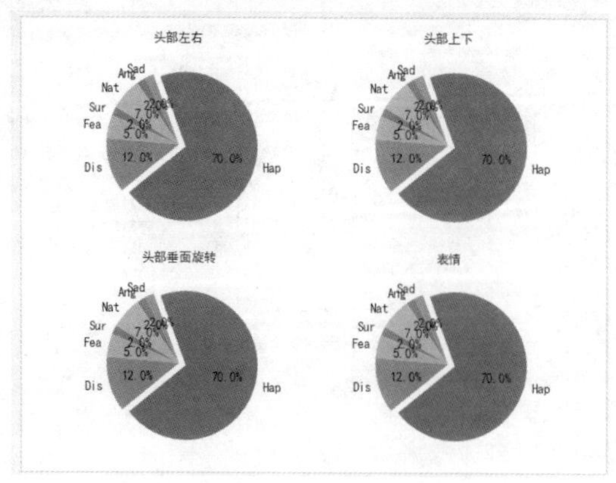

图 4-6　执行效果

（2）请看下面的实例文件 pie07.py，功能是使用 matplotlib 同时绘制 3 个并排的饼状图。

```
import matplotlib.pyplot as plt

labels = 'Comments rated 1', 'Comments rated 2', 'Comments rated 3', 'Comments
rated 4', 'Comments rated 5'
sizes = [38, 16, 18, 54, 107]
sizes1 = [84, 45, 59, 132, 314]
sizes2 = [205, 172, 273, 561, 1954]
explode = (0, 0, 0, 0, 0.025)  # 设置各部分距离圆心的距离
fig1 = plt.figure(facecolor='white',figsize=(16,8))
ax1=plt.subplot(1,3,1)
ax1.pie(sizes, explode=explode,autopct='%1.1f%%',
        shadow=False, startangle=90)
ax1.axis('equal')
ax1.legend(labels)

ax1=plt.subplot(1,3,2)
ax1.pie(sizes1, explode=explode,autopct='%1.1f%%',
        shadow=False, startangle=90)
ax1.axis('equal')
ax1.legend(labels)

ax1=plt.subplot(1,3,3)
ax1.pie(sizes2, explode=explode,autopct='%1.1f%%',
        shadow=False, startangle=90)
ax1.axis('equal')
ax1.legend(labels)

plt.tight_layout()
plt.show()
fig1.savefig('饼状图.jpg',dpi=400)
```

执行效果如图 4-7 所示。

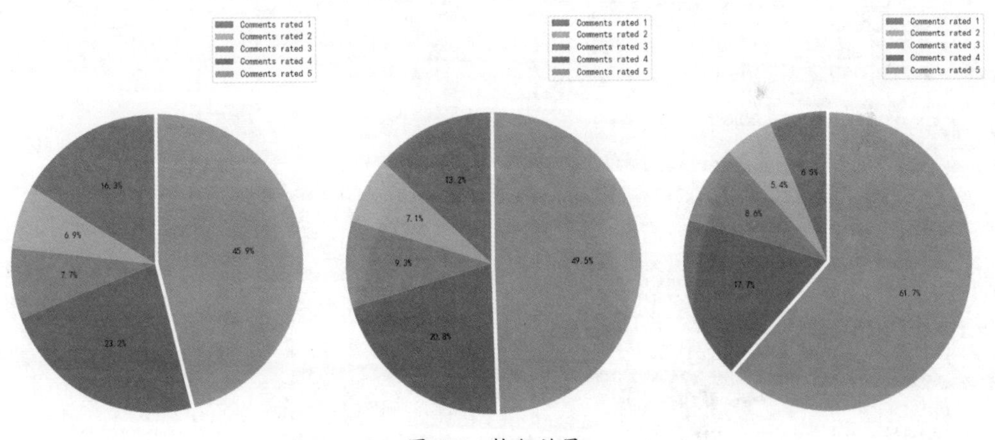

图 4-7　执行效果

（3）请看下面的实例文件 pie08.py，功能是使用 matplotlib 绘制一个 3 行 2 列的饼状图。

```
import numpy as np
import matplotlib.pyplot as plt
import pylab as pl
#画出 3 行 2 列的饼图

labels = ['Flat','Reduce','Raise']
# 321 > 3行2列第1个
fig1 = pl.subplot(321)
pl.pie([12,1,7],labels=labels,autopct='%1.1f%%',shadow=False,startangle=90)
plt.axis('equal')
```

```
    plt.title("group A_20min")

    # 322 > 3行2列第2个
    fig2 = pl.subplot(322)
    pl.pie([2,2,1],labels=labels,autopct='%1.1f%%',shadow=False,startangle=90)
    plt.axis('equal')
    plt.title("group B_20min")

    # 323 > 3行2列第3个
    fig3 = pl.subplot(323)
    pl.pie([8,2,10],labels=labels,autopct='%1.1f%%',shadow=False,startangle=90)
    plt.axis('equal')
    plt.title("group A_60min")

    # 324 > 3行2列第4个
    fig4 = pl.subplot(324)
    pl.pie([2,1,2],labels=labels,autopct='%1.1f%%',shadow=False,startangle=90)
    plt.axis('equal')
    plt.title("group B_60min")

    # 325 > 3行2列第5个
    fig5 = pl.subplot(325)
    pl.pie([4,1,15],labels=labels,autopct='%1.1f%%',shadow=False,startangle=90)#sta
rtangle表示饼图的起始角度
    plt.axis('equal') #这行代码加入饼图不会画成椭圆
    plt.title("group A_70min")

    # 326 > 3行2列第6个
    fig6 = pl.subplot(326)
    pl.pie([3,1,1],labels=labels,autopct='%1.1f%%',shadow=False,startangle=90)
    plt.axis('equal')
    plt.title("group B_70min")

    pl.tight_layout() #布局方法
    pl.savefig('vc4.jpg',dpi = 500) #dpi实参改变图像的分辨率
    pl.show() #显示方法
```

执行效果如图 4-8 所示。

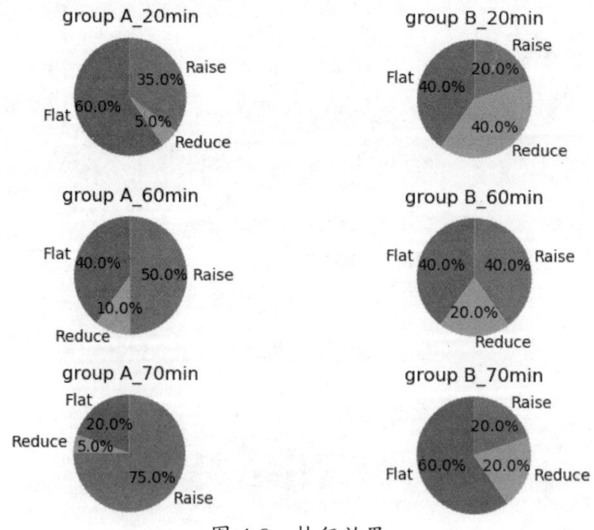

图 4-8　执行效果

4.2 实践案例：可视化分析热门电影信息

在电影行业中，大数据和数据分析技术变得愈发重要。通过挖掘数据，可以及时地发现我们的观众在哪里，然后就可以把相应的电影推荐给他们，实现精准化营销。另外，还可以分析热门电影的数据，了解观众的喜好和发展趋势。在本节的内容中，将通过具体实例讲解可视化分析热门电影信息的过程。

↑扫码看视频（本节视频课程时间：2 分 22 秒）

本实例的基本思路：首先使用爬虫技术抓取某电影网的热门电影信息，然后将抓取的电影信息保存到 MySQL 数据库中，最后使用 matplotlib 绘制电影信息的饼状统计图。

4.2.1 创建 MySQL 数据库

编写文件 myPymysql.py，功能是使用 pymysql 建立和指定 MySQL 数据库的连接，并创建指定选项的数据库表。文件 myPymysql.py 的主要实现代码如下所示：

```
# 获取 logger 的实例
logger = logging.getLogger("myPymysql")
# 指定 logger 的输出格式
formatter = logging.Formatter('%(asctime)s %(levelname)s %(message)s')
# 文件日志，终端日志
file_handler = logging.FileHandler("myPymysql")
file_handler.setFormatter(formatter)

# 设置默认的级别
logger.setLevel(logging.INFO)
logger.addHandler(file_handler)

class DBHelper:
  def __init__(self, host="127.0.0.1", user='root',
               pwd='66688888',db='testdb',port=3306,
               charset='utf-8'):
      self.host = host
      self.user = user
      self.port = port
      self.passwd = pwd
      self.db = db
      self.charset = charset
      self.conn = None
      self.cur = None

  def connectDataBase(self):
      """
      连接数据库
      """
      try:
        self.conn =pymysql.connect(host="127.0.0.1",
          user='root',password="66688888",db="testdb",charset="utf8")

      except:
        logger.error("connectDataBase Error")
        return False

      self.cur = self.conn.cursor()
      return True

  def execute(self, sql, params=None):
```

```
        """
        执行一般的 sq 语句
        """
        if self.connectDataBase() == False:
          return False

        try:
          if self.conn and self.cur:
            self.cur.execute(sql, params)
            self.conn.commit()
        except:
          logger.error("execute"+sql)
          logger.error("params",params)
          return False
        return True

      def fetchCount(self, sql, params=None):
        if self.connectDataBase() == False:
          return -1
        self.execute(sql, params)
        return self.cur.fetchone() # 返回操作数据库操作得到一条结果数据

      def myClose(self):
        if self.cur:
          self.cur.close()
        if self.conn:
          self.conn.close()
        return True
    if __name__ == '__main__':
      dbhelper = DBHelper()

       sql  =  "create  table  maoyan(title  varchar(50),actor  varchar(200),time
varchar(100));"
      result = dbhelper.execute(sql, None)
      if result == True:
        print("创建表成功")
      else:
        print("创建表失败")
      dbhelper.myClose()
      logger.removeHandler(file_handler)
```

执行后会在名为“testdb”的数据库中创建名为“maoyan”的数据库表，如图 4-9 所示。

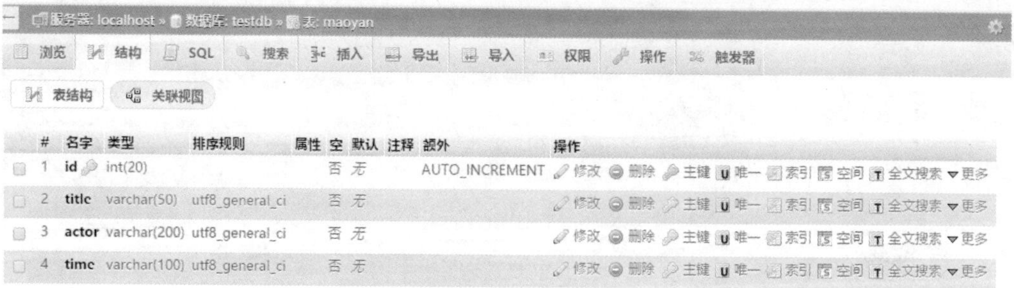

服务器: localhost » 数据库: testdb » 表: maoyan

浏览 结构 SQL 搜索 插入 导出 导入 权限 操作 触发器

表结构 关联视图

#	名字	类型	排序规则	属性	空	默认	注释	额外	操作
1	id	int(20)			否	无		AUTO_INCREMENT	修改 删除 主键 唯一 索引 空间 全文搜索 更多
2	title	varchar(50)	utf8_general_ci		否	无			修改 删除 主键 唯一 索引 空间 全文搜索 更多
3	actor	varchar(200)	utf8_general_ci		否	无			修改 删除 主键 唯一 索引 空间 全文搜索 更多
4	time	varchar(100)	utf8_general_ci		否	无			修改 删除 主键 唯一 索引 空间 全文搜索 更多

图 4-9　创建的 MySQL 数据库

4.2.2　抓取并分析电影数据

编写文件 maoyan.py，功能是抓取指定网页的电影信息，将抓取到的数据添加到 MySQL 数据库中；然后建立和 MySQL 数据库的连接，并使用 matplotlib 将数据库中的电影数据绘制国别类别的统计饼状图。文件 maoyan.py 的主要实现代码如下所示：

```
import logging

# 获取 logger 的实例
logger = logging.getLogger("maoyan")
# 指定 logger 的输出格式
formatter = logging.Formatter('%(asctime)s %(levelname)s %(message)s')
# 文件日志，终端日志
file_handler = logging.FileHandler(" 域名 .txt")
file_handler.setFormatter(formatter)

# 设置默认的级别
logger.setLevel(logging.INFO)
logger.addHandler(file_handler)

def get_one_page(url):
    """
    发起 Http 请求，获取 Response 的响应结果
    """
     ua_headers = {"User-Agent":"Mozilla/4.0 (Macintosh; U; Intel Mac OS X 
10_6_8; en-us) AppleWebKit/534.50 (KHTML, like Gecko) Version/4.1 Safari/534.50"}
    reponse = requests.get(url,headers=ua_headers)
    if reponse.status_code == 200: #ok
        return reponse.text
    return None

def write_to_sql(item):
    """
    把数据写入数据库
    """
    dbhelper = myPymysql.DBHelper()
    title_data = item['title']
    actor_data = item['actor']
    time_data = item['time']
    sql = "INSERT INTO testdb.maoyan(title,actor,time) VALUES (%s,%s,%s);"
    params = (title_data, actor_data, time_data)
    result = dbhelper.execute(sql, params)
    if result == True:
        print(" 插入成功 ")
    else:
        logger.error("execute: "+sql)
        logger.error("params: ",params)
        logger.error(" 插入失败 ")
        print(" 插入失败 ")

def parse_one_page(html):
    """
    从获取到的 html 页面中提取真实想要存储的数据：
    电影名，主演，上映时间
    """
     pattern = re.compile('<p class="name">.*?title="([\s\S]*?)"[\s\S]*?<p 
class="star">([\s\S]*?)</p>[\s\S]*?<p class="releasetime">([\s\S]*?)</p>')
    items = re.findall(pattern,html)

    # yield 在返回的时候会保存当前的函数执行状态
    for item in items:
        yield {
                'title':item[0].strip(),
                'actor':item[1].strip(),
                'time':item[2].strip()
    }
```

```
import matplotlib.pyplot as plt

def analysisCounry():
    # 从数据库表中查询出每个国家的电影数量来进行分析
    dbhelper = myPymysql.DBHelper()
    # fetchCount
    Total = dbhelper.fetchCount("SELECT count(*) FROM `testdb`.`maoyan`;")
    Am = dbhelper.fetchCount('SELECT count(*) FROM `testdb`.`maoyan` WHERE time
like "% America %";')
    Ch = dbhelper.fetchCount('SELECT count(*) FROM `testdb`.`maoyan` WHERE time
like "% China %";')
    Jp = dbhelper.fetchCount('SELECT count(*) FROM `testdb`.`maoyan` WHERE time
like "% Japan %";')
    Other = Total[0] - Am[0] - Ch[0] - Jp[0]
    sizes = Am[0], Ch[0], Jp[0], Other
    labels = 'America','China','Japan','Others'
    colors = 'Yellow','Red','Black','Green'
    explode = 0,0,0,0
    # 画出统计图表的饼状图
    plt.pie(sizes,explode=explode,labels=labels,
        colors=colors, autopct="%1.1f%%", shadow=True)
    plt.show()


def CrawlMovieInfo(lock, offset):
    """
    抓取电影的电影名，主演，上映时间
    """
    url = 'http:// 域名 .com/board/4?offset='+str(offset)
    # 抓取当前的页面
    html = get_one_page(url)
    #print(html)

    # 这里的 for
    for item in parse_one_page(html):
        lock.acquire()
        #write_to_file(item)
        write_to_sql(item)
        lock.release()

    # 每次下载完一个页面，随机等待 1~3 秒再次去抓取下一个页面
    #time.sleep(random.randint(1,3))


if __name__ == "__main__":
    analysisCounry()
    # 把页面做 10 次的抓取，每一个页面都是一个独立的入口
    from multiprocessing import Manager
    #from multiprocessing import Lock 进程池中不能用这个 lock

    # 进程池之间的 lock 需要用 Manager 中的 lock
    manager = Manager()
    lock = manager.Lock()

    # 使用 functools.partial 对函数做一层包装，从而把这把锁传递进进程池
    # 这样进程池内就有一把锁可以控制执行流程
    partial_CrawlMovieInfo = functools.partial(CrawlMovieInfo, lock)
    pool = Pool()
    pool.map(partial_CrawlMovieInfo, [i*10 for i in range(10)])
    pool.close()
    pool.join()
    logger.removeHandler(file_handler)
```

执行后会将抓取的电影信息添加到数据库中，会根据数据库数据绘制饼状统计图，如图 4-10 所示。

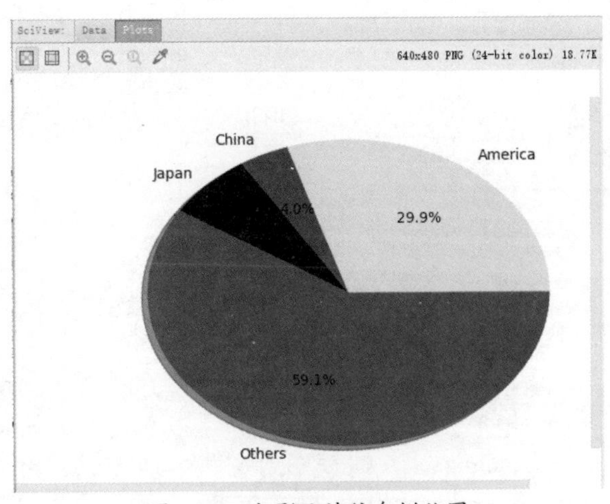

图 4-10　电影统计信息饼状图

4.3　实践案例：可视化展示名著《西游记》中出现频率最多的文字

在数据可视化分析领域中，文字识别和统计处理是最常用的一种应用情形之一。小到文字识别，大到海量数据检索和统计分析，甚至包括语言翻译。在本节的内容中，将通过一个具体实例讲解文字识别和统计处理的过程。

↑扫码看视频（本节视频课程时间：4 分 07 秒）

本节实例的基本思路：在记事本文件中保存四大名著之一的《西游记》电子书，然后使用 Python 技术对这个记事本文件进行文字计数测试，最后使用饼状图展示出现频率最多的文字。

4.3.1　单元测试文件

单元测试（unit testing）是指对软件中的最小可测试单元进行检查和验证。总的来说，单元就是人为规定的最小的被测功能模块。单元测试是在软件开发过程中要进行的最低级别的测试活动，软件的独立单元将在与程序的其他部分相隔离的情况下进行测试。在本项目中的“test”目录下保存了单元测试文件，各个文件的主要功能如下表 4-1 所示。

表 4-1

文件名称	说明
test_arc.py	功能是实现 matplotlib.pyplot 绘图测试
test_embedded_tk.py	功能是实现 tkinter 窗体内嵌 pyplot 测试
test_thread.py	功能是实现多线程绘图测试
wordCount.py	功能是实现文本文件计数测试
wordGenerate.py	功能是实现生成测试所用文本文件

4.3.2 GUI 界面

本项目是使用 Python GUI 界面库 Tkinter 实现的，为了提高统计小说文字的效率，使用内置库 threading 实现多线程处理。编写程序文件 MainWindow.py 实现本项目的 UI 界面，具体实现流程可分为 5 个步骤。

（1）编写类 MainWindow 实现主窗体，在顶部设计一个“打开文件”菜单。对应的实现代码如下所示：

```
class MainWindow:
    def __init__(self):
        self.root = tk.Tk()  # 创建主窗体
        self.root.geometry("600x500")
        menubar = tk.Menu(self.root)
        self.canvas = tk.Canvas(self.root)
        menubar.add_command(label='打开文件', command=self.open_file_dialog)
        self.root['menu'] = menubar
        self.expected_percentage = 0.8
```

（2）编写函数 open_file_dialog(self)，单击“打开文件”菜单后会弹出一个打开文件对话框。函数 open_file_dialog(self) 的具体实现代码如下所示：

```
    def open_file_dialog(self):
         file = tk.filedialog.askopenfilename(title='打开文件', filetypes=[('text',
'*.txt'), ('All Files', '*')])
        self.dialog = InputDialog.InputDialog(self)
        self.dialog.mainloop()
            load_thread = threading.Thread(target=self.create_matplotlib,
args=(file,))
        load_thread.start()
```

（3）编写函数 create_matplotlib(self, f)，功能是使用 matplotlib 绘制饼状图，首先设置字体为“SimHei”，然后使用函数 open() 打开要读取的文件，通过循环统计记事本文件中的所有文字，最终根据统计结果绘制饼状图。为了提高效率，使用了以下两种机制：

- 利用 re 模块进行正则匹配筛选文本，简化代码编写量；
- 利用 threading 模块实现多线程读取文件并进行文字统计，从而不影响窗体循环，使得用户可以继续进行操作。

函数 create_matplotlib(self, f) 的具体实现代码如下所示：

```
    def create_matplotlib(self, f):
        plt.rcParams['font.sans-serif'] = ['SimHei']
        plt.rcParams['axes.unicode_minus'] = False
        plt.style.use("ggplot")

        file = open(f, 'r', encoding='utf-8')
        word_list = list(file.read())
        word_count = {}
        total_words = 0
        for word in word_list:
            if re.match('\\w', word):
                total_words += 1
                if word in word_count:
                    word_count[word] += 1
                else:
                    word_count[word] = 1
        max_list = []
        max_count = []
        now_percentage = 0
        if self.expected_percentage >= 1:
```

```
            while len(word_count) != 0:
                max_count.append(0)
                max_list.append("")
                for word, count in word_count.items():
                    if max_count[-1] < count:
                        max_count[-1] = count
                        max_list[-1] = word
                # if max_list[-1] in word_count:
                word_count.pop(max_list[-1])
        else:
            while now_percentage < self.expected_percentage:
                max_count.append(0)
                max_list.append("")
                for word, count in word_count.items():
                    if max_count[-1] < count:
                        max_count[-1] = count
                        max_list[-1] = word
                # if max_list[-1] in word_count:
                word_count.pop(max_list[-1])
                now_percentage = 0
                for val in max_count:
                    now_percentage += val / total_words

        print(max_list, [val / total_words for val in max_count], sep="\n")

        fig = plt.figure(figsize=(32, 32))
        ax1 = fig.add_subplot(111)
        tuple_builder = [0.0 for index in range(len(max_count))]
        tuple_builder[0] = 0.1
        explode = tuple(tuple_builder)
        ax1.pie(max_count, labels=max_list, autopct='%1.1f%%', explode=explode,
textprops={'size': 'larger'})
        ax1.axis('equal')
        ax1.set_title('文字统计')

        self.create_form(fig)
```

（4）编写函数 create_form(self, figure)，使用 Tkinter 的 canvas 绘制 pyplot，然后利用库 Matplotlib 提供的 NavigationToolbar2Tk 将 pyplot 的导航栏内嵌到 tkinter 窗体中。函数 create_form(self, figure) 的具体实现代码如下所示：

```
    def create_form(self, figure):
        self.clear()
        figure_canvas = FigureCanvasTkAgg(figure, self.root)
        self.canvas = figure_canvas.get_tk_widget()

        self.toolbar = NavigationToolbar2Tk(figure_canvas, self.root)
        self.toolbar.update()
        self.canvas.pack(side=tk.TOP, fill=tk.BOTH, expand=1)
```

（5）编写函数 clear(self) 销毁创建的工具栏，在主函数中启动 GUI 界面。函数 clear(self) 的具体实现代码如下所示：

```
    def clear(self):
        self.canvas.destroy()
        if hasattr(self, 'toolbar'):
            self.toolbar.destroy()


if __name__ == "__main__":
    mainWindow = MainWindow()
    mainWindow.root.mainloop()
```

执行文件 MainWindow.py 后会启动 GUI 主界面，执行效果如图 4-11 所示。单击 GUI 主界面菜单中的“打开文件”后会弹出一个“打开文件”对话框，如图 4-12 所示。

图 4-11　GUI 主界面

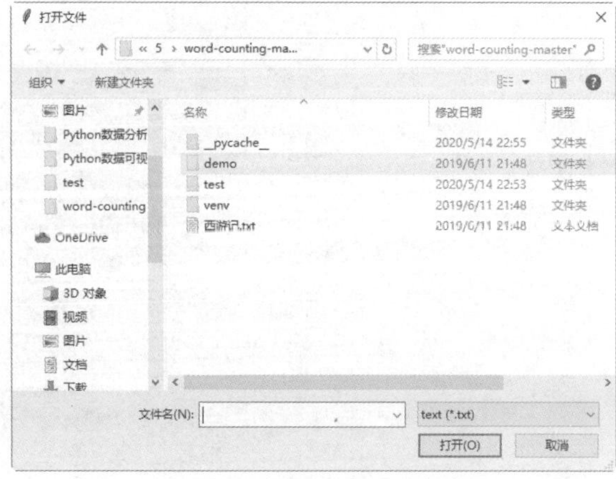

图 4-12　“打开文件”对话框

4.3.3　设置所需显示的出现频率

编写文件 InputDialog.py，功能是当在“打开文件”对话框中选择一个要统计的小说文件后会弹出一个输入对话框，提示用户输入所需显示的出现频率，然后根据用户输入的频率值统计文字。文件 InputDialog.py 的具体实现代码如下所示：

```
import tkinter as tk

class InputDialog(tk.Tk):

    def __init__(self, main_window):
        tk.Tk.__init__(self)
        self.main_window = main_window
        tk.Label(self, text="输入所需显示的出现频率百分数：").grid(row=0, column=0)
        self.input = tk.Entry(self, bd=5)
        self.input.insert(0, "20%")
        self.input.grid(row=0, column=1)
        tk.Button(self, text="Commit", command=self.__commit).grid(row=1,
column=0, columnspan=2, padx=5, pady=5)

    def __commit(self):
        self.main_window.expected_percentage = float(self.input.get().
strip("%"))/100
        self.quit()
        self.withdraw()


if __name__ == '__main__':
    window = MainWindow.MainWindow()
    window.root.mainloop()
```

执行后的效果如图 4-13 所示，这说明本项目默认的统计频率值是 20%。单击“Commit”按钮后会在“输出”界面打印显示下面的统计结果，并统计当前文件中频率为 20% 的文字的统计饼状图，如图 4-14 所示。

```
['道', '不', '一', '了', '那', '我', '是', '来', '他', '个', '行', '你', '的',
'者', '有', '大', '得', '这', '上', '去']
```

```
    [0.018718106708697443, 0.014997293644660065, 0.013463692285695805,
0.013073909032418393, 0.012735189087430597, 0.012141152600994713,
0.010975207061715115, 0.010103726600238977, 0.009727560229172127,
0.009656071597566663, 0.00958968929679016, 0.009308841101197264,
0.009182884940749541, 0.008502040830221309, 0.007553965406310744,
0.007249287666684936, 0.006478231711676136, 0.006445891616426045,
0.006292701691557193, 0.006214404618846446]
```

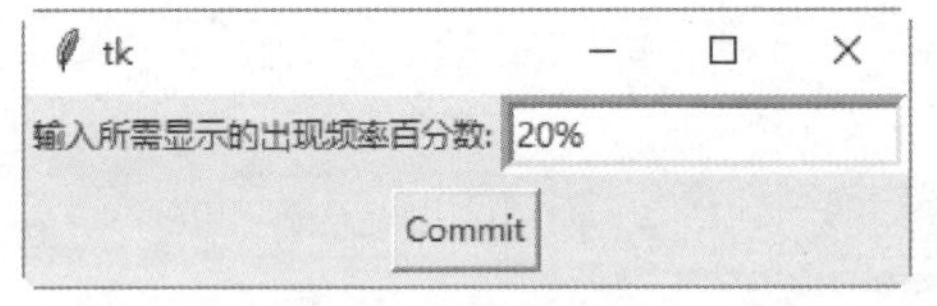

图 4-13　执行结果

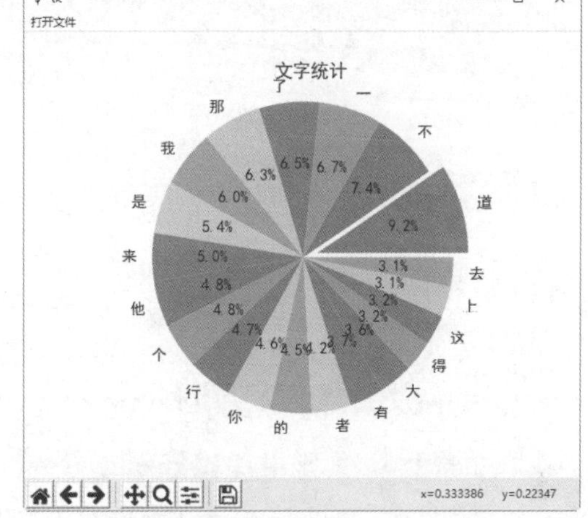

图 4-14　统计饼状图

4.4　绘制雷达图

雷达图是以从同一点开始的轴上表示的三个或更多个常量或变量的二维图表的形式，这样可以显示多变量数据的图形。在数据可视化分析工作中，经常需要绘制和数据统计相关的雷达图。在本节的内容中，将详细讲解在 Python 程序中使用各种库绘制雷达图的知识。

↑扫码看视频（本节视频课程时间：3 分 34 秒）

4.4.1　创建极坐标图

在 Python 程序中，可以使用库 Matplotlib 绘制雷达图，在绘制之前需要使用极坐标体系辅助实现。在库 Matplotlib 的 pyplot 子库中提供了绘制极坐标图的方法 subplot()，在通过此方法创建子图时通过设置 projection='polar' 便可创建一个极坐标子图，然后调用方法 plot() 在极坐标子图中绘图。请看下面的实例文件 lei01.py，功能是使用库 Matplotlib 分别创建一个极坐标图和一个直角坐标子图进行对比。

```
import matplotlib.pyplot as plt
import numpy as np
theta=np.arange(0,2*np.pi,0.02)

ax1 = plt.subplot(121, projection='polar')
ax2 = plt.subplot(122)
ax1.plot(theta,theta/6,'--',lw=2)
ax2.plot(theta,theta/6,'--',lw=2)
plt.show()
```

执行效果如图 4-15 所示。

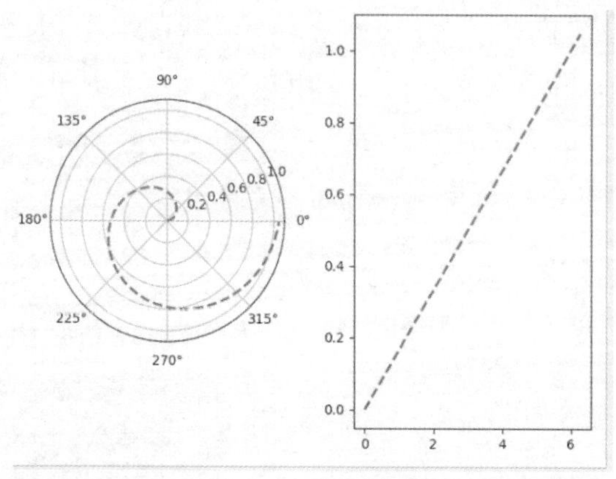

图 4-15　执行效果

4.4.2　设置极坐标的正方向

在 Python 程序中使用库 matplotlib 绘制极坐标图时，函数 set_theta_direction() 用于设置极坐标的正方向。当函数 set_theta_direction() 的参数值为 1、'counterclockwise' 或者是 'anticlockwise' 的时候，正方向为逆时针；当函数 set_theta_direction() 的参数值为 -1 或者是 'clockwise' 的时候，正方向为顺时针。请看下面的实例文件 lei02.py，功能是使用库 matplotlib 创建极坐标图，并且设置极坐标的正方向。

```
import matplotlib.pyplot as plt
import numpy as np
theta=np.arange(0,2*np.pi,0.02)
ax1= plt.subplot(121, projection='polar')
ax2= plt.subplot(122, projection='polar')
ax2.set_theta_direction(-1)
ax1.plot(theta,theta/6,'--',lw=2)
ax2.plot(theta,theta/6,'--',lw=2)
plt.show()
```

执行效果如图 4-16 所示。

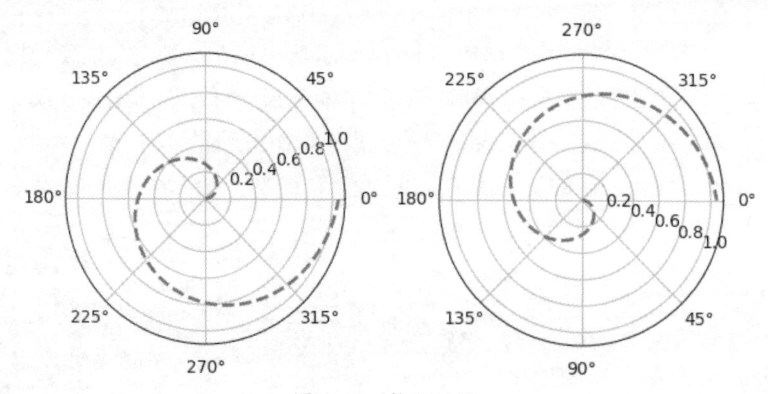

图 4-16　执行效果

4.4.3　绘制一个基本的雷达图

请看下面的实例文件 lei03.py，功能是使用库 matplotlib 绘制一个基本的雷达图。首先使用内置函数 np.random.randint() 随机生成了 3 组整数型数据，然后使用这三组整形数据绘制雷达图。

```
import matplotlib.pyplot as plt
import numpy as np

# 雷达图 1 - 极坐标的折线图 / 填图 - plt.plot()

plt.figure(figsize=(8,4))
ax1= plt.subplot(111, projection='polar')
ax1.set_title('radar map\n')  # 创建标题
ax1.set_rlim(0,12)

data1 = np.random.randint(1,10,10)
data2 = np.random.randint(1,10,10)
data3 = np.random.randint(1,10,10)
theta=np.arange(0, 2*np.pi, 2*np.pi/10)
# 创建数据

ax1.plot(theta,data1,'.--',label='data1')
ax1.fill(theta,data1,alpha=0.2)
ax1.plot(theta,data2,'.--',label='data2')
ax1.fill(theta,data2,alpha=0.2)
ax1.plot(theta,data3,'.--',label='data3')
ax1.fill(theta,data3,alpha=0.2)
# 绘制雷达线
plt.show()
```

执行效果如图 4-17 所示。

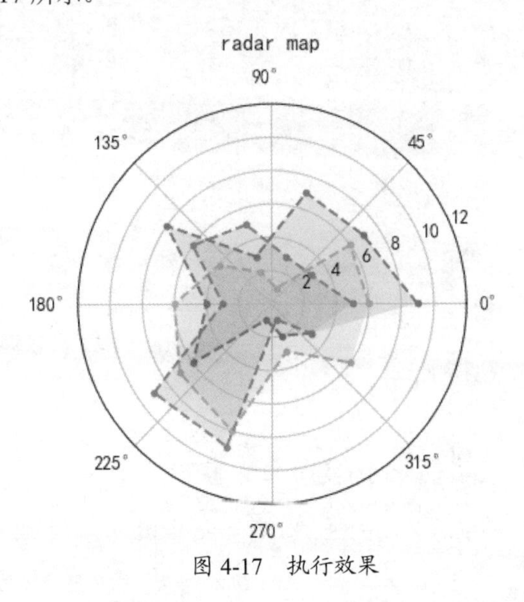

图 4-17　执行效果

4.4.4　绘制汽车性能雷达图

请看下面的实例文件 lei06.py，功能是使用库 matplotlib 绘制一个雷达图，从耐撞、加速、集气、转向、喷射和漂移等 5 个方面可视化展示某款汽车的性能。

```
import numpy as np                                       # 导入科学计算基础包(已安装)
import matplotlib.pyplot as plt                          # 导入绘图库（已安装）
from matplotlib.font_manager import FontProperties       # 导入下载的 matplotlib 下
可用的中文字体

# 从文件路径下选择可用的中文字体
font_set=FontProperties(fname="C:\Windows\Fonts\simhei.ttf", size=15)
# 将属性标签放入数组
label=np.array(['耐撞','加速','集气','转向','喷射','漂移'])
#将各属性得分放入数组
data=np.array([3.5,4.5,3.8,5,4,4.5])

# 将2π分为六部分，放入一个数组
angles=np.linspace(0,2*np.pi,len(label),endpoint=False)
data=np.concatenate((data,[data[0]]))
angles=np.concatenate((angles,[angles[0]]))

fig=plt.figure()
ax=fig.add_subplot(111,polar=True)

ax.plot(angles,data,'bo',linewidth=2)
ax.fill(angles,data,facecolor='b',alpha=0.25)
ax.set_thetagrids(angles*180/np.pi,label,fontproperties=font_set)
ax.set_rlim(0,5)
ax.grid(True)

plt.show()
```

执行效果如图 4-18 所示。

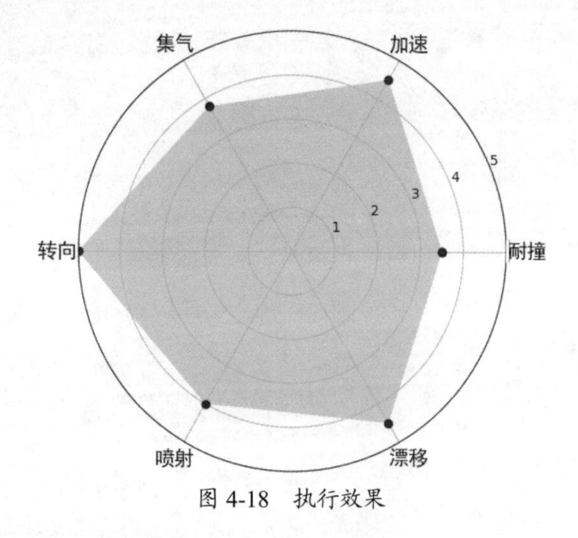

图 4-18　执行效果

4.4.5　使用雷达图比较两名研发部同事的能力

请看下面的实例文件 lei05.py，功能是使用库 matplotlib 分别绘制两个雷达图，从 5 个方面（编程能力、沟通技能、专业知识、团队协作、工具掌握）可视化比较两名研发部同事的能力。

```
import numpy as np
from matplotlib import pyplot as plt
fig=plt.figure(figsize=(10,5))
ax1=fig.add_subplot(1,2,1,polar=True) #设置第一个坐标轴为极坐标体系
ax2=fig.add_subplot(1,2,2,polar=True) #设置第二个坐标轴为极坐标体系
fig.subplots_adjust(wspace=0.4) #设置子图间的间距，为子图宽度的 40%
```

```
    p1={"编程能力":60,"沟通技能":70,"专业知识":65,"团体协作":75,"工具掌握":80}  #创建第一个人的数据
    p2={"编程能力":70,"沟通技能":60,"专业知识":75,"团体协作":65,"工具掌握":70}  #创建第二个人的数据

    data1=np.array([i for i in p1.values()]).astype(int) #提取第一个人的信息
    data2=np.array([i for i in p2.values()]).astype(int) #提取第二个人的信息
    label=np.array([j for j in p1.keys()]) #提取标签

    angle = np.linspace(0, 2*np.pi, len(data1), endpoint=False) #data 里有几个数据，就把整圆 360°分成几份
    angles = np.concatenate((angle, [angle[0]])) #增加第一个 angle 到所有 angle 里，以实现闭合
    data1 = np.concatenate((data1, [data1[0]])) #增加第一个人的第一个 data 到第一个人所有的 data 里，以实现闭合
    data2 = np.concatenate((data2, [data2[0]])) #增加第二个人的第一个 data 到第二个人所有的 data 里，以实现闭合

    #设置第一个坐标轴
    ax1.set_thetagrids(angles*180/np.pi, label, fontproperties="Microsoft Yahei")
    #设置网格标签
    ax1.plot(angles,data1,"o-")
    ax1.set_theta_zero_location('NW') #设置极坐标 0°位置
    ax1.set_rlim(0,100) #设置显示的极径范围
    ax1.fill(angles,data1,facecolor='g', alpha=0.2) #填充颜色
    ax1.set_rlabel_position(255) #设置极径标签位置
    ax1.set_title("同事甲",fontproperties="SimHei",fontsize=16) #设置标题

    #设置第二个坐标轴
    ax2.set_thetagrids(angles*180/np.pi, label, fontproperties="Microsoft Yahei")
    #设置网格标签
    ax2.plot(angles,data2,"o-")
    ax2.set_theta_zero_location('NW') #设置极坐标 0°位置
    ax2.set_rlim(0,100) #设置显示的极径范围
    ax2.fill(angles,data2,facecolor='g', alpha=0.2) #填充颜色
    ax2.set_rlabel_position(255) #设置极径标签位置
    ax2.set_title("同事乙",fontproperties="SimHei",fontsize=16) #设置标题

    plt.show()
```

执行效果如图 4-19 所示。

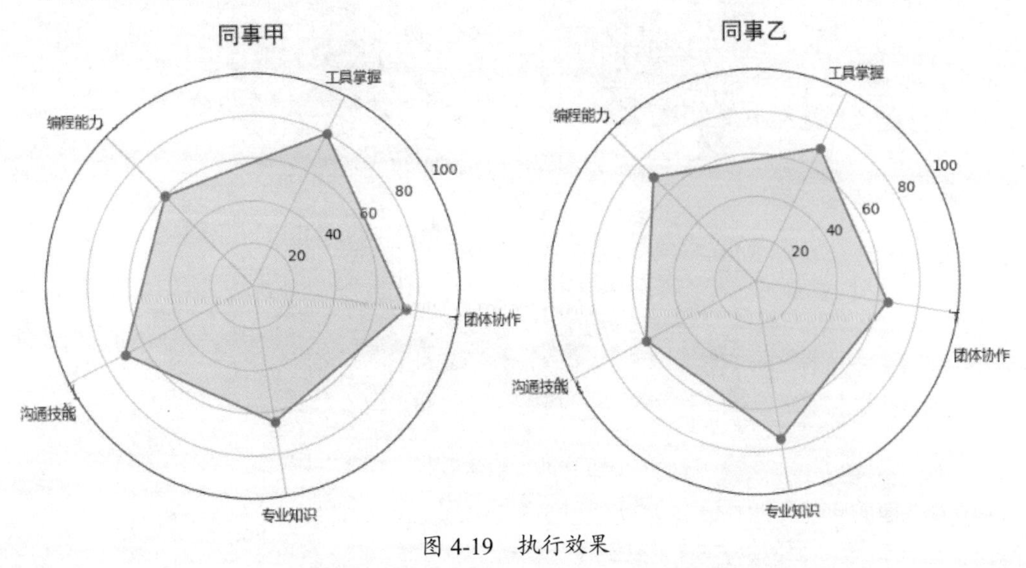

图 4-19　执行效果

4.5 绘制热力图

热力图是指以特殊高亮的形式显示访客热衷的页面区域和访客所在的地理区域的图示。在现实应用中，通常使用热力图展示不同区域某件事情的热度，例如交通热力图等。在本节的内容中，将详细讲解在 Python 程序中使用各种库绘制热力图的知识。

↑扫码看视频（本节视频课程时间：3 分 02 秒）

4.5.1 使用库 matplotlib 绘制基本的热力图

在 Python 程序中，可以使用库 matplotlib 绘制热力图。请看下面的实例文件 re02.py，功能是使用库 matplotlib 绘制一个基本的热力图。首先使用内置函数 numpy.meshgrid() 生成网格点坐标矩阵并赋值给变量 x 和 y，然后为 z 赋值为 x**2 + y**2 的平方根。

```
import numpy as np
import matplotlib.pyplot as plt

points = np.arange(-5,5,0.01)

x,y = np.meshgrid(points,points)
z = np.sqrt(x**2 + y**2)

cmaps = [plt.cm.jet, plt.cm.gray, plt.cm.cool, plt.cm.hot]

fig, axes = plt.subplots(2, 2)

for i, ax in enumerate(axes.ravel()):
    ax.imshow(z,cmap=cmaps[i])

plt.show()
```

执行效果如图 4-20 所示。

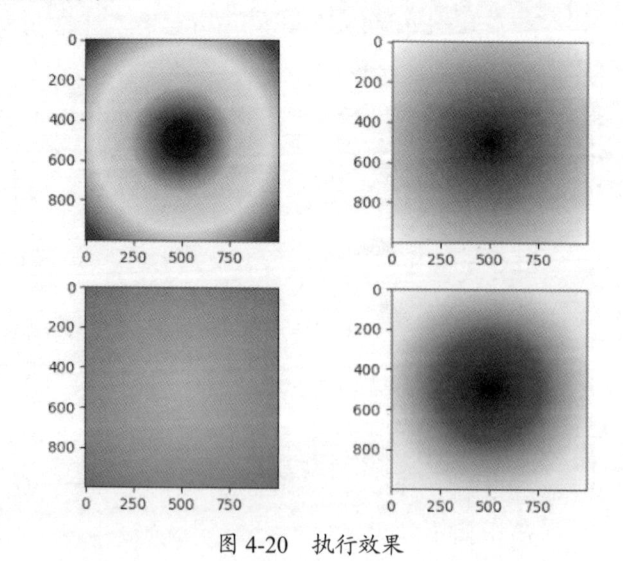

图 4-20　执行效果

请看下面的实例文件 re03.py，功能也是使用库 matplotlib 绘制一个热力图。

```
import matplotlib.pyplot as plt
```

```
import matplotlib.cm as cm
from matplotlib.colors import LogNorm
import numpy as np
x, y = np.random.rand(10), np.random.rand(10)
z = (np.random.rand(9000000)+np.linspace(0,1, 9000000)).reshape(3000, 3000)
plt.imshow(z+10, extent=(np.amin(x), np.amax(x), np.amin(y), np.amax(y)),
    cmap=cm.hot, norm=LogNorm())
plt.colorbar()
plt.show()
```

执行效果如图 4-21 所示。

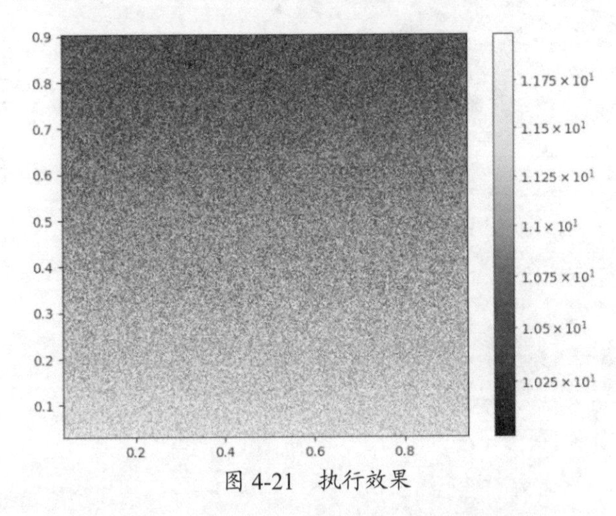

图 4-21　执行效果

4.5.2　将 Excel 文件中的地址信息可视化为交通热力图

请看下面的实例项目，功能是在一个 Excel 文件中保存一些地址名称。然后将这些文字格式的地址信息转换为坐标信息，并在地图中展示这些地址的实时热力信息。

（1）将地址转换为 JS 格式

在 Excel 文件 address.xlsx 中保存了文字格式的地址信息，编写 Python 文件 index_address.py 将文件 address.xlsx 中的地址保存到 JS 文件 address.js 中。文件 index_address.py 的具体实现代码如下所示：

```
import xlrd
import os
data = xlrd.open_workbook('address.xlsx')
table = data.sheets()[0]
nrows = table.nrows #行数
ccols = table.col_values(0)
address = []

for rownum in range(1,nrows):
    adr = table.cell(rownum,0).value
    address.append({'name':adr,'address':adr})
print (len(address))

# 保存到文件
address = str(address).encode('utf-8').decode("unicode_escape")
address = address.replace("u", "")
address = '// 自动生成\n' + 'var address = ' + address
with open('./docs/js/data/address.js','a',encoding='utf-8') as jsname:
    jsname.write(address)
```

（2）将 JS 地址转换为坐标

编写文件 amap.js 将 JS 文件 address.js 中的地址转换为坐标格式，具体实现代码如下所示：

```
var map = new AMap.Map('container',{
    resizeEnable: true,
    zoom: 13
});
map.setCity('北京市');

var geocoder
AMap.plugin(['AMap.ToolBar', 'AMap.Geocoder'],function(){
    geocoder = new AMap.Geocoder({
        city: "010"//城市，默认："全国"
    });
    map.addControl(new AMap.ToolBar());
    map.addControl(geocoder);
});

function getMarker (map, model, markers) {
    var address = model.address;
    geocoder.getLocation(address, function(status,result){
        queryNum++
        var marker
        if(status=='complete'&&result.geocodes.length){
            marker = new AMap.Marker({
                map: map,
                position: result.geocodes[0].location
            });
            marker.model = model
              marker.setLabel({//label默认蓝框白底左上角显示，样式className为：amap-
marker-label
                offset: new AMap.Pixel(20, 20),//修改label相对于maker的位置
                content: model.name
            });
            markers.push(marker);

            position.push({
                lng: result.geocodes[0].location.lng,
                lat: result.geocodes[0].location.lat,
                count: position.length + 1,
                ...model
            })
        }else{
            console.log('获取位置失败', address);
        }
        return marker
    })
}
var markers = []
var position = []
var locationCount = 0
var queryNum = 0
function modelsToMap (map, models) {
    markers = []
    locationCount= models ? models.length : 0
    queryNum = 0
    var model
    for (model of models) {
        getMarker(map, model, markers)
    }
}

function load () {
```

```
        // address 从 address.js 中获取
        // address 数据格式
         // var address = [{'name': '天安门','address':'xxx号'},{'name': '水立方','address':'yyy号'}]
        modelsToMap(map, address)
    }
    function downloadPostions() {
        if (queryNum == locationCount) {
            // 调用 download.js 中的方法
              download('position' + position.length, '// amsp.js 生成\n' + 'var postions = ' +JSON.stringify(position))
        }else{
            alert('还没处理完.请稍后.....')
        }
    }
```

（3）在地图中显示地址的热力信息

申请高德地图 API 的 key，编写文件 index_address.html，在高德地图中可视化展示文件 address.xlsx 中各个地址的热力信息。文件 index_address.html 的具体实现代码如下所示：

```
          <link rel="stylesheet" href="http://cache.amap.com/lbs/static/main1119.css"/>
          <script src="http://webapi.amap.com/maps?v=1.4.0&key=您申请的key值"></script>
          <script type="text/javascript" src="http://cache.amap.com/lbs/static/addToolbar.js"></script>
        <script type="text/javascript" src="./js/data/postions.js"></script>

    </head>
    <body>
    <div id="container"></div>
    <div class="button-group">
          <input type="button" class="button" value="显示热力图" onclick="heatmap.show()"/>
          <input type="button" class="button" value="关闭热力图" onclick="heatmap.hide()"/>
    </div>
    </body>
    <script type="text/javascript" src="./js/reli/amap.js"></script>
```

执行后会显示文件 address.xlsx 中各个地址的热力信息，如图 4-22 所示。

图 4-22　执行效果

4.6 绘制词云图

“词云”就是通过形成“关键词云层”或“关键词渲染”，对网络文本中出现频率较高的“关键词”形成视觉上的突出。在本节的内容中，将通过两个实例讲解使用 Python 语言创建词云图的方法。

↑扫码看视频（本节视频课程时间：1 分 24 秒）

4.6.1 绘制 B 站词云图

编写实例文件 BiliBili.py，功能抓取 B 站排行榜的信息，然后提取排行榜前 50 名视频的所有标签信息，最后将提取的标签文本添加到词云图中。文件 BiliBili.py 的具体实现代码如下所示：

```
    sys.stdout = io.TextIOWrapper( sys.stdout.buffer, encoding='gb18030')#编码
    url='https://www.bilibili.com/ranking'#B 站排行榜链接
    response=requests.get(url)
    html=response.text
    video_list=re.findall(r'<a href="(.*?)" target="_blank">', html)#B 站排行榜链接列表
    label_list=[]
    video_name=re.findall(r'target="_blank"  class="title">(.*?)</a><!----><div
class="detail"><span class="data-box">',html)#B 站排行榜视频名
    video_play=re.findall(r'<i class="b-icon play"></i>(.*?)</span>', html)#播放数
    video_view=re.findall(r'<i  class="b-icon  view"></i>(.*?)</span><a  target="_
blank"',html)#评论数
    video_up=re.findall(r'<i class="b-icon author"></i>(.*?)</span></a>', html)#UP 主
    for i in range (0, 100):
        print('%d.'%(i+1), end='')
        print('%-65s'%video_name[i],end='')
        print('up 主 : %-15s'%video_up[i], end='')
        print(' 播放数 : %-8s'%video_play[i], end='')
        print(' 评论数 : %s'%video_view[i])#循环输出视频名、 UP 主、 播放数、 评论数
    for video in video_list:
        video_response=requests.get(video)
        video_html=video_response.text
        video_label=re.findall(r'target="_blank">(.*?)</a>', video_html)
        for label in video_label:
                label_list.append(label)#把排行榜视频的所有标签添加进 label_list
    label_string=" ".join(label_list)#把 label_list 转为 string 型
    w = wordcloud.WordCloud(width=1000,
                            height=700,
                            background_color='white',
                            font_path='msyh.ttc')
    w.generate(label_string)
    w.to_file('BiliBili.png')
```

执行后会创建一幅名为“BiliBili.png”的词云图，效果如图 4-23 所示。

图 4-23　词云图效果

4.6.2　绘制知乎词云图

编写实例文件 zhihu.py，抓取知乎热榜的信息，然后提取排行榜前 50 名信息的所有标签信息，最后将提取的标签文本添加到词云图中。文件 zhihu.py 的具体实现代码如下所示：

```
    import matplotlib.pyplot as plt
    sys.stdout = io.TextIOWrapper( sys.stdout.buffer, encoding='gb18030')#编码
    headers={
        'User-Agent':'Mozilla/5.0  (X11;  Linux  x86_64)  AppleWebKit/537.36  (KHTML,
like Gecko) Chrome/66.0.3359.181 Safari/537.36',
        'Referer':'https://www.zhihu.com/',
        'Cookie':'_zap=d39ea994-dcf8-444d-bbb0-2cda0cafc120;
    response=requests.get("https://www.zhihu.com/hot",headers=headers)#知乎热榜链接
    html=response.text
    html_list=re.findall(r'</div></div><div  class="HotItem-content"><a  href="(.*?)"
title="', html)#抓取所有热榜话题链接
    print(html_list)
    print(len(html_list))
    label_list=[]
    for i in range(0, 50):
        hot_response=requests.get(html_list[i], headers=headers)#必须要有 headers，要不
然无法访问
        hot_html=hot_response.text
        hot_label=re.findall(r'keywords"  content="(.*?)"/><meta
itemProp="answerCount"', hot_html)#抓取所有热词
        hot_name=re.findall(r'><title  data-react-helmet="true">(.*?) ?  -   知  乎 </
title><meta name="viewport"', hot_html)#抓取标题
        print('%d.'%(i+1), end='')
        for i in range(len(hot_label)):
              label_list.append(hot_label[i])
    print(label_list)
    label_string=" ".join(label_list)#转换为 string 型
    print(label_string)
    w = wordcloud.WordCloud(width=1000,
                            height=700,
                            background_color='white',
                            font_path='msyh.ttc')
    w.generate(label_string)
    w.to_file('zhihu.png')#生成词云图片
```

执行后会创建一幅名为“zhihu.png”的词云图，效果如图 4-24 所示。

图 4-24　词云图效果

4.7　实践案例：使用热力图可视化展示某城市的房价信息

在本项目实例中，将某城市某个时间点的房价信息保存在数据库 db.sqlite3 中，然后使用库 Django 开发一个 Web 项目，在网页版百度地图中使用热力图可视化展示这些房价信息。

↑扫码看视频（本节视频课程时间：1 分 07 秒）

4.7.1　准备数据

在数据库 db.sqlite3 中保存某城市某个时间点的房价信息，有关数据库表的具体设计请参考 Django 文件 models.py，文件 models.py 的具体实现代码如下所示：

```
from django.db import models
from django.contrib.auth.models import User

class taizhou(models.Model):
    id = models.BigAutoField(primary_key=True)
    name = models.CharField(max_length=100)
    cityid = models.CharField(max_length=10)
    info = models.TextField(blank=True)
    mi2=models.BigIntegerField()
    tel = models.TextField(blank=True)
    avg=models.BigIntegerField(null=True)
    howsell= models.TextField(null=True)
    getdate = models.DateTimeField()
    GPS_lat = models.TextField(null=True)
    GPS_lng = models.TextField(null=True)
```

4.7.2　使用热力图可视化展示信息

编写 Django 项目的模板文件 hot.html，使用百度地图的 API 展示此城市的房价热力图信息，文件 hot.html 的具体实现代码如下所示：

```
{% load static %}
<!DOCTYPE html>
<html lang="en">
```

```
    <head>
    <!DOCTYPE html>
    <html>
    <head>
        <meta http-equiv="Content-Type" content="text/html; charset=utf-8" />
        <meta name="viewport" content="initial-scale=1.0, user-scalable=no" />
        <script type="text/javascript" src="http://api.map.baidu.com/api?v=2.0&ak=Rpu35NRbsPN7gkRWOhT0hTBdYY1BLBp8"></script>
        <script type="text/javascript" src="http://api.map.baidu.com/library/Heatmap/2.0/src/Heatmap_min.js"></script>
        <title>台州市房产分布热力图</title>
        <style type="text/css">
              ul,li{list-style: none;margin:0;padding:0;float:left;}
              html{height:100%}
              body{height:100%;margin:0px;padding:0px;font-family:"微软雅黑";}
              #container{height:100%;width:100%;}
              #r-result{width:100%;}
        </style>
    </head>
    <body>
       <div id="container"></div>
       <div id="r-result" style="display:none">
              <input type="button"  onclick="openHeatmap();" value="显示热力图"/><input type="button"  onclick="closeHeatmap();" value="关闭热力图"/>
       </div>
    </body>
    </html>
    <script type="text/javascript">
        var map = new BMap.Map("container");           // 创建地图实例

        var point = new BMap.Point(121.267705,28.655381);
        map.centerAndZoom(point, 14); // 初始化地图，设置中心点坐标和地图级别
        map.setCurrentCity("台州");  //设置当前显示城市
        map.enableScrollWheelZoom(); // 允许滚轮缩放


        var points =[

    {% for taizhou in maplist %}
      {"lng":"{{taizhou.GPS_lng}}","lat":"{{taizhou.GPS_lat}}","count":"{{taizhou.avg}}"},
    {% empty %}
      No Data
    {% endfor %}

    ];//这里面添加经纬度


        if(!isSupportCanvas()){
            alert('热力图目前只支持有canvas支持的浏览器,您所使用的浏览器不能使用热力图功能~')
        }
        //详细的参数,可以查看heatmap.js的文档 https://github.com/pa7/heatmap.js/blob/master/README.md
        //参数说明如下:
        /* visible 热力图是否显示,默认为true
         * opacity 热力的透明度,1-100
         * radius 势力图的每个点的半径大小
         * gradient  {JSON} 热力图的渐变区间 . gradient如下所示
         *  {
               .2:'rgb(0, 255, 255)',
               .5:'rgb(0, 110, 255)',
               .8:'rgb(100, 0, 255)'
           }
```

```
        其中 key 表示插值的位置， 0~1.
            value 为颜色值 .
     */
    heatmapOverlay = new BMapLib.HeatmapOverlay({"radius":100,"visible":true});
    map.addOverlay(heatmapOverlay);
    heatmapOverlay.setDataSet({data:points,max:20000});

    //closeHeatmap();



    //判断浏览区是否支持canvas
    function isSupportCanvas(){
        var elem = document.createElement('canvas');
        return !!(elem.getContext && elem.getContext('2d'));
    }

    function setGradient(){
        /*格式如下所示：
        {
            0:'rgb(102, 255, 0)',
            .5:'rgb(255, 170, 0)',
            1:'rgb(255, 0, 0)'
        }*/
        var gradient = {};
        var colors = document.querySelectorAll("input[type='color']");
        colors = [].slice.call(colors,0);
        colors.forEach(function(ele){
            gradient[ele.getAttribute("data-key")] = ele.value;
        });
        heatmapOverlay.setOptions({"gradient":gradient});
    }

    function openHeatmap(){
        heatmapOverlay.show();
    }

    function closeHeatmap(){
        heatmapOverlay.hide();
    }
</script>
</body>
</html>
```

通过如下命令启动 Django Web 项目。

```
python manage.py runserver
```

此时会在命令行界面显示以下信息。

```
System check identified no issues (0 silenced).
May 16, 2020 - 21:15:29
Django version 3.0.5, using settings 'hotmap.settings'
Starting development server at http://127.0.0.1:8000/
Quit the server with CTRL-BREAK.
```

在浏览器中输入 http://127.0.0.1:8000/ 后会在网页中显示对应的可视化热力图效果，执行效果如图 4-25 所示。

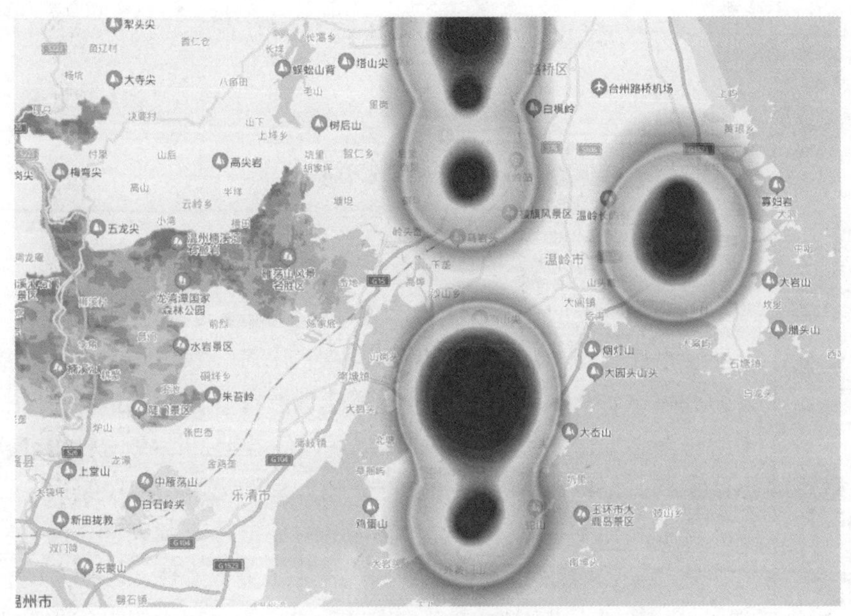
犁头尖
大寺尖
蜈蚣山背
塔山尖
树后山
梅窝尖
高尖岩
五龙尖
龙湾潭国家森林公园
水岩景区
朱苔岭
上岩山
中雁荡山
白石岭尖
新田拔教
东蒙山
乐清市
温州市
鸡蛋山
台州路桥机场
白枫岭
温岭市
大岩山
烟灯山
大园头山头
大岙山
玉环市大鹿岛景区

图 4-25　执行效果

第 5 章

可视化 CSV 文件数据

CSV 是逗号分隔值（Comma-Separated Values）的缩写，其文件以纯文本形式存储表格数据（数字和文本）。有时也被称为字符分隔值，因为分隔字符也可以不是逗号。纯文本意味着该文件是一个字符序列，不用包含必须像二进制数字那样被解读的数据。CSV 文件由任意数目的记录组成，记录间以某种换行符分隔。每条记录由字段组成，字段间的分隔符是其他字符或字符串，最常见的是逗号或制表符。通常所有的记录都有完全相同的字段序列，一般都是纯文本文件。在本章的内容中，将详细讲解使用 Python 语言处理 CSV 文件的知识，为读者步入本书后面知识的学习打下基础。

5.1 内置 csv 模块介绍

在 Python 程序中，建议使用内置模块 csv 来处理 CSV 文件。在本节的内容中，将详细讲解 Python 语言中 csv 模块的内置成员以及对 csv 文件的基本应用。

↑扫码看视频（本节视频课程时间：13 分 44 秒）

5.1.1 内置成员

在 Python 语言的内置模块 csv 中，存在内置方法、类、常量、dialect 和格式化参数、Reader 对象以及 writer 对象等 6 类。内置成员。

1. 内置方法

内置 csz 模块中包含的内置方法如表 5-1 所示。

表 5-1

方法名称	说　明
csv.reader(csvfile, dialect='excel', **fmtparams)	返回一个读取器对象，将在给定的 csvfile 文件中进行迭代
参数 csvfile	可以是任何支持 iterator 协议的对象，并且每次调用 __next__() 方法时返回一个字符串。参数 csvfile 可以是文件对象或列表对象，如果 csvfile 是文件对象，则应使用 newline='' 打开它
参数 dialect	表示编码风格，默认为 excel 的风格，用逗号 “,” 分隔。dialect 方式也支持自定义格式，通过调用 register_dialect 方法来注册

续上表

方法名称	说　明
参数 fmtparam	这是一个格式化参数，用来覆盖 dialect 参数指定的编码风格
csv.writer(csvfile, dialect='excel', **fmtparams)	返回一个 writer 对象，负责将用户的数据转换为给定类文件对象上的分隔字符串。各个参数的具体含义同前面的方法 reader() 相同
csv.register_dialect(name[, dialect[, **fmtparams]])	功能是定义一个 dialect 编码风格的名字。参数“name”表示所自定义的 dialect 的名字，比如默认的是“excel”，也可以定义为“mydialect”
csv.unregister_dialect(name)	功能是从注册表中删除名为“name”的 dialect。如果名称不是注册的 dialect 名称，则会引发错误
csv.get_dialect(name)	返回名称为“name”的 dialect。如果名称不是注册的 dialect 名称，则会引发错误
csv.list_dialects()	返回所有注册 dialect 的名称
csv.field_size_limit([new_limit])	返回解析器允许的当前最大字段大小，如果给出了参数 new_limit，则将成为新的限制大小

2. 类

在 Python 语言的内置模块 csv 中，包含了表 5-2 所示的类。

表 5-2

类名称	说　明
csv.DictReader(csvfile, fieldnames=None, restkey=None, restval=None, dialect='excel', *args, **kwds)	功能是创建一个对象，其操作类似于普通读取器，但将读取的信息映射到一个字典中，其中的键由可选参数 fieldnames 给出。参数 fieldnames 是一个序列，其元素按顺序与输入数据的字段相关联，这些元素将成为结果字典的键。如果省略 fieldnames 参数，则 csvfile 的第一行中的值将用作字段名称。如果读取的行具有比字段名序列更多的字段，则剩余数据将作为键值为 restkey 的序列添加。如果读取的行具有比字段名序列少的字段，则剩余的键使用可选的 restval 参数的值。任何其他可选或关键字参数都传递给底层的 reader 实例
csv.DictWriter(csvfile, fieldnames, restval='', extrasaction='raise', dialect='excel', *args, **kwds)	功能是创建一个类似于常规 writer 的对象，但是将字典映射到输出行。参数 fieldnames 是一个序列，用于标识传递给 writerow() 方法的字典中的值被写入 csvfile。如果字典在 fieldnames 中缺少键，则可选的 restval 参数指定要写入的值。如果传递给 writerow() 方法的字典包含 fieldnames 中未找到的键，则可选的 extrasaction 参数指示要执行的操作。如果将“extrasaction”设置为“raise”则会引发 ValueError 错误，如果设置为“ignore”则会忽略字典中的额外值。任何其他可选或关键字参数都传递给底层的 writer 实例
注意：类 ignore 与类 DictReader 不同，类 DictWriter 中的参数 fieldnames 不是可选的。因为 Python 中的 dict 对象没有排序，所以没有足够的信息来推断将该行写入 csvfile 的顺序	
csv.Dialect	主要依赖于其属性的容器类，用于定义特定 reader 或 writer 实例的参数
csv.excel	定义 Excel 生成的 CSV 文件的常用属性，用 dialect 名称 'excel' 注册
csv.excel_tab	定义了 Excel 生成的 TAB 分隔文件的常用属性，用 dialect 名称 'excel-tab' 注册
csv.unix_dialect	定义在 UNIX 系统上生成的 CSV 文件的常用属性，即使用 '\n' 作为行终止符并引用所有字段，用 dialect 名称 'unix' 注册
csv.Sniffer	用于推导 CSV 文件的格式。在类 Sniffer 中提供了如下所示的两个方法：
sniff(sample, delimiters=None)	分析给定的 sample 示例，并返回一个参数对应的 Dialect 子类。如果给出了可选的分隔符参数，它将被解释为包含所有可能的有效分隔符字符的字符串
has_header(sample)	分析示例文本（假定为 CSV 格式），如果第一行显示为一系列列标题，则返回 True

3. 常量

在 Python 语言的内置模块 csv 中，包含了表 5-3 所示的常量成员。

表 5-3

常量名称	说　明
csv.QUOTE_ALL	设置 writer 对象引用所有字段
csv.QUOTE_MINIMAL	设置 writer 对象仅引用包含特殊字符（如分隔符，quotechar 或 lineterminator t5>
csv.QUOTE_NONNUMERIC	设置 writer 对象引用所有非数字字段，指示读者将所有未引用的字段转换为 float
csv.QUOTE_NONE	设置 writer 对象从不引用字段。当输出数据中出现当前定界符时，其前面为当前 escapechar 字符。如果未设置 escapechar，如果遇到需要转义的字符，则写入程序时会引发 Error，指示 reader 不对引号字符执行特殊处理
csv.Error	检测到错误时由任何函数引发的错误常量

4. dialect 和格式化参数

为了更容易地设置输入和输出记录的格式，特定的格式化参数被分组到 dialect 中。dialect 是具有一组特定方法和单个 validate() 方法的 Dialect 类的子类。当创建 reader 或 writer 对象时，程序员可以指定类Dialect的字符串或子类作为 dialect 参数。类Dialects 支持如下表 5-4 所示的属性。

表 5-4

属性名称	说　明
Dialect.delimiter	用于分隔字段的单字符字符串，默认为逗号 “,”
Dialect.doublequote	控制在字段中出现的 quotechar 实例本身应如何引用。当为 True 时，字符加倍。当为 False 时，escapechar 用作 quotechar 的前缀。默认为 True
Dialect.lineterminator	用于终止由 writer 生成的行的字符串，默认为 “\r\n”
Dialect.quotechar	用于引用包含特殊字符（例如分隔符或 quotechar）或包含换行字符的字段的单字符字符串，它默认为单个双引号
Dialect.quoting	控制何时由作者生成并由读者识别，可以接受任何 QUOTE_* 常量，默认为 QUOTE_MINIMAL
Dialect.skipinitialspace	当为 True 时，紧跟分隔符后的空格将被忽略。默认值为 False
Dialect.strict	当为 True 时，在错误的 CSV 输入上引发异常 Error。默认值为 False

5. Reader 对象

在 Reader 对象（DictReader 实例和 reader() 函数返回的对象）中只包含一个公共方法 csvreader.__next__()，其功能是返回读者的可迭代对象的下一行作为列表，根据当前 Dialect 进行解析。通常应该把它叫作 next(reader)。

在 Reader 对象中包含如下所示的公共属性：

- csvreader.dialect：解析器使用的 Dialect 的只读描述；
- csvreader.line_num：从源迭代器读取的行数。这与返回的记录数不同，因为记录可以跨越多行。

6. Writer 对象

在 Writer 对象（DictWriter 实例和由 writer() 返回的对象）中包含如下所示的公共方法：

- csvwriter.writerow(row)：将行参数写入文件对象，根据当前 Dialect 进行格式化处理；
- csvwriter.writerows(rows)：将所有行参数写入作者的文件对象，根据当前 Dialect 格式化。

在 Writer 对象中包含了公共属性 csvwriter.dialect，其功能是使用规定格式来读取或写入 CSV。在 DictWriter 对象中包含了公共方法 DictWriter.writeheader()，其功能是用字段名写入一行（在构造函数中指定）。

5.1.2　操作 CSV 文件

假设存在一个名为 sample.csv 的 CSV 文件，在里面保存了 Title、Release Date 和 Director 三种数据，具体内容如下所示：

```
Title,Release Date,Director
And Now For Something Completely Different,1971,Ian MacNaughton
Monty Python And The Holy Grail,1975,Terry Gilliam and Terry Jones
Monty Python's Life Of Brian,1979,Terry Jones
Monty Python Live At The Hollywood Bowl,1982,Terry Hughes
Monty Python's The Meaning Of Life,1983,Terry Jones
```

接下来，我们通过 8 个实例文件讲述一下对 CSV 文件的基本操作，最后实现可视化。

通过如下所示的实例文件 001.py, 可以打印输出文件 sample.csv 中的日期和标题内容。

```
for line in open("sample.csv"):
    title, year, director = line.split(",")
    print(year, title)
```

执行后会输出结果如下：

```
Release Date Title
1971 And Now For Something Completely Different
1975 Monty Python And The Holy Grail
1979 Monty Python's Life Of Brian
1982 Monty Python Live At The Hollywood Bowl
1983 Monty Python's The Meaning Of Life
```

在下面的实例文件 002.py 中，演示了使用 csv 模块打印输出文件 sample.csv 中的日期和标题内容的过程。

```
import csv
reader = csv.reader(open("sample.csv"))
for title, year, director in reader:
    print(year, title)
```

执行后会输出结果如下：

```
Release Date Title
1971 And Now For Something Completely Different
1975 Monty Python And The Holy Grail
1979 Monty Python's Life Of Brian
1982 Monty Python Live At The Hollywood Bowl
1983 Monty Python's The Meaning Of Life
```

在下面的实例文件 003.py 中，演示了将数据保存为 CSV 格式的过程。

```
import csv
import sys

data = [
    ("And Now For Something Completely Different", 1971, "Ian MacNaughton"),
    ("Monty Python And The Holy Grail", 1975, "Terry Gilliam, Terry Jones"),
    ("Monty Python's Life Of Brian", 1979, "Terry Jones"),
    ("Monty Python Live At The Hollywood Bowl", 1982, "Terry Hughes"),
    ("Monty Python's The Meaning Of Life", 1983, "Terry Jones")
]

writer = csv.writer(sys.stdout)

for item in data:
    writer.writerow(item)
```

在上述代码中，通过 csv.writer() 方法生成 csv 文件。执行后输出结果如下：

```
And Now For Something Completely Different,1971,Ian MacNaughton
Monty Python And The Holy Grail,1975,"Terry Gilliam, Terry Jones"
Monty Python's Life Of Brian,1979,Terry Jones
Monty Python Live At The Hollywood Bowl,1982,Terry Hughes
Monty Python's The Meaning Of Life,1983,Terry Jones
```

在下面的实例文件 004.py 中，演示了读取指定 CSV 文件的文件头内容的过程。

```
import csv

# Get dates, high, and low temperatures from file.
filename = '2018.csv'
with open(filename) as f:
    reader = csv.reader(f)
    header_row = next(reader)
    print(header_row)
```

执行上述代码后，将会打印输出文件 2018.csv 的文件头的内容，执行后输出结果如下：

```
    ['AKDT', 'Max TemperatureF', 'Mean TemperatureF', 'Min TemperatureF', 'Max
Dew PointF', 'MeanDew PointF', 'Min DewpointF', 'Max Humidity', ' Mean Humidity',
' Min Humidity', ' Max Sea Level PressureIn', ' Mean Sea Level PressureIn', ' Min
Sea Level PressureIn', ' Max VisibilityMiles', ' Mean VisibilityMiles', ' Min
VisibilityMiles', ' Max Wind SpeedMPH', ' Mean Wind SpeedMPH', ' Max Gust SpeedMPH',
'PrecipitationIn', ' CloudCover', ' Events', ' WindDirDegrees']
```

在下面的实例文件 005.py 中，演示了打印输出指定 CSV 文件的文件头和对应位置中的过程。

```
import csv

# Get dates, high, and low temperatures from file.
filename = '2018.csv'
with open(filename) as f:
    reader = csv.reader(f)
    header_row = next(reader)

    for index,column_header in enumerate(header_row):
        print(index,column_header)
```

执行后输出结果如下：

```
0 AKDT
1 Max TemperatureF
2 Mean TemperatureF
3 Min TemperatureF
4 Max Dew PointF
5 MeanDew PointF
6 Min DewpointF
7 Max Humidity
8  Mean Humidity
9  Min Humidity
10  Max Sea Level PressureIn
11  Mean Sea Level PressureIn
12  Min Sea Level PressureIn
13  Max VisibilityMiles
14  Mean VisibilityMiles
15  Min VisibilityMiles
16  Max Wind SpeedMPH
17  Mean Wind SpeedMPH
18  Max Gust SpeedMPH
19 PrecipitationIn
20  CloudCover
21  Events
```

```
22  WindDirDegrees
```

在下面的实例文件 006.py 中，演示了打印输出指定 CSV 文件中每天最高气温的过程。

```
import csv
filename = '2018.csv'
with open(filename) as f:
    reader = csv.reader(f)
    header_row = next(reader)

    highs = []
    for row in reader:
            highs.append(row[1])
    print(highs)
```

执行后输出结果如下：

```
['64', '71', '64', '59', '69', '62', '61', '55', '57', '61', '57', '59', '57',
'61', '64', '61', '59', '63', '60', '57', '69', '63', '62', '59', '57', '57', '61',
'59', '61', '61', '66']
```

在下面的实例文件 007.py 中，演示了数据分析指定 CSV 文件中的内容，并绘制某地 2018 年每天最高温度和最低温度图表的过程。

```
import csv
from datetime import datetime

from matplotlib import pyplot as plt
# Get dates, high, and low temperatures from file.
filename = '2018.csv'
with open(filename) as f:
    reader = csv.reader(f)
    header_row = next(reader)

    dates, highs, lows = [], [], []
    for row in reader:
        try:
            current_date = datetime.strptime(row[0], "%Y-%m-%d")
            high = int(row[1])
            low = int(row[3])
        except ValueError:
            print(current_date, 'missing data')
        else:
            dates.append(current_date)
            highs.append(high)
            lows.append(low)

# Plot data.
fig = plt.figure(dpi=128, figsize=(10, 6))
plt.plot(dates, highs, c='red', alpha=0.5)
plt.plot(dates, lows, c='blue', alpha=0.5)
plt.fill_between(dates, highs, lows, facecolor='blue', alpha=0.1)

# Format plot.
title = "wendu - 2018\nDeath Valley, CN"
plt.title(title, fontsize=20)
plt.xlabel('', fontsize=16)
fig.autofmt_xdate()
plt.ylabel("wendu(F)", fontsize=16)
plt.tick_params(axis='both', which='major', labelsize=16)

plt.show()
```

执行后的效果如图 5-1 所示。

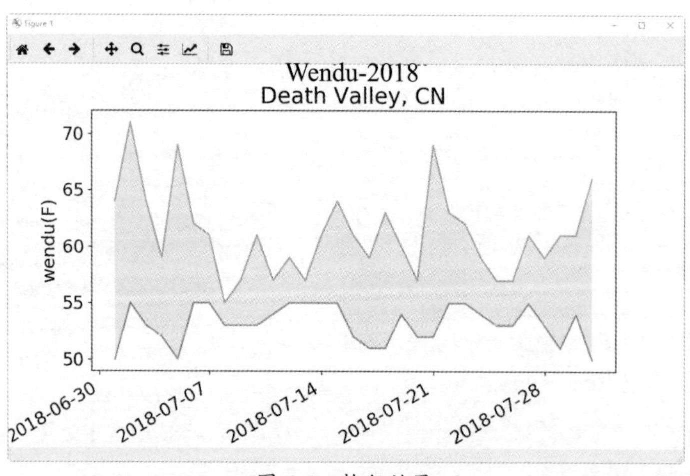

图 5-1　执行效果

在下面的实例文件 008.py 中，演示了数据分析指定 CSV 文件中的内容，并根据 CSV 文件 test.csv 的内容绘制统计曲线的过程。

```
import csv
from matplotlib import pyplot as plt
from datetime import datetime

# 从 csv 中获取数据绘制图表
filename = 'test.csv'
with open(filename) as f:
    reader = csv.reader(f)
    header_row = next(reader)
    print(header_row)

    header_dict = dict(zip(list(range(1, len(header_row) + 1)), header_row))

    for index, value in header_dict.items():
        print(str(index) + ":" + value)

    highs = []
    for row in reader:
        highs.append(row[13])

    # 相当于做了错误数据检查，把不合法的数据转换为合法数据。也可以用 try，遇到不合法数据跳过
    print('highs::', highs)
    for index, v in enumerate(highs):
        if v == '':
            highs[index] = '0'

    print('highs::', highs)

figure = plt.figure(figsize=(10, 6))
plt.plot(list(range(0, 20)), highs[0:20], c='red')
plt.title("Highs", fontsize='8')
plt.xlabel("date", fontsize='16')
plt.ylabel("H")
# 可以让 x 轴的值斜着显示
figure.autofmt_xdate()
plt.tick_params(axis='both')

plt.show()

# datetime 用法
```

```
data = datetime.strptime('2015-11-11', '%Y-%m-%d')
print(data)
# 2015-11-11 00:00:00

# 给两条折线之间填充颜色
plt.fill_between(dates,highs,lows,facecolor='blue',alpha=0.1)
```

执行后会输出 CSV 文件 test.csv 中的数据：

```
['AIANHH', 'STATE', 'COUNTY', 'COUSUBCE', 'RT', 'CODE', 'POP', 'VAPOP', 'VACIT',
'VACLANG', 'VACLEP', 'ILLIT', 'CILLIT', 'LEPPCT', 'ILLRAT', 'FENG5I', 'FENG10I',
'NAME1', 'STABRV', 'NAME21', 'RACEGP']
1:AIANHH
2:STATE
3:COUNTY
4:COUSUBCE
5:RT
6:CODE
7:POP
8:VAPOP
9:VACIT
10:VACLANG
11:VACLEP
12:ILLIT
13:CILLIT
14:LEPPCT
15:ILLRAT
16:FENG5I
17:FENG10I
18:NAME1
19:STABRV
20:NAME21
21:RACEGP
highs:: ['418', '2702', '2082', '1364', '1032', '831', '558', '757', '626',
'1593', '1186', '1167', '798', '1872', '1524', '2744', '2652', '699', '536', '890',
'674', '595', '552', '419', '270', '1717', '1158', '562', '397', '3128', '2968',
'934', '823', '1015', '502', '980', '245', '315', '217', '', '173', '097', '245',
'114', '108', '034', '991', '831', '602', '214', '134', '046', '166', '072', '1003',
'721', '605', '462', '172', '053', '2314', '2164', '241', '163', '781', '604',
'860', '636', '212', '123', '388', '187', '1580', '858', '192', '070', '025', '085',
'061', '755', '613', '322', '095', '072', '042', '282', '182', '1170', '773', '216',
'097', '296', '140', '1082', '780', '475', '293', '114', '032', '1050', '395',
'091', '075', '280', '228', '090', '072', '028', '731', '539', '598', '
##### 此处省略很多输出
```

并且同时会根据 CSV 文件 test.csv 中的数据绘制统计曲线图，如图 5-2 所示。

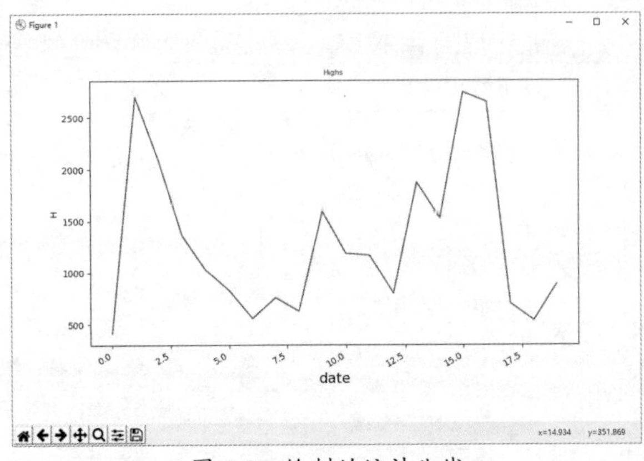

图 5-2　绘制的统计曲线

5.1.3 提取 CSV 数据并保存到 MySQL 数据库

请看下面的实例，功能是提取 3 个 CSV 文件中的数据，并将提取的数据保存到 MySQL 数据库中。

（1）首先提供 3 个 CSV 文件：shop_info.csv、user_pay.csv 和 user_view.csv。

（2）创建一个名为“tianchi_1”的 MySQL 数据库，然后分别创建 3 个数据表（表 5-5~表 5-7）。

表 5-5

Field	Sample	Description
shop_id	000001	商家 id
city_name	北京	市名
location_id	001	所在位置编号，位置接近的商家具有相同的编号
per_pay	3	人均消费（数值越大消费越高）
score	1	评分（数值越大评分越高）
comment_cnt	2	评论数（数值越大评论数越多）
shop_level	1	门店等级（数值越大门店等级越高）
cate_1_name	美食	一级品类名称
cate_2_name	小吃	二级分类名称
cate_3_name	其他小吃	三级分类名称

表 5-6

Field	Sample	Description
user_id	0000000001	用户 id
shop_id	000001	商家 id，与 shop_info 对应
time_stamp	2015-10-10 11:00:00	支付时间

表 5-7

Field	Sample	Description
user_id	0000000001	用户 id
shop_id	000001	商家 id，与 shop_info 对应
time_stamp	2015-10-10 10:00:00	浏览时间

（3）提取数据。在下面的实例文件 psql.py 中，演示了使用 pandas 提取上述 3 个 CSV 文件中数据的过程。

```
import pandas

def Init():
    print('正在提取商家数据……')
     shop_info = pandas.read_csv(r'shop_info.csv',header=None,names=['shop_
id','city_name','location_id','per_pay','score','comment_cnt','shop_level','cate_1_
name','cate_2_name','cate_3_name'])
    print(shop_info.head(5))
```

```
        print('正在提取支付数据……')
        user_pay = pandas.read_csv(r'user_pay.csv', iterator=True,header=None,names
=['user_id','shop_id','time_stamp'])
        try:
            df = user_pay.get_chunk(5)
        except StopIteration:
            print("Iteration is stopped.")
        print(df)

        print("正在提取浏览数据……")
        user_view = pandas.read_csv(r'user_viewcsv',header=None,names=['user_
id','shop_id','time_stamp'])
        print(user_view.head(5))

    if __name__=='__main__':
        Init()
```

在上述代码中有如下 3 个重要的参数，我们重点讲解一下：

- header：指定某一行为列名，默认 header=0，即指定第一行的所有元素名对应为每一列的列名。若 header=None，则不指定列名行；
- names：与 header 配合使用，若 header=None，则可以使用该参数手动指定列名；
- iterator：返回一个 TextFileReader 对象，以便逐块处理文件。默认值为 False。

执行后输出结果如下：

```
    正在提取商家数据……
            shop_id  city_name  location_id          per_pay  score  comment_cnt  \
    0   000001，湖州，885          8              4   12，2，美食，休闲茶饮，饮品      NaN
NaN
    1   000002，广州，885          8              4   12，2，美食，休闲茶饮，饮品      NaN
NaN

      shop_level  cate_1_name  cate_2_name  cate_3_name
    0         NaN          NaN          NaN          NaN
    1         NaN          NaN          NaN          NaN
    正在提取支付数据……
      user_id                          shop_id  time_stamp
    0  1，00001  00002，2018-10-10 11:00:00          NaN
    1  2，00003  00004，2018-10-17 11:00:00          NaN
    正在提取浏览数据……
                                          user_id  shop_id  time_stamp
    0  0000000001，000001，2018-10-17 11:00:00       NaN       NaN
    1  0000000002，000002，2018-10-17 11:00:00       NaN       NaN
```

（4）提取数据并添加到数据库。在下面的实例文件 009.py 中，演示了使用 pandas 和 pymysql 提取数据上述 3 个 CSV 文件中数据，并将提取的数据添加到数据库“tianchi_1”中的过程。

```
    import pandas
    import pymysql

    def Init():
        # 连接数据库
        conn= pymysql.connect(
                host='localhost',
                port = 3306,
                user='root',
                passwd='66688888',
                db ='tianchi_1',
                charset = 'utf8',          # 不声明编码导入的数据会显示出错
                )
```

```
        cur = conn.cursor()

        print(" 正在提取商家数据……")
         shop_info = pandas.read_csv(r'shop_info.csv', iterator=True,chunksize=1,h
eader=None,names=['shop_id','city_name','location_id','per_pay','score','comment_
cnt','shop_level','cate_1_name','cate_2_name','cate_3_name'])
        print(" 正在将数据导入数据库……")
        for i,shop in enumerate(shop_info):
            # 用 -1 或者 '' 代替空值 NAN
            shop = shop.fillna({'cate_1_name':'','cate_2_name':'','cate_3_name':''})
# 替换字符串空值
            shop = shop.fillna(-1)      # 替换整数空值
            shop = shop.values[0]       # Series 类型转换成列表类型
            #print shop
             sql ="insert into shop_info ('shop_id','city_name','location_id','per_
pay','score','comment_cnt','shop_level','cate_1_name','cate_2_name','cate_3_name')
values('%s','%s','%s','%s','%d','%s','%s','%s','%s','%s')"\
                 %(shop[0],shop[1],shop[2],shop[3],shop[4],shop[5],shop[6],shop[7]
,shop[8],shop[9])
            cur.execute(sql)
            print('%d / 2000'%(i+1))
        conn.commit()

        print(' 正在提取支付数据……')
         user_pay = pandas.read_csv(r'user_pay.csv', iterator=True,chunksize=1,heade
r=None,names=['user_id','shop_id','time_stamp'])
        print(' 正在将数据导入数据库……')
        for i,user in enumerate(user_pay):
            # 用 -1 代替空值 NAN
            user = user.fillna(-1)      # 替换整数空值
            user = user.values[0]       # Series 类型转换成列表类型
            #print user
                sql ="insert into user_pay ('user_id','shop_id','time_stamp')
values('%s','%s','%s')"\
                 %(user[0],user[1],user[2])
            cur.execute(sql)
            print('%d'%(i+1))
        conn.commit()

        print(' 正在提取浏览数据……')
         user_view = pandas.read_csv(r'user_view.csv', iterator=True,chunksize=1,hea
der=None,names=['user_id','shop_id','time_stamp'])
        print(' 正在将数据导入数据库……')
        for i,user in enumerate(user_view):
            # 用 -1 代替空值 NAN
            user = user.fillna(-1)      # 替换整数空值
            user = user.values[0]       # Series 类型转换成列表类型
            #print user
                sql ="insert into user_view ('user_id','shop_id','time_stamp')
values('%s','%s','%s')"\
                 %(user[0],user[1],user[2])
            cur.execute(sql)
            print('%d'%(i+1))
        conn.commit()

    if __name__=='__main__':
        Init()
```

在前面的文件 008.py 中，在提取 user_pay 的数据时使用了迭代提取法，这是因为如果 user_pay 的 csv 文件太大（例如好几个 G 大小）。这时如果使用的是 Windows 32bit Python，内存大小会有限制，无法一次性读取这么大的数据集（会提示 MemoryError）。所以在上述代码中，也是采用了迭代的方式将提取的数据都保存到数据库，在迭代的过程中执行 SQL

语句将数据插入数据库表中。执行后会成功在数据库“tianchi_1”中添加 CSV 文件中的数据，执行效果如图 5-3 所示。

浏览 结构 SQL 搜索 插入 导出 导入 权限 操作 触发

	id	city_name	location_id	per_pay	score	comment_cnt	shop_level
编辑 复制 删除	4	8	4	12	-1	-1	-1
编辑 复制 删除	5	8	4	12	-1	-1	-1
编辑 复制 删除	6	8	4	12	-1	-1	-1
编辑 复制 删除	7	8	4	12	-1	-1	-1
编辑 复制 删除	8	8	4	12	-1	-1	-1
编辑 复制 删除	9	8	4	12	-1	-1	-1
编辑 复制 删除	10	8	4	12	-1	-1	-1
编辑 复制 删除	11	8	4	12	-1	-1	-1
编辑 复制 删除	12	8	4	12	-1	-1	-1
编辑 复制 删除	13	8	4	12	-1	-1	-1
编辑 复制 删除	14	8	4	12	-1	-1	-1
编辑 复制 删除	15	8	4	12	-1	-1	-1
编辑 复制 删除	16	8	4	12	-1	-1	-1

图 5-3　执行效果

5.1.4　提取 CSV 数据并保存到 SQLite 数据库

请看下面的实例，功能是提取两个 CSV 文件中的数据，并将提取的数据保存到 SQLite 数据库中。

（1）首先提供两个 CSV 文件：courses.csv 和 peeps.csv。

（2）在下面的实例文件 db_builder.py 中，提取 CSV 文件中的数据，并将提取的数据添加到 SQLite 数据库中。文件 db_builder.py 的具体实现代码如下所示：

```
    import sqlite3
    import csv

    f="database.db"
    db = sqlite3.connect(f) #如果数据库存在则打开，否则将创建
    c = db.cursor()     #创建游标
    #c.execute('.open database.db')
    #==========================================================
    #在这个区域插入你的填充代码，打开两个 CSV 文件
    coursesfile = open('courses.csv','rU')
    studentsfile = open('peeps.csv','rU')
    coursedict = csv.DictReader(coursesfile)
    studentdict = csv.DictReader(studentsfile)
    c.execute('CREATE TABLE courses (code TEXT, mark NUMERIC, id NUMERIC);')
    c.execute('CREATE TABLE students (name TEXT, age NUMERIC, id NUMERIC PRIMARY
KEY);')
    for row in coursedict:
        code = row['code']
        #print code
        mark = row['mark']
        #print mark
        idnum = row['id']
        #print idnum
        filler = repr(code) + ',' +str(mark) + ',' + str(idnum)
        #print filler
        c.execute('INSERT INTO courses VALUES ('+ filler +');')
```

```
    for row in studentdict:
        name = row['name']
        age = row['age']
        idnum = row['id']
        filler = repr(name) + ',' +str(age) + ',' + str(idnum)
        c.execute('INSERT INTO students VALUES ('+ filler +');')


c.execute('SELECT * FROM courses;')
print(c.fetchone())
print('\n')
c.execute('SELECT * FROM students;')
print(c.fetchall())

#=============================================================
db.commit() #保存修改
db.close()  #关闭连接
```

执行后会打印输出 SQLite 数据库中的数据信息，SQLite 数据库中的数据是从 CSV 文件中提取出来的。

```
('systems', 75, 1)


[('kruder', 44, 1), ('dorfmeister', 33, 2), ('sasha', 22, 3), ('digweed', 11, 4), ('tiesto', 99, 5), ('bassnectar', 13, 6), ('TOKiMONSTA', 972, 7), ('jphlip', 27, 8), ('tINI', 23, 9), ('alison', 23, 10)]
```

5.2 实践案例：爬取图书信息并保存为 CSV 文件

在数据可视化分析应用中，经常将海量的数据信息先保存到 CSV 文件中，然后进行下一步处理。例如在爬虫程序中，会先将爬取到的数据保存到 CSV 文件，然后使用 Pandas 或 Matplotlib 绘制可视化程序。在本节的内容中，将通过一个具体实例的实现过程，详细讲解将爬虫抓取的图书信息保存为 CSV 文件的方法。

↑扫码看视频（本节视频课程时间：2 分 55 秒）

5.2.1 实例介绍

本实例是抓取某知名图书商城网站主页中每一个热门标签的第一页的图书，抓取的信息有图书名字、图书出版社信息、作者、评论数等，然后把信息存储到本地的一个 CSV 文件。本实例用到了两个如下所示的第三方库：

- requests：是用 Python 语言基于 urllib 编写的，采用的是 Apache2 Licensed 开源协议的 HTTP 库，Requests 会比 urllib 更加方便，可以提高程序员的开发效率；
- bs4：全名是 BeautifulSoup，是编写 python 爬虫常用库之一，主要用来解析 HTML 标签。

5.2.2 具体实现

编写实例文件 DouBanSpider.py，抓取指定目标网页中的图书标签信息和对应的第一页的图书信息。首先获取指定 URL 中图书信息的分类标签，然后根据标签的内容和链接抓取图书信息。

（1）使用 import 导入需要的库，对应代码如下所示：

```
import ssl
import bs4
import re
import requests
import csv
import codecs
import time

from urllib import request, error

context = ssl._create_unverified_context()
```

（2）因为有一些网站不喜欢被爬虫程序访问，所以会检测连接对象，如果是爬虫程序，也就是非人点击访问，它就不会让你继续访问，所以为了要让程序可以正常运行，需要隐藏自己的爬虫程序的身份。此时，我们就可以通过设置 User Agent 的来达到隐藏身份的目的，User Agent 的中文名为用户代理，简称 UA。在本实例中，在初始化方法 __init__ 中设置 User Agent 浏览器代理，具体实现代码如下所示：

```
class DouBanSpider:
    def __init__(self):
        self.userAgent = "Mozilla/5.0 (Macintosh; Intel Mac OS X 10_12_6)
AppleWebKit/537.36 (KHTML, like Gecko) Chrome/59.0.3071.115 Safari/537.36"
        self.headers = {"User-Agent": self.userAgent}
```

（3）通过方法 getBookCategroies(self) 获取指定网站中图书的分类标签，具体实现代码如下所示：

```
    def getBookCategroies(self):
        try:
            url = "https://book.douban.com/tag/?view=type&icn=index-sorttags-
all"
            response = request.urlopen(url, context=context)
            content = response.read().decode("utf-8")
            return content
        except error.HTTPError as identifier:
            print("errorCode: " + identifier.code + "errrorReason: " + identifier.
reason)
            return None
```

（4）通过方法 getCategroiesContent(self) 获取每个标签的内容，具体实现代码如下所示：

```
    def getCategroiesContent(self):
        content = self.getBookCategroies()
        if not content:
            print("页面抓取失败...")
            return None
        soup = bs4.BeautifulSoup(content, "lxml")
        categroyMatch = re.compile(r"^/tag/*")
        categroies = []
        for categroy in soup.find_all("a", {"href": categroyMatch}):
            if categroy:
                categroies.append(categroy.string)
        return categroies
```

（5）通过方法 getCategroyLink(self) 获取每个标签的链接，具体实现代码如下所示：

```
    def getCategroyLink(self):
        categroies = self.getCategroiesContent()
        categroyLinks = []
        for item in categroies:
            link = "https://book.douban.com/tag/" + str(item)
```

```
            categroyLinks.append(link)
        return categroyLinks
```

（6）通过方法 getBookInfo 获取图书的详细信息，包括书名、出版社信息、作者和评论数等，具体实现代码如下所示：

```
    def getBookInfo(self, categroyLinks):
        self.setCsvTitle()
        categroies = categroyLinks
        try:
            for link in categroies:
                print("正在爬取：" + link)
                bookList = []
                response = requests.get(link)
                soup = bs4.BeautifulSoup(response.text, 'lxml')
                bookCategroy = soup.h1.string
                for book in soup.find_all("li", {"class": "subject-item"}):
                    bookSoup = bs4.BeautifulSoup(str(book), "lxml")
                    bookTitle = bookSoup.h2.a["title"]
                    bookAuthor = bookSoup.find("div", {"class": "pub"})
                    bookComment = bookSoup.find("span", {"class": "pl"})
                    bookContent = bookSoup.li.p
                    # print(bookContent)
                    if bookTitle and bookAuthor and bookComment and bookContent:
                        bookList.append([bookCategroy.strip(), bookTitle.strip(),
bookAuthor.string.strip(), bookComment.string.strip(), bookContent.string.strip()])
                self.saveBookInfo(bookList)
                time.sleep(3)

            print("爬取结束....")

        except error.HTTPError as identifier:
            print("errorCode: " + identifier.code + "errrorReason: " + identifier.
reason)
            return None

    def setCsvTitle(self):
        csvFile = codecs.open("./test.csv", 'a', 'utf_8_sig')
        try:
            writer = csv.writer(csvFile)
            writer.writerow(['图书分类', '图书名字', '图书信息', '图书评论数', '图
书内容'])
        finally:
            csvFile.close()
```

（7）通过方法 saveBookInfo 将抓取的图书信息保存为 CSV 文件 test.csv，具体实现代码如下所示：

```
    def saveBookInfo(self, bookList):
        bookList = bookList
        csvFile = codecs.open("./test.csv", 'a', 'utf_8_sig')
        try:
            writer = csv.writer(csvFile)
            for book in bookList:
                writer.writerow(book)
        finally:
            csvFile.close()
```

（8）开始运行爬虫程序，具体实现代码如下所示：

```
    def start(self):
        categroyLink = self.getCategroyLink()
        self.getBookInfo(categroyLink)
```

```
douBanSpider = DouBanSpider()
douBanSpider.start()
```

在 IDLE 中运行后会显示爬虫过程，如图 5-4 所示。

```
Python 3.6.2 Shell
File Edit Shell Debug Options Window Help
正在爬取：https://book.douban.com/tag/外国文学
正在爬取：https://book.douban.com/tag/文学
正在爬取：https://book.douban.com/tag/随笔
正在爬取：https://book.douban.com/tag/中国文学
正在爬取：https://book.douban.com/tag/经典
正在爬取：https://book.douban.com/tag/日本文学
正在爬取：https://book.douban.com/tag/散文
正在爬取：https://book.douban.com/tag/村上春树
正在爬取：https://book.douban.com/tag/诗歌
正在爬取：https://book.douban.com/tag/童话
正在爬取：https://book.douban.com/tag/儿童文学
正在爬取：https://book.douban.com/tag/古典文学
正在爬取：https://book.douban.com/tag/名著
正在爬取：https://book.douban.com/tag/王小波
正在爬取：https://book.douban.com/tag/杂文
正在爬取：https://book.douban.com/tag/余华
正在爬取：https://book.douban.com/tag/张爱玲
正在爬取：https://book.douban.com/tag/当代文学
正在爬取：https://book.douban.com/tag/钱钟书
正在爬取：https://book.douban.com/tag/外国名著
正在爬取：https://book.douban.com/tag/鲁迅
正在爬取：https://book.douban.com/tag/诗词
正在爬取：https://book.douban.com/tag/茨威格
正在爬取：https://book.douban.com/tag/米兰·昆德拉
正在爬取：https://book.douban.com/tag/杜拉斯
正在爬取：https://book.douban.com/tag/港台
正在爬取：https://book.douban.com/tag/漫画
正在爬取：https://book.douban.com/tag/推理
正在爬取：https://book.douban.com/tag/绘本
正在爬取：https://book.douban.com/tag/青春
正在爬取：https://book.douban.com/tag/东野圭吾
正在爬取：https://book.douban.com/tag/科幻
正在爬取：https://book.douban.com/tag/悬疑
正在爬取：https://book.douban.com/tag/言情
正在爬取：https://book.douban.com/tag/奇幻
正在爬取：https://book.douban.com/tag/武侠
正在爬取：https://book.douban.com/tag/日本漫画
正在爬取：https://book.douban.com/tag/推理小说
正在爬取：https://book.douban.com/tag/耽美
正在爬取：https://book.douban.com/tag/韩寒
Ln: 16 Col: 0
```

图 5-4　爬虫过程

爬虫结束后会将抓取的图书信息保存为 CSV 文件 test.csv，在笔者电脑执行后抓取了 2 829 条图书信息，如图 5-5 所示。

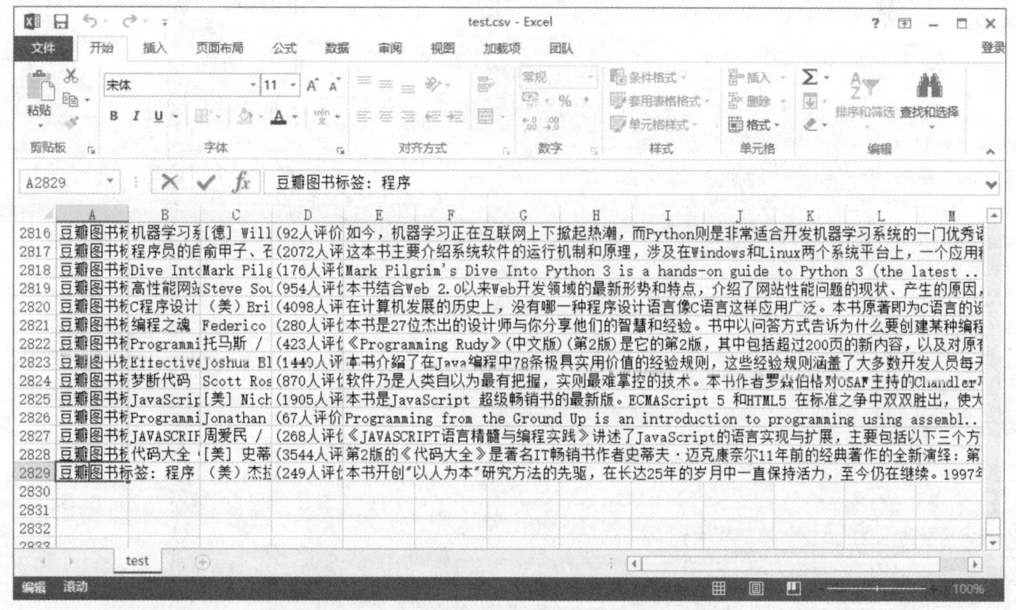

	A	B	C	D	E
2816	豆瓣图书标	机器学习系	[德] Will	(92人评价	如今，机器学习正在互联网上下掀起热潮，而Python则是非常适合开发机器学习系统的一门优秀语
2817	豆瓣图书标	程序员的自	俞甲子、石	(2072人评	这本书主要介绍系统软件的运行机制和原理，涉及在Windows和Linux两个系统平台上，一个应用程
2818	豆瓣图书标	Dive Into	Mark Pilg	(176人评价	Mark Pilgrim's Dive Into Python 3 is a hands-on guide to Python 3 (the latest ..
2819	豆瓣图书标	高性能网站	Steve Sou	(954人评价	本书结合Web 2.0以来Web开发领域的最新形势和特点，介绍了网站性能问题的现状、产生的原因，
2820	豆瓣图书标	C程序设计	（美）Bri	(4098人评	在计算机发展的历史上，没有哪一种程序设计语言像C语言这样应用广泛。本书原著即为C语言的设
2821	豆瓣图书标	编程之魂	Federico	(280人评价	本书是27位杰出的设计师与你分享他们的智慧和经验。书中以问答方式告诉为什么要创建某种编程
2822	豆瓣图书标	Programmi	托马斯 /	(423人评价	《Programming Rudy》(中文版)(第2版)是它的第2版，其中包括超过200页的新内容，以及对原有
2823	豆瓣图书标	Effective	Joshua Bl	(1440人评	本书介绍了在Java编程中78条极具实用价值的经验规则，这些经验规则涵盖了大多数开发人员每天
2824	豆瓣图书标	梦断代码	Scott Ros	(870人评价	软件乃是人类自以为最有把握，实则最难掌控的技术。本书作者罗森伯格对OSAF主持的ChandlerJ
2825	豆瓣图书标	JavaScrip	[美] Nich	(1905人评	本书是JavaScript 超级畅销书的最新版。ECMAScript 5 和HTML5 在标准之争中双双胜出，使大
2826	豆瓣图书标	Programmi	Jonathan	(67人评价	Programming from the Ground Up is an introduction to programming using assembl..
2827	豆瓣图书标	JAVASCRIP	周爱民 /	(268人评价	《JAVASCRIPT语言精髓与编程实践》讲述了JavaScript的语言实现与扩展，主要包括以下三个方
2828	豆瓣图书标	代码大全	[美] 史蒂	(3544人评	第2版的《代码大全》是著名IT畅销书作者史蒂夫·迈克康奈尔11年前的经典著作的全新演绎：第
2829	豆瓣图书标签：程序		（美）杰拉	(249人评价	本书开创"以人为本"研究方法的先驱，在长达25年的岁月中一直保持活力，至今仍在继续。1997年
2830					
2831					
2832					

图 5-5　在 CSV 文件中保存抓取的图书信息

5.3 实践案例：使用 CSV 文件保存 Scrapy 抓取的数据

Scrapy 是使用纯 Python 语言编写的基于异步处理的爬虫框架，开发者可以轻松地实现一个爬虫应用，用来帮助开发者快速抓取网页内容以及各种图片。在本节的内容中，将通过一个具体实例的实现过程，详细讲解使用 Scrapy 抓取信息的方法，介绍将抓取的信息保存为 CSV 文件的过程。

↑扫码看视频（本节视频课程时间：1 分 41 秒）

5.3.1 准备 Scrapy 环境

因为爬虫应用程序的需求日益高涨，所以在市面中诞生了很多第三方开源爬虫框架，其中 Scrapy 是一个为了爬取网站数据，提取结构性数据而编写的应用框架。Scrapy 框架的用途十分广泛，可以用于数据挖掘、数据监测和自动化测试等工作。

在本地计算机安装 Python 后，可以使用“pip”命令或“easy_install”命令来安装 Scrapy，具体命令格式如下所示：

```
pip install scrapy
easy_install scrapy
```

另外还需要确保已经安装了“win32api”模块，在安装此模块时必须安装和本地 Python 版本相对应的版本和位数（32 位或 64 位）。读者可以登录 https://www.lfd.uci.edu/~gohlke/pythonlibs/ 找到需要的版本。

5.3.2 具体实现

在自动生成的 Scrapy 项目的基础上编写代码，首先设置要爬取的目标 URL 和要爬取的内容，然后设置保存爬虫数据的 CSV 文件名。

（1）编写文件 shushan.py 抓取指定网址中的条目信息，具体实现代码如下所示：

```
import scrapy
from scrapy import Request
from scrapy.spiders import Spider
from test100.items import ShushanItem #需要根据自己的项目名称和 Item 类修改，可不修改

class ShushanSpider(Spider):
    name = "shushan" #爬虫命名，可按照自己要求修改，可不修改
    headers = {
            'User-Agent': 'Mozilla/5.0 (Windows NT 5.1; Win64; x64)
AppleWebKit/537.36 (KHTML, like Gecko) Chrome/53.0.2785.143 Safari/537.36',
    } #代理设置，可参考本人设置
    def start_requests(self):
#访问抓取的初始地址，一般为自己要爬取的网页的第一页地址
            url = 'http://www.shushan.net.cn/index.php?_m=mod_product&_
a=prdlist&cap_id=69'
        yield Request(url, headers=self.headers)
    def parse(self, response):
        item = ShushanItem()        #参考 Item.py 里面的设置，根据爬取的内容项取的名字，可
随意取，但此处必须引用
#获取的是所有 item 的上层元素位置
            row_all = response.xpath('//div[@class="prod_list_con"]/div[@
class="prod_list_list"]')
        for row in row_all:                #定义单个行或块
```

```
                item['name'] = row.xpath(
                       './/div[@class="prod_list_name"]/a/text()').extract()[0]# 获取
name 项的元素位置，并用 extract()[0] 提取值
                item['category'] = row.xpath(
                       './/div[@class="prod_list_type"]/a/text()').extract()[0]# 获取
category 项的元素位置，并用 extract()[0] 提取值
                yield item
              next_url = response.xpath('//a[contains(text()," 下 一 页 ")]/@href').
extract()# 获取下一页的网址，根据初始访问地址页面底部，下一页元素的 href 设置
          if next_url:                                  # 如果 next_url 不空
                 next_url = 'http://www.shushan.net.cn/index.php?_m=mod_product&_
a=prdlist&cap_id=69' + next_url[0]# 获取下一页的访问地址
                yield Request(next_url, headers=self.headers) # 循环访问下一页的数据，
获取 name 项和 category 项的值
```

（2）修改文件 items.py，在类 ShushanItem 中设置只包含需要下载的数据项的分类和名称。开发者可以根据自己要抓取的网页内容添加项目名称，有多少要抓取的条目就添加多少项名称。文件 items.py 的具体实现代码如下所示：

```
    class ShushanItem(scrapy.Item):
        # 分类
        category = scrapy.Field()
        # 名称
        name = scrapy.Field()
```

将 CMD 命令行来到 spider 目录，然后运行如下所示的命令会生成 CSV 文件 shushan.csv。

```
    scrapy crawl shushan -o shushan.csv
```

运行过程如图 5-6 所示，在文件 shushan.csv 中会保存抓取的数据信息。如图 5-7 所示。

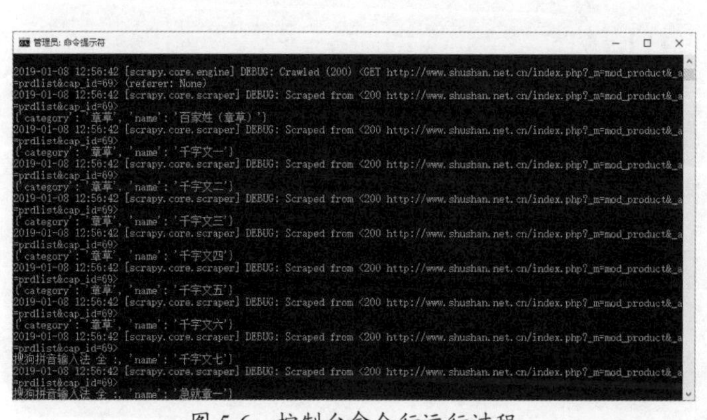

图 5-6　控制台命令行运行过程

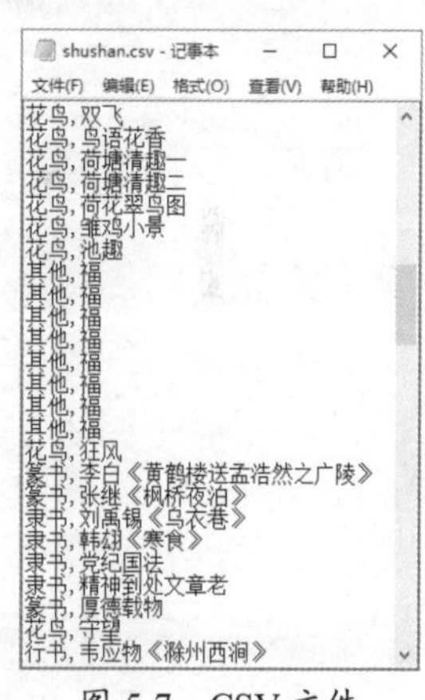

图 5-7　CSV 文件

5.4　实践案例：抓取电子书信息并实现一个 Web 下载系统

在本节内容中，我们会详细讲解抓取指定电子书网站图书信息的方法；通过本实例的学习，将进一步了解 Python 爬虫技术的精髓，掌握将爬虫数据保存为 CSV 文件并存储到 SQLite 数据库中的方法。

↑扫码看视频（本节视频课程时间：3 分 53 秒）

5.4.1 抓取信息并保存到 CSV 文件

编写文件 allitebooks.py，首先设置要抓取的目标 URL，然后使用循环抓取每一个 URL 分页中的图书信息，最后将抓取到的信息分别保存到 CSV 文件和 SQLite 数据库中。

（1）定义变量 FILE_NAME 表示保存的 CSV 文件名，定义变量 ZIP_NAME 表示下载的压缩文件名，通过变量 html 设置要抓取的网页地址。对应代码如下所示：

```
FILE_NAME = 'books_list.csv'
ZIP_NAME= 'allitebooks_ebooks.zip'

html= get('http://www.allitebooks.com/')
html= fromstring(html.text)
```

（2）定义函数 index()，功能是抓取目标网址的电子书信息，然后将抓取到的信息保存到文件 books_list.csv。具体实现代码如下所示：

```
def index():
    with open(FILE_NAME, 'w',encoding='utf-8') as csvfile:
        write = writer(csvfile)
          write.writerow(['Category', 'Title', 'Author(s)', 'ISBN', 'Year',
'Cover_Image', 'Description',
                        'Pages', 'Categories', 'Language', 'File_Size', 'File_
Format', 'Download_Link'])

        categories_title = html.xpath('//*[@id="side-content"]/ul/li/a/text()')
        categories_link  = html.xpath('//*[@id="side-content"]/ul/li/a/@href')
        category         = dict(zip(categories_title, categories_link))

        for c_title, link in tqdm(category.items(), 'Categories'):
            category_html = get(link)
            category_html = fromstring(category_html.text)
              pages = int(category_html.xpath('//*[@id="main-content"]/div/div/
a[last()]/text()')[0])

            for page in tqdm(range(pages), c_title):
                books = get(link + 'page/{}/'.format(page+1))
                books = fromstring(books.text)
                  books_link = books.xpath('//*[@id="main-content"]//header/h2/
a/@href')

                for each_book in tqdm(books_link, 'Page - {}'.format(page+1)):
                    book = get(each_book)
                    book = fromstring(book.text)

                    def get_value(name, link = None):
                          header_details = book.xpath('//*[@id="main-content"]/
div/article/header')[0]
                        if name is 'title': return (header_details.xpath('./h1/
text()') or [''])[0]
                              if name is 'dlink': return (book.xpath('//span[@
class="download-links"]/a[contains\
                                                       (@href, "file.allitebooks.
com")]/@href') or [''])[0]
                          if name is 'cover': return (header_details.xpath('.//
div/a/img/@src') or [''])[0]
                        if name is 'descr':
                              desc = book.xpath('//div[@class="entry-content"]//
text()')
                                return ' '.join([x for x in [x.strip() for x in
desc]])
```

```
                    if link: return ', '.join([x for x in (header_details.
xpath('.//*[text() = "{}"]\
                                                /following-sibling::dd[1]/a/
text()'.format(name)) or [''])])
                    return (header_details.xpath('.//*[text() = "{}"]/
following-sibling::dd[1]//text()'.format(name)) or [''])[0]

                category      = c_title
                book_name     = get_value('title')
                cover_img     = get_value('cover')
                authors       = get_value('Author:', link = True)
                isbn          = get_value('ISBN-10:')
                year          = get_value('Year:')
                pages         = get_value('Pages:')
                description   = get_value('descr')
                language      = get_value('Language:')
                file_size     = get_value('File size:')
                file_format   = get_value('File format:')
                categories    = get_value('Category:', link = True)
                download_link = get_value('dlink')

                write.writerow([category, book_name, authors, isbn, year,
cover_img, description,
                                pages, categories, language, file_size, file_
format, download_link])
```

（3）编写函数 download()，功能是下载文件中 books_list.csv 中的所有链接的电子书，具体实现代码如下所示：

```
def download():
    try:
        files = ZipFile(ZIP_NAME).namelist()
    except: files = []
    with ZipFile(ZIP_NAME, 'a', ZIP_DEFLATED) as output:
        with open(FILE_NAME, 'r',encoding='utf-8') as csvfile:
            reader = DictReader(csvfile)
            for row in reader:
                b_name    = row['Download_Link']
                response  = get(b_name)
                file_name = b_name.split('/')[-1]
                if file_name not in files:
                    print('\rDownloading : {}'.format(file_name), end='')
                    output.writestr(b_name.split('/')[-1], response.content)
                else:
                    print('\rSkipping : {}'.format(file_name), end='')
```

（4）提示用户选择一个操作，根据用户的输入执行对应操作，具体实现代码如下所示：

```
if __name__ == '__main__':
    print('请选择要进行的操作：')
    print('1 - 用下载链接列出 csv 文件中的所有电子书）\
        \n2 - Scrape 全部（索引和下载压缩文件）\n')
    choice = int(input('输入你要进行的操作：'))
    if isinstance(choice, int) and choice in [1, 2]:
        if choice == 1:
            index()
        else:
            index()
            download()
    else:
        raise RuntimeError('请输入有效选项！！！')
```

执行后输出结果如下：

```
请选择要进行的操作：
```

```
1 - 用下载链接列出 csv 文件中的所有电子书）
2 - Scrape 全部（索引和下载压缩文件）

输入你要进行的操作 : 1
Categories:   0%|          | 0/15 [00:00<?, ?it/s]
Web Development:   0%|          | 0/187 [00:00<?, ?it/s]

Page - 1:   0%|          | 0/10 [00:00<?, ?it/s]

Page - 1:  10%|█         | 1/10 [00:01<00:14,  1.61s/it]

Page - 1:  20%|██        | 2/10 [00:04<00:15,  1.93s/it]

Page - 1:  30%|███       | 3/10 [00:10<00:22,  3.27s/it]

Page - 1:  40%|████      | 4/10 [00:12<00:16,  2.81s/it]

Page - 1:  50%|█████     | 5/10 [00:15<00:13,  2.78s/it]

Page - 1:  60%|██████    | 6/10 [00:16<00:09,  2.36s/it]

Page - 1:  70%|███████   | 7/10 [00:17<00:06,  2.10s/it]

Page - 1:  80%|████████  | 8/10 [00:19<00:03,  2.00s/it]

Page - 1:  90%|█████████ | 9/10 [00:26<00:03,  3.43s/it]

Page - 1: 100%|██████████| 10/10 [00:27<00:00,  2.71s/it]

Web Development:   1%|          | 1/187 [00:30<1:33:01, 30.01s/it]

Page - 2:   0%|          | 0/10 [00:00<?, ?it/s]

Page - 2:  10%|█         | 1/10 [00:01<00:17,  1.98s/it]

Page - 2:  20%|██        | 2/10 [00:03<00:14,  1.86s/it]
### 省略后面的抓取过程
```

抓取的数据被保存到文件 books_list.csv 中，如图 5-8 所示。

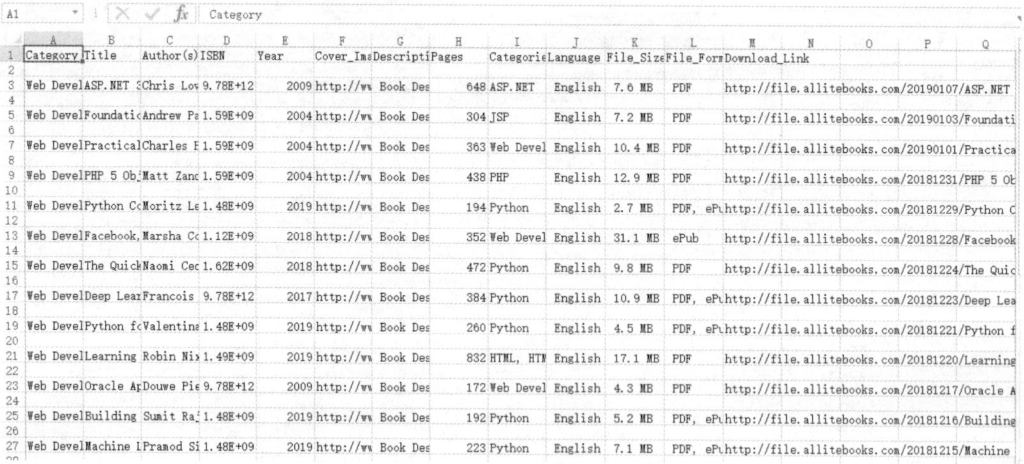

A1 | Category

	Category	Title	Author(s)	ISBN	Year	Cover_Ima	Descripti	Pages	Categorie	Language	File_Size	File_Form	Download_Link
2													
3	Web Devel	ASP.NET 3	Chris Lov	9.78E+12	2009	http://ww	Book Des	648	ASP.NET	English	7.6 MB	PDF	http://file.allitebooks.com/20190107/ASP.NET
4													
5	Web Devel	Foundatic	Andrew Pa	1.59E+09	2004	http://ww	Book Des	304	JSP	English	7.2 MB	PDF	http://file.allitebooks.com/20190103/Foundati
6													
7	Web Devel	Practical	Charles F	1.59E+09	2004	http://ww	Book Des	363	Web Devel	English	10.4 MB	PDF	http://file.allitebooks.com/20190101/Practica
8													
9	Web Devel	PHP 5 Obj	Matt Zanc	1.59E+09	2004	http://ww	Book Des	438	PHP	English	12.9 MB	PDF	http://file.allitebooks.com/20181231/PHP 5 Ob
10													
11	Web Devel	Python Cc	Moritz Le	1.48E+09	2019	http://ww	Book Des	194	Python	English	2.7 MB	PDF, ePu	http://file.allitebooks.com/20181229/Python C
12													
13	Web Devel	Facebook,	Marsha Cc	1.12E+09	2018	http://ww	Book Des	352	Web Devel	English	31.1 MB	ePub	http://file.allitebooks.com/20181228/Facebook
14													
15	Web Devel	The Quick	Naomi Cec	1.62E+09	2018	http://ww	Book Des	472	Python	English	9.8 MB	PDF	http://file.allitebooks.com/20181224/The Quic
16													
17	Web Devel	Deep Lear	Francois	9.78E+12	2017	http://ww	Book Des	384	Python	English	10.9 MB	PDF, ePu	http://file.allitebooks.com/20181223/Deep Lea
18													
19	Web Devel	Python fc	Valentina	1.48E+09	2019	http://ww	Book Des	260	Python	English	4.5 MB	PDF, ePu	http://file.allitebooks.com/20181221/Python f
20													
21	Web Devel	Learning	Robin Nix	1.49E+09	2019	http://ww	Book Des	832	HTML, HTM	English	17.1 MB	PDF	http://file.allitebooks.com/20181220/Learning
22													
23	Web Devel	Oracle Ap	Douwe Pie	9.78E+12	2009	http://ww	Book Des	172	Web Devel	English	4.3 MB	PDF	http://file.allitebooks.com/20181217/Oracle A
24													
25	Web Devel	Building	Sumit Raj	1.48E+09	2019	http://ww	Book Des	192	Python	English	5.2 MB	PDF, ePu	http://file.allitebooks.com/20181216/Building
26													
27	Web Devel	Machine L	Pramod Si	1.48E+09	2019	http://ww	Book Des	223	Python	English	7.1 MB	PDF, ePu	http://file.allitebooks.com/20181215/Machine

图 5-8　文件 books_list.csv

5.4.2 抓取信息并保存到 SQLite 数据库

编写文件 allitebooks_db.py，功能是爬虫抓取指定电子书网站图书信息的方法，并将抓取到的信息保存到 SQLite 数据库。文件 allitebooks_db.py 的主要实现代码如下所示：

```
    db_schema = """
        CREATE TABLE IF NOT EXISTS ebooks_index (
            category        CHAR NOT NULL,
            book_name       CHAR NOT NULL,
            cover_img       CHAR NOT NULL,
            authors         CHAR,
            isbn            CHAR,
            year            CHAR,
            pages           CHAR,
            description     CHAR,
            language        CHAR,
            file_size      CHAR,
            file_format    CHAR,
            categories      CHAR,
            download_link CHAR NOT NULL UNIQUE,
            download_file BLOB );
    """

    html = get('http://www.allitebooks.com/')
    html = fromstring(html.text)

    conn = connect('allitebooks.db')
    curs = conn.cursor()
    curs.executescript(db_schema)
    categories_title = html.xpath('//*[@id="side-content"]/ul/li/a/text()')
    categories_link  = html.xpath('//*[@id="side-content"]/ul/li/a/@href')
    category         = dict(zip(categories_title, categories_link))

    for c_title, link in tqdm(category.items(), 'Categories'):
        category_html = get(link)
        category_html = fromstring(category_html.text)
         pages = int(category_html.xpath('//*[@id="main-content"]/div/div/a[last()]/
text()')[0])

        for page in tqdm(range(pages), c_title):
            books = get(link + 'page/{}/'.format(page+1))
            books = fromstring(books.text)
            books_link = books.xpath('//*[@id="main-content"]//header/h2/a/@href')

            for each_book in tqdm(books_link, 'Page - {}'.format(page+1)):
                book = get(each_book)
                book = fromstring(book.text)

                def get_value(name, link = None):
                        header_details = book.xpath('//*[@id="main-content"]/div/
article/header')[0]
                   if name is 'title': return (header_details.xpath('./h1/text()')
or [''])[0]
                   if name is 'dlink': return (book.xpath('//span[@class="download-
links"]/a[contains\
                                                (@href, "file.allitebooks.com")]/@
href') or [''])[0]
                       if name is 'cover': return (header_details.xpath('.//div/a/
img/@src') or [''])[0]
                   if name is 'descr':
                       desc = book.xpath('//div[@class="entry-content"]//text()')
                       return ' '.join([x for x in [x.strip() for x in desc]])
```

```
                        if link: return ', '.join([x for x in (header_details.
xpath('.//*[text() = "{}"]\
                                            /following-sibling::dd[1]/a/text()'.
format(name)) or [''])])
                        return (header_details.xpath('.//*[text() = "{}"]/following-
sibling::dd[1]//text()'.format(name)) or [''])[0]

                down = get(get_value('dlink'))
                curs.execute("INSERT OR REPLACE INTO ebooks_index \
                            (category, book_name, cover_img, authors, isbn, year,
pages, description, \
                            language, file_size, file_format, categories, download_
link, download_file) VALUES \
                            (?, ?, ?, ?, ?, ?, ?, ?, ?, ?, ?, ?, ?, ?)", (c_title,
get_value('title'), \
                            get_value('cover'), get_value('Author:', link = True),
get_value('ISBN-10:'),\
                                 get_value('Year:'), get_value('Pages:'), get_
value('descr'), get_value('Language:'),\
                              get_value('File size:'), get_value('File format:'),
get_value('Category:', link = True),\
                           get_value('dlink'), down.content))
                conn.commit()
    conn.close()
```

执行后会创建数据库 allitebooks.db，在里面保存抓取到的数据信息，如图 5-9 所示。

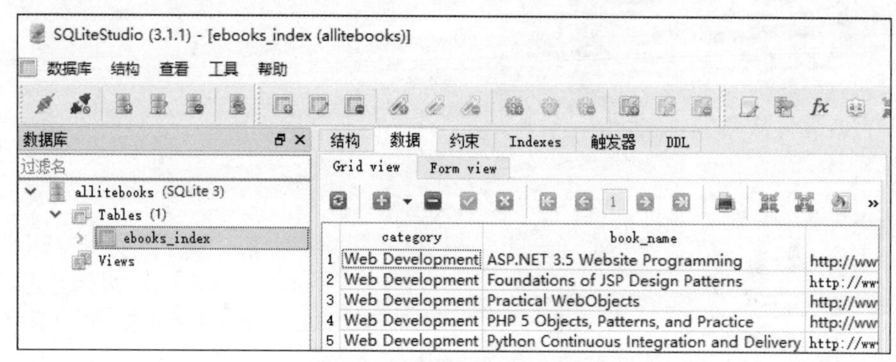

图 5-9　数据库 allitebooks.db

5.4.3　利用爬虫数据建立自己的电子书下载系统

接下来创建一个 Flask Web 项目，读取保存爬虫数据的数据库文件 allitebooks.db，建立一个人自己的在线电子书下载系统。

（1）创建一个 Flask Web 项目，启动文件 web.py 的具体实现代码如下所示：

```
    from flask    import Flask, render_template
    from sqlite3 import connect
    from os       import path

    path = path.dirname(path.realpath(__file__))
    db   = path + '/allitebooks.db'
    app = Flask(__name__)

    connection = connect(db)
    cursor     = connection.cursor()
    data          = cursor.execute("SELECT category, book_name, cover_img, authors,
pages, \
```

```
                                        description, file_size, download_link FROM ebooks_
index LIMIT 100").fetchall()

    @app.route('/')
    def index():
        return render_template('index.html', result = data)

    if __name__ == '__main__':
            app.run()
```

（2）通过模板文件 index.html 显示数据库保存的电子书信息，具体实现代码如下所示：

```
    <!doctype html>
    <html lang="en">
      <head>
        <meta charset="utf-8">
         <meta name="viewport" content="width=device-width, initial-scale=1.0,
minimum-scale=1.0">
        <title>ALLITEBOOKS</title>

         <link rel="stylesheet" href="https://fonts.googleapis.com/css?family=Roboto
:regular,bold,italic,thin,light,bolditalic,black,medium&amp;lang=en">
          <link rel="stylesheet" href="https://fonts.googleapis.com/
icon?family=Material+Icons">
         <link rel="stylesheet" href="https://code.getmdl.io/1.3.0/material.deep_
purple-pink.min.css">
        <link rel="stylesheet" href="{{ url_for('static', filename='styles.css') }}">
        <style>
        #view-source {
          position: fixed;
          display: block;
          right: 0;
          bottom: 0;
          margin-right: 40px;
          margin-bottom: 40px;
          z-index: 900;
        }
        </style>
      </head>
      <body class="mdl-demo mdl-color--grey-100 mdl-color-text--grey-700 mdl-base">
        <div class="mdl-layout mdl-js-layout mdl-layout--fixed-header">
          <main class="mdl-layout__content">
            <div class="mdl-layout__tab-panel is-active" id="overview">

              {% for data in result %}
                <section class="section--center mdl-grid mdl-grid--no-spacing mdl-
shadow--2dp">
                  <header class="section__play-btn mdl-cell mdl-cell--3-col-desktop mdl-
cell--2-col-tablet mdl-cell--5-col-phone mdl-color--teal-100 mdl-color-text--white">
                    <img width="100%" src="{{ data[2] }}">
                  </header>
                  <div class="mdl-card mdl-cell mdl-cell--9-col-desktop mdl-cell--5-
col-tablet mdl-cell--5-col-phone">
                    <div class="mdl-card__supporting-text">
                      <h4>{{ data[1] }}</h4>
                      <h6>By: {{ data[3] }}</h6>
                      <h6>Category : {{ data[0] }}</h6>
                       {{ data[5]|truncate(500) }}
                    </div>
                    <div class="mdl-card__actions">
                            <a href="{{ data[7] }}" class="mdl-button" target="_
blank">DOWNLOAD</a>
                      <a href="#" class="mdl-button">File size : {{ data[6] }}</a>
                      <a href="#" class="mdl-button">Pages : {{ data[4] }}</a>
```

```
            </div>
          </div>
        </section>
        {% endfor %}

      </div>
      <div class="mdl-layout__tab-panel" id="features">
      </div>
    </main>
  </div>
  <script src="https://code.getmdl.io/1.3.0/material.min.js"></script>
</body>
</html>
```

在本地浏览器中输入“http://127.0.0.1:5000/”后会显示抓取到的电子书信息，如图 5-10 所示。

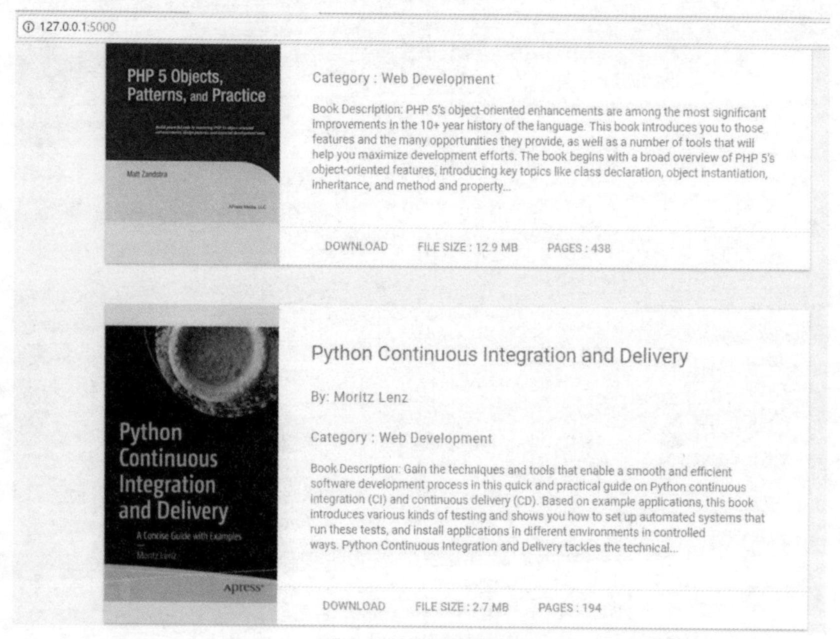

图 5-10　执行效果

5.5 实践案例：某网店口罩销量数据的可视化

假设有一个在天猫平台销售口罩的网店，在一个 CSV 文件中保存了 2020 年第一季度各类口罩的销量。我们可以编写一个 Python 程序，使用 Matplotlib 绘制可视化展示各类口罩销量的统计柱状图。

↑扫码看视频（本节视频课程时间：2 分 04 秒）

5.5.1 准备 CSV 文件

准备 CSV 文件 fall-2020.csv，在里面保存了各类口罩的销量数据，文件 fall-2020.csv 的具体内容如下所示：

```
7
普通口罩,1244
普通医用口罩,1142
医用外科口罩,4250
N90 口罩,1754
N95 口罩,1569
KN94 口罩,3763
N99 口罩,1149
```

5.5.2　可视化 CSV 文件中的数据

编写 Python 程序文件 n95.py，功能是读取 CSV 文件“fall-2020.csv”的内容，并根据读取的数据绘制柱状图，展示一个类口罩商品的销量情况。程序文件 n95.py 的具体实现代码如下所示：

```
import numpy as np
import matplotlib.pyplot as plt
plt.rcParams['font.sans-serif'] = ['SimHei'] # 指定默认字体

def read_file(file_name): # 读取 CSV 文件的内容
    i = 0
    data = open(file_name, 'r', encoding='UTF-8')
    for line in data:
        sections = int(line[0])
         college_array = np.empty(sections, dtype = object) # 为不同口罩类型创建空
数组
        enrollment_array = np.zeros(sections) # 为销量创建空数组
        break

    # 使用逗号分隔数据，然后使用 for 循环将其添加到分隔数组中
    for line in data:
        line = line.split(',')
        line[1] = line[1][:-1]
        college_names = line[0]
        college_array[i] = college_names
        enrollment_array[i] = line[1]
        i +=1
     return college_array, enrollment_array # 分别返回 college department 和 statistic 数
组的元组


def main(file_name):
    i = 0
    read_file(file_name) # 读取 CSV 文件的内容
     college_names, college_enrollments = read_file(file_name) #gets the results
from the read_file method
    print(college_names)
    print(college_enrollments)

    # 使用 matplotlib 库绘制图形并以可读的方式显示信息
    plt.figure("2020 某网店的口罩销量统计图 ")
     plt.bar(college_names, college_enrollments, width = 0.8, color = ["blue",
"gold"])
    plt.yticks(ticks = np.arange(0,4800, 400))
    plt.ylim(0,4400)

    plt.xlabel(" 口罩类型 ")
    plt.ylabel(" 销量 ")

    plt.show() # 显示可视化统计图形

main("fall-2020.csv")
```

执行效果如图 5-11 所示。

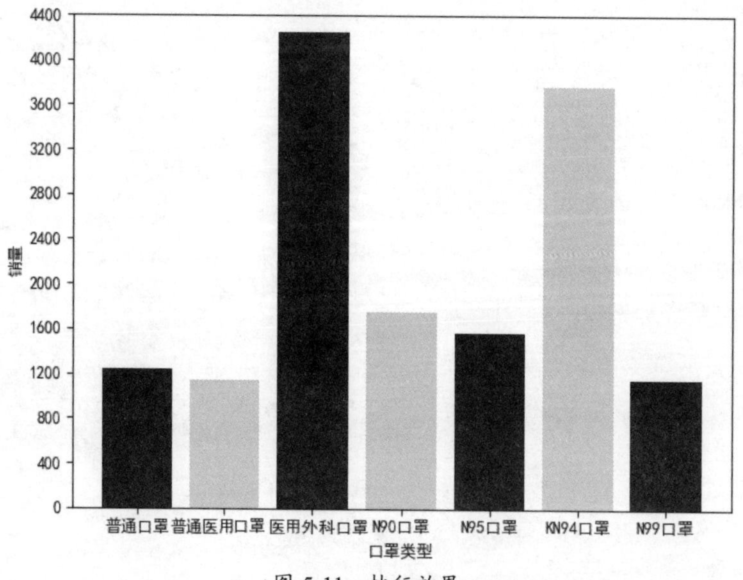

图 5-11　执行效果

5.6　实践案例：根据 CSV 文件绘制可视化 3D 图形

假设现在有一个 CSV 文件，在里面保存了绘制三维 matplotlib 图形所需要的数据。要求编写一个 Python 程序，调用这个 CSV 文件中的数据使用 matplotlib 绘制可视化 3D 图形。根据实例描述和要求可知，我们只需提取 CSV 文件中的数据，然后利用提取到的数据绘制 3D 图形。

↑扫码看视频（本节视频课程时间：1 分 14 秒）

5.6.1　准备 CSV 文件

创建文件夹“csv_files”，在这个文件夹中准备 CSV 文件 useasimilar.csv，在里面保存了绘制三维 matplotlib 图形所需要的数据，文件 useasimilar.csv 的具体内容如下所示：

```
Organization,Recreation,Education,Diversity_and_inclusivity
American Esports,0,0,-1
Brass City Gamers,1,.5,-1
Evolve Youth Esports,1,-1,-1
Minnesota Esports Club,1,0,-1
US Esports Foundation,-1,1,1
```

5.6.2　绘制可视化 3D 图形

编写 Python 程序文件 pltr.py，功能提示用户输入要读取的 CSV 文件名，然后要求用户分别输入 3D 图的标题、x 轴标识、y 轴标识和 z 轴标识，最后根据用户输入的信息和 CSV 文件中的信息绘制可视化 3D 图形。程序文件 pltr.py 的具体实现代码如下所示：

```
import csv
import matplotlib.pyplot as plt
```

```
from mpl_toolkits.mplot3d import Axes3D
import numpy as np
from os import chdir


# 查找的文件夹
chdir("csv_files")
filename = input("file name:  ")

xr, yr, zr = [], [], []

# 从 csv 中提取数据
with open(filename, newline="") as csvfile:
    reader = csv.reader(csvfile, delimiter=",")
    header = next(reader)
    if header != None:
            for i in reader:
                    xr.append(float(i[1]))
                    yr.append(float(i[2]))
                    zr.append(float(i[3]))

data_raw, data_unique = [], []
# 获取绘图列表
for i in zip(xr, yr, zr):
    data_raw.append(i)
for i in data_raw:
    if i not in data_unique:
            data_unique.append(i)

point_frequency = []
# 获取频率值
for i in data_unique:
    c = 0
    for j in data_raw:
            if i == j:
                    c += 1
    point_frequency.append(c)
    #TODO 将比例因子添加到点频率

xu, yu, zu = [], [], []
# x、y、z 轴的刻度
for i in data_unique:
    xu.append(i[0])
    yu.append(i[1])
    zu.append(i[2])

#指定自定义点标记权重
# n = len(data_raw)

# 制作散点图
fig = plt.figure()
ax = Axes3D(fig)
ax.scatter(xu, yu, zu, c=point_frequency)

"""
for x, y, z, i in zip(xu, yu, zu, point_frequency):
    ax.text(x, y, z, i, size=15)
"""

# 设置轴范围
ax.set_xlim(-1, 1)
ax.set_ylim(-1, 1)
ax.set_zlim(-1, 1)
```

```
# 设置轴刻度
ax.set_xticks(np.arange(-1, 1, step=.5))
ax.set_yticks(np.arange(-1, 1, step=.5))
ax.set_zticks(np.arange(-1, 1, step=.5))

# 设置标题和轴名称
ax.set_title(input("plot title:  "))
ax.set_xlabel(input("x axis:  "))
ax.set_ylabel(input("y axis:  "))
ax.set_zlabel(input("z axis:  "))
```

执行后会成功绘制可视化 3D 图形，如图 5-12 所示。

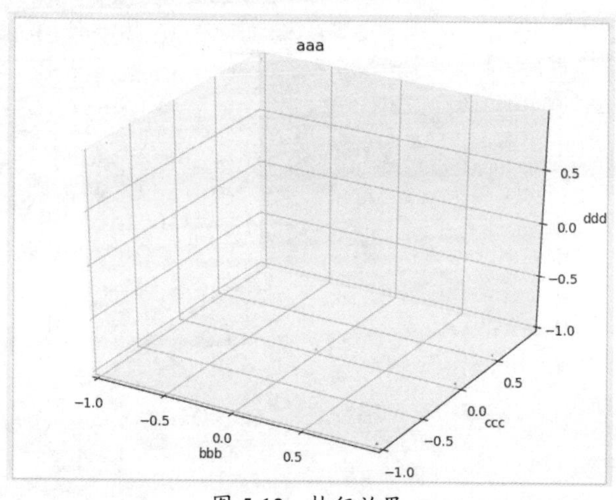

图 5-12　执行效果

5.7　数据挖掘：可视化处理文本情感分析数据

情感分析是自然语言处理和数据挖掘领域中的重要内容，也是机器学习开发者们乐于研究的领域。文本情感分析又被称为意见挖掘、倾向性分析等，是对带有情感色彩的主观性文本进行分析、处理、归纳和推理的过程。在互联网（例如博客和论坛以及社会服务网络如大众点评）中产生了大量的用户参与的、对于诸如人物、事件、产品等有价值的评论信息。我们可以编写一个 Python 程序，使用库 matplotlib 绘制可视化处理文本情感分类数据的柱状图。

↑扫码看视频（本节视频课程时间：3 分 35 秒）

5.7.1　准备 CSV 文件

预先准备两部电视剧的剧本文件，这些文件的格式是记事本格式“.txt”，在这些剧本文件中有大量的英文单词。将整个剧本文件分为两类：a 和 b，分别表示电视剧 a 和电视剧 b 的剧本文件。在 CSV 文件 sentiment_lex.csv 中保存了 4 798 个情感分析单词的值。如图 5-13 所示。

图 5-13　准备的文件

5.7.2　可视化两个剧本的情感分析数据

编写 Python 程序文件 main.py，预先设置我们要在柱状图中统计显示的 5 种情感类型，然后遍历所有第 1 部和第 2 部电视剧剧本对应的记事本文件，并分割每一个单词的内容。

（1）打入需要的库，设置 font.sans-serif 保证能够在绘制的柱状图中显示中文。对应的实现代码如下所示：

```
import numpy as np
import matplotlib.pyplot as mplot
from glob import glob
mplot.rcParams['font.sans-serif'] = ['SimHei'] # 指定默认字体
t=0
t2=0
```

（2）准备要处理的文件，设置第 1 部电视剧剧本的对应的记事本文件名以字母“a”开头，第 2 部电视剧剧本的对应的记事本文件名以字母“b”开头。对应的实现代码如下所示：

```
afilenames = glob('a*.txt')
allawords=[]

bfilenames = glob('b*.txt')
allbwords=[]

file=[]

lex={}
```

（3）设置我们要在柱状图中统计显示的 5 种情感类型 neg（Negative，阴性）、wneg（Weakly Negative，弱阴性）、neu（Neutral，中性）、wpos（Weakly Positive，弱阳性）和 pos（Positive，阳性）。对应的实现代码如下所示：

```
neg=0
wneg=0
neu=0
wpos=0
pos=0
```

（4）遍历所有第 1 部电视剧剧本的对应的记事本文件，分割每一个单词的内容。对应的实现代码如下所示：

```
for i in range(0,len(afilenames)):
    data=open(afilenames[i], "r")
    data=data.read()
```

```
        data=data.replace("\n", " ")
        data=data.replace(".", "")
        data=data.replace("?", "")
        data=data.replace(",", "")
        data=data.replace("!", "")
        data=data.replace("[", "")
        data=data.replace("]", "")
        data=data.replace("(", "")
        data=data.replace(")", "")
        data=data.replace(">", "")
        data=data.replace("<", "")
        data=data.split(" ")
        for j in range(0, len(data)):
            if len(data[j]) > 0:
                if (data[j]) != (data[j].upper()):
                    allawords.append(data[j].lower())
```

（5）遍历所有第 2 部电视剧剧本对应的记事本文件，分割每一个单词的内容。对应的实现代码如下所示：

```
    for i in range(0,len(bfilenames)):
        data=open(bfilenames[i], "r")
        data=data.read()
        data=data.replace("\n", " ")
        data=data.replace(".", "")
        data=data.replace("?", "")
        data=data.replace(",", "")
        data=data.replace("!", "")
        data=data.replace("[", "")
        data=data.replace("]", "")
        data=data.replace("(", "")
        data=data.replace(")", "")
        data=data.split(" ")
        for j in range(0, len(data)):
            if len(data[j]) > 0:
                if (data[j]) != (data[j].upper()):
                    allbwords.append(data[j].lower())
```

（6）打开预先设置的 CSV 文件 sentiment_lex.csv，然后读取里面的情感分析值。对应的实现代码如下所示：

```
    file=open("sentiment_lex.csv", "r")
    sent=file.read()
    sent=sent.split("\n")


    for i in range(0,len(sent)-2):
        sent[i]=sent[i].split(",")
        sent[i][1]= np.float64(sent[i][1])
        lex[sent[i][0]]=sent[i][1]
```

（7）提示用户输入要分析哪一部电视剧的剧本，如果用户输入“a”，则分析第 1 部电视剧的剧本内容，并使用库 matplotlib 绘制情感分析统计柱状图。对应的实现代码如下所示：

```
    choice = input("分析哪一部电视剧的剧本？ ")
    if choice == "a":
        for i in range(0, len(allawords)):
            if allawords[i] in lex:
                if lex[allawords[i]]>=-1 and lex[allawords[i]]<-.6:
                    neg+=1
                if lex[allawords[i]]>=-.6 and lex[allawords[i]]<-.2:
                    wneg+=1
                if lex[allawords[i]]>=-.2 and lex[allawords[i]]<=.2:
```

```
                neu+=1
            if lex[allawords[i]]>.2 and lex[allawords[i]]<=.6:
                wpos+=1
            if lex[allawords[i]]>.6 and lex[allawords[i]]<=1:
                pos+=1

    y = [neg, wneg, neu, wpos, pos]
    y=np.log10(y)
    x = ["Neg", "W.Neg", "Neu", "W.Pos", "Pos"]

    mplot.title("a 类情感分析 ")
    mplot.xlabel(" 情感 ")
    mplot.ylabel("log10")
    mplot.bar(x,y)
    mplot.show()
```

（8）如果用户输入“b”，则分析第 2 部电视剧的剧本内容。并使用库 matplotlib 绘制情感分析统计柱状图。对应的实现代码如下所示：

```
elif choice == "b":
    for i in range(0, len(allbwords)):
        if allbwords[i] in lex:
            if lex[allbwords[i]]>=-1 and lex[allbwords[i]]<-.6:
                neg+=1
            if lex[allbwords[i]]>=-.6 and lex[allbwords[i]]<-.2:
                wneg+=1
            if lex[allbwords[i]]>=-.2 and lex[allbwords[i]]<=.2:
                neu+=1
            if lex[allbwords[i]]>.2 and lex[allbwords[i]]<=.6:
                wpos+=1
            if lex[allbwords[i]]>.6 and lex[allbwords[i]]<=1:
                pos+=1


    y = [neg, wneg, neu, wpos, pos]
    y=np.log10(y)
    x = ["Neg", "W.Neg", "Neu", "W.Pos", "Pos"]

    mplot.title("b 类情感分析 ")
    mplot.xlabel(" 情感 ")
    mplot.ylabel("log10")
    mplot.bar(x,y)
    mplot.show()
```

执行后会提示用户输入要分析哪一部电视剧的剧本，如果用户输入“a”，则分析第 1 部电视剧的剧本内容，并使用库 matplotlib 绘制情感分析统计柱状图。如图 5-14 所示。

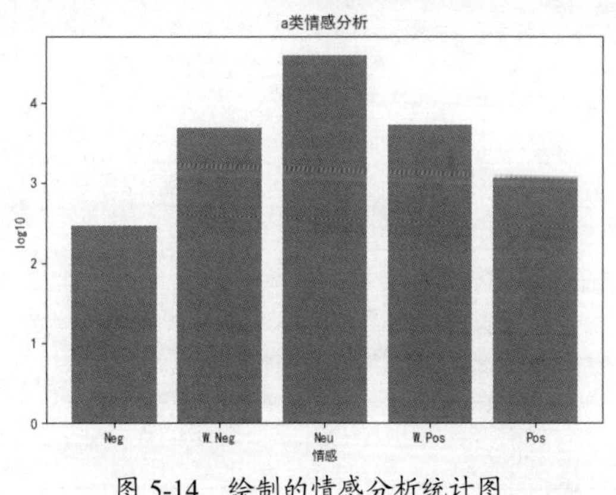

图 5-14　绘制的情感分析统计图

第 6 章 可视化处理 JSON 数据

JSON（JavaScript Object Notation）是基于 ECMAScript（可以理解为是 JavaScript 的一个标准）的一个子集；它是一种轻量级的数据交换格式，可以被几乎所有的编程语言兼容。在本节的内容中，将详细讲解使用 Python 语言解析并可视化 JSON 数据的知识，为读者步入本书后面知识的学习打下基础。

6.1 Python 内置的 JSON 处理模块

对于 Python 语言来说，处理 JSON 数据非常方便，因为 Python 语言提供了内置的数据类型和专业的内置模块来处理 JSON 数据。在本节的内容中，将简要讲解使用 Python 内置的模块处理 JSON 数据的方法。

↑扫码看视频（本节视频课程时间：8 分 28 秒）

6.1.1 内置的类型转换

在 JSON 的编码和解码过程中，Python 的原始类型与 JSON 类型会相互转换。Python 编码为 JSON 类型的转换对应表如表 6-1 所示。

表 6-1

Python	JSON
dict	object
list, tuple	array
str	string
int, float, int- & float-derived Enums	number
True	true
False	false
None	null

JSON 解码为 Python 类型的转换对应表如表 6-2 所示。

表 6-2

JSON	Python
object	dict
array	list

续表

JSON	Python
string	str
number (int)	int
number (real)	float
true	True
false	False
null	None

6.1.2　使用内置 json 模块实现编码和解码

在 Python 程序中，可以使用 json 模块对 JSON 数据进行编码和解码操作。例如在下面的实例文件 js.py 中，演示了将 Python 字典类型转换为 JSON 对象的过程。

```
import json
# 将字典类型转换为 JSON 对象
data = {
    'no' : 1,
    'name' : 'laoguan',
    'url' : 'http://www.toppr.net'
}
json_str = json.dumps(data)
print ("Python 原始数据：", repr(data))
print ("JSON 对象：", json_str)
```

执行后的效果如图 6-1 所示。通过输出结果可以看出，简单类型通过编码后跟其原始的 repr() 输出结果非常相似。

```
Python 原始数据： {'url': 'http://www.toppr.net', 'no': 1, 'name': 'laoguan'}
JSON 对象： {"url": "http://www.toppr.net", "no": 1, "name": "laoguan"}
```

图 6-1　执行效果

在下面的实例文件 fan.py 中，可以将一个 JSON 编码的字符串转换回一个 Python 数据结构。

```
import json
# 将字典类型转换为 JSON 对象
data1 = {
    'no' : 1,
    'name' : 'laoguan',
    'url' : 'http://www.toppr.net'
}
json_str = json.dumps(data1)
print ("Python 原始数据：", repr(data1))
print ("JSON 对象：", json_str)
# 将 JSON 对象转换为字典
data2 = json.loads(json_str)
print ("data2['name']: ", data2['name'])
print ("data2['url']: ", data2['url'])
```

执行后的效果如图 6-2 所示。

```
Python 原始数据： {'name': 'laoguan', 'url': 'http://www.toppr.net', 'no': 1}
JSON 对象： {"name": "laoguan", "url": "http://www.toppr.net", "no": 1}
data2['name']:  laoguan
data2['url']:  http://www.toppr.net
```

图 6-2　执行效果

在 Python 程序中，如果要处理的是 JSON 文件而不是字符串，那么可以使用函数 json.

dump() 和函数 json.load() 来编码和解码 JSON 数据。例如下面的演示代码：

```
# 写入 JSON 数据
with open('data.json', 'w') as f:
json.dump(data, f)
# 读取数据
with open('data.json', 'r') as f:
data = json.load(f)
```

例如在下面的实例文件 jsonparser.py 中，首先自定义编写了解析 JSON 数据的类 jsonparser，在此类中创建了如表 6-3 所示的几个功能函数。

表 6-3

函数名称	说明
parse(self)	跳过空白、换行或 tab 解析数据
parse_string(self)	找出两个双引号中的 string
parse_value(self)	解析 value 值，包含字符串、数字、bool 变量和 null
parse_array(self)	解析数组
parse_object(self)	解析对象实例
to_str(pv)	把 Python 变量转换成 string 字符串

最后调用类 jsonparser 中的上述功能函数，解析指定的 JSON 文件的内容。

```
import sys
from imp import reload
reload(sys)
import json

def txt2str(file='jsondata2.txt'):
    '''
    打开指定的 json 文件
    '''
    fp=open(file,encoding='UTF-8')
    allLines = fp.readlines()
    fp.close()
    str=""
    for eachLine in allLines:
            #eachLine=ConvertCN(eachLine)

            # 转换成字符串
            for i in range(0,len(eachLine)):
                    #if eachLine[i]!= ' ' and eachLine[i]!= ' ' and eachLine[i]!='\n':
# 删除空格和换行符，但是 json 双引号中的内容空格不能删除
                    str+=eachLine[i]
    return str


class jsonparser:

    def __init__(self, str=None):
            self._str = str
            self._index=0

    def _skipBlank(self):
            '''
            跳过空白、换行或 tab: \n\t\r
            '''
            while self._index<len(self._str) and self._str[self._index] in '
\n\t\r':
                    self._index=self._index+1
```

```
    def parse(self):
        '''
        进行解析的主要函数
        '''
        self._skipBlank()
        if self._str[self._index]=='{':
            self._index+=1
            return self._parse_object()
        elif self._str[self._index] == '[':
            self._index+=1
            return self._parse_array()
        else:
            print("Json format error!")
    def _parse_string(self):
        '''
        找出两个双引号中的string
        '''
        begin = end =self._index
        #找到string的范围
        while self._str[end]!='"':
            if self._str[end]=='\\': #表明其后面的是配合\的转义符号，如
\",\t,\r，主要针对\"的情况
                end+=1
                if self._str[end] not in '"\\/bfnrtu':
                    print
            end+=1
        self._index = end+1
        return self._str[begin:end]

    def _parse_number(self):
        '''
        数值没有双引号
        '''
        begin = end = self._index
        end_str=' \n\t\r,}]' #数字结束的字符串
        while self._str[end] not in end_str:
            end += 1
        number = self._str[begin:end]

        #进行转换
        if '.' in number or 'e' in number or 'E' in number :
            res = float(number)
        else:
            res = int(number)
        self._index = end
        return res

    def _parse_value(self):
        '''
        解析值，包括string，数字
        '''
        c = self._str[self._index]

        #解析对象
        if c == '{':
            self._index+=1
            self._skipBlank()
            return self._parse_object()
        #解析数组
        elif c == '[':
            #array
            self._index+=1
            self._skipBlank()
```

```
                return self._parse_array()
        # 解析 string
        elif c == '"':
                #string
                self._index += 1
                self._skipBlank()
                return self._parse_string()
        # 解析 null
        elif c=='n' and self._str[self._index:self._index+4] == 'null':
                #null
                self._index+=4
                return None
        # 解析 bool 变量 true
        elif c=='t' and self._str[self._index:self._index+4] == 'true':
                #true
                self._index+=4
                return True
        # 解析 bool 变量 false
        elif c=='f' and self._str[self._index:self._index+5] == 'false':
                #false
                self._index+=5
                return False
        # 剩下的情况为 number
        else:
                return self._parse_number()

    def _parse_array(self):
        '''
        解析数组
        '''
        arr=[]
        self._skipBlank()
        # 空数组
        if self._str[self._index]==']':
                self._index +=1
                return arr
        while True:
                val = self._parse_value() # 获取数组中的值，可能是 string，obj 等
                arr.append(val)           # 添加到数组中
                self._skipBlank()         # 跳过空白
                if self._str[self._index] == ',':
                        self._index += 1
                        self._skipBlank()
                elif self._str[self._index] ==']':
                        self._index += 1
                        return arr
                else:
                        print("array parse error!")
                        return None

    def _parse_object(self):
        '''
        解析对象
        '''
        obj={}
        self._skipBlank()
        # 空 object
        if self._str[self._index]=='}':
                self._index +=1
                return obj
        #elif self._str[self._index] !='"':
                # 报错
```

```
            self._index+=1 #跳过当前的双引号
            while True:
                    key = self._parse_string() #获取 key 值
                    self._skipBlank()

                    self._index = self._index+1#跳过冒号：
                    self._skipBlank()

                    #self._index = self._index+1#跳过双引号
                    #self._skipBlank()
                    #获取 value 值，目前假设只有 string 的 value 和数字
                    obj[key]= self._parse_value()
                    self._skipBlank()
                    #print key,":",obj[key]
                    #对象结束了，break
                    if self._str[self._index]=='}':
                            self._index +=1
                            break
                    elif self._str[self._index]==',':
                            self._index +=1
                            self._skipBlank()
                    self._index +=1#跳过下一个对象的第一个双引号
            return obj#返回对象

    def display(self):
            displayStr=""
            self._skipBlank()
            while self._index<len(self._str):
                    displayStr=displayStr+self._str[self._index]
                    self._index=self._index+1
                    self._skipBlank()
            print(displayStr)


def _to_str(pv):
    '''把python变量转换成string'''
    _str=''
    if type(pv) == type({}):
            #处理对象
            _str+='{'
            _noNull = False
            for key in pv.keys():
                    if type(key) == type(''):
                            _noNull = True #对象非空
                            _str+='"'+key+'":'+_to_str(pv[key])+','
            if _noNull:
                    _str = _str[:-1] #把最后的逗号去掉
            _str +='}'

    elif type(pv) == type([]):
            #处理数组
            _str+='['
            if len(pv) >0: #数组不为空，方便后续格式合并
                    _str += _to_str(pv[0])
            for i in range(1,len(pv)):
                    _str+=','+_to_str(pv[i])#因为已经合并了第一个，所以可以加逗号
            _str+=']'

    elif type(pv) == type(''):
            #字符串
            _str = '"'+pv+'"'
    elif pv == True:
            _str+='true'
```

```
        elif pv == False:
                _str+='false'
        elif pv == None:
                _str+='null'
        else:
                _str = str(pv)
        return _str


#main 函数
if __name__ == '__main__':
    print("test")
    '''
    jsonInstance=jsonparser(txt2str())
    jsonTmp = jsonInstance.parse()
    print jsonTmp
    print jsonTmp['obj1']['family']['father']
    print jsonTmp['obj1']['family']['sister']

    print ' '
    jsonInstance=jsonparser(txt2str('jsondataArray.txt'))
    jsonTmp = jsonInstance.parse()
    print jsonTmp
    print ' '
    '''
    jsonInstance=jsonparser(txt2str('jsonTestFile.txt'))
    jsonTmp = jsonInstance.parse()
    print(jsonTmp)
    print(_to_str(jsonTmp))

    print(' ')
    jsonInstance=jsonparser(txt2str('json.txt'))
    jsonTmp = jsonInstance.parse()
    print(jsonTmp)

    print(_to_str(jsonTmp))
```

在上述代码中，提及的 JSON 主体内容主要是指两大类：对象 object 和数组 array。因为一个 json 格式的字符串不是一个 object 就是一个 array。所以在上述编写的类 jsonparser 中，有 _parse_object 和 _parse_array 两个函数。首先通过 parse 函数直接判断开始的符号为大括号“{”还是中括号“[”，进而决定调用 _parse_object 还是 _parse_array。在标准 json 格式中的 key 是 string 类型，使用双引号包括，其中上面的 _parse_string() 函数专门用来解析 key。而 json 中的 value 则相对复杂一些，类型可以是 object、array、string、数字、true、false 和 null。

通过上述代码，设置如果遇到的字符为大括号“{”则调用 _parse_object() 函数。如果遇到中括号“[”则调用 _parse_array 函数，并且把 value 解析统一封装到函数 _parse_value() 中。

本实例执行后会输出：

```
    test
    ['JSON Test Pattern pass1', {'object with 1 member': ['array with 1
element']}, {}, [], -42, True, False, None, {'integer': 1234567890, 'real':
-9876.54321, 'e': 1.23456789e-13, 'E': 1.23456789e+34, '': -inf, 'zero': 0, 'one':
1, 'space': '', 'singlequote': '\'\\"', 'singlequote2': "'", 'quote': '\\"',
'backslash': '\\\\', 'controls': '\\b\\f\\n\\r\\t', 'slash': '/ & \\/', 'alpha':
'abcdefghijklmnopqrstuvwyz', 'ALPHA': 'ABCDEFGHIJKLMNOPQRSTUVWYZ', 'digit':
'0123456789', 'special': "`1~!@#$%^&*()_+-={':[,]}|;.</>?", 'hex': '\\u0123\\u4567\\
u89AB\\uCDEF\\uabcd\\uef4A', 'true': True, 'false': False, 'null': None, 'array':
[], 'object': {}, 'address': '50 St. James Street', 'url': 'http://www.JSON.org/',
'comment': '// /* <!-- --', '# -- --> */': '', ' s p a c e d ': [1, 2, 3, 4, 5, 6, 7],
```

```
'compact': [1, 2, 3, 4, 5, 6, 7], 'jsontext': '{\\"object with 1 member\\":[\\"array
with 1 element\\"]}', '\\/\\\\\\"\\uCAFE\\uBABE\\uAB98\\uFCDE\\ubcda\\uef4A\\b\\
f\\n\\r\\t`1~!@#$%^&*()_+-=[]{}|;:\',./<>?': 'A key can be any string'}, 0.5, 98.6,
99.44, 1066, 'rosebud']
    ["JSON Test Pattern pass1",{"object with 1 member":["array with
1 element"]},{},[],-42,true,false,null,{"integer":1234567890,"real":-
9876.54321,"e":1.23456789e-13,"E":1.23456789e+34,"":-inf,"zero":false,"one":true,
"space":"","singlequote":"'\"","singlequote2":"'","quote":"\"","backslash":"\\","
controls":"\b\f\n\r\t","slash":"/ & \/","alpha":"abcdefghijklmnopqrstuvwyz","ALP
HA":"ABCDEFGHIJKLMNOPQRSTUVWYZ","digit":"0123456789","special":"`1~!@#$%^&*()_+-
={':[,]}|;.</>?","hex":"\u0123\u4567\u89AB\uCDEF\uabcd\uef4A","true":true
,"false":false,"null":null,"array":[],"object":{},"address":"50 St. James
Street","url":"http://www.JSON.org/","comment":"// /* <!-- --","# -- --> */":""," s
p a c e d ":[true,2,3,4,5,6,7],"compact":[true,2,3,4,5,6,7],"jsontext":"{\"object
with 1 member\":[\"array with 1 element\"]}","\/\\\"\uCAFE\uBABE\uAB98\uFCDE\ubcda\
uef4A\b\f\n\r\t`1~!@#$%^&*()_+-=[]{}|;:',./<>?":"A key can be any string"},0.5,98.6
,99.44,1066,"rosebud"]

    Json format error!
    None
    null
```

6.2　实践案例：可视化分析世界人口数据

现在已经存在世界人口地图的 JSON 统计文件 population_data.json，在里面保存了 2010 年世界人口的详细信息。在本节的内容中，将详细讲解分析这个 JSON 文件并可视化的过程。

↑扫码看视频（本节视频课程时间：1 分 17 秒）

6.2.1　输出每个国家 2010 年的人口数量

文件 population_data.json 的部分内容如下所示：

```
{
    "Country Name": "Arab World",
    "Country Code": "ARB",
    "Year": "1960",
    "Value": "96388069"
  },
  {
    "Country Name": "Arab World",
    "Country Code": "ARB",
    "Year": "1961",
    "Value": "98882541.4"
  },
```

编写文件 fan01.py，功能是输出国别码以及相对应国家在 2010 年的人口数量，首先加载 JSON 文件 population_data.json，然后提取 Year 值为“2010”的数据。具体实现代码如下所示：

```
import json
#1. 将数据加载到一个列表中
filename='population_data.json'
with open(filename)as f:
    pop_data=json.load(f)
#2. 打印每个国家 2010 年的人口数量
for pop_dict in pop_data:
    if pop_dict['Year']=='2010':
```

```
        country_name = pop_dict["Country Name"]
        population = int(float(pop_dict["Value"]))
        print(country_name + ":" + str(population))
```

执行后输出结果如下：

```
Arab World:357868000
Caribbean small states:6880000
East Asia & Pacific (all income levels):2201536674
East Asia & Pacific (developing only):1961558757
Euro area:331766000
Europe & Central Asia (all income levels):890424544
Europe & Central Asia (developing only):405204000
European Union:502125000
Heavily indebted poor countries (HIPC):635663000
High income:1127437398
High income: nonOECD:94204398
High income: OECD:1033233000
Latin America & Caribbean (all income levels):589011025
Latin America & Caribbean (developing only):582551688
Least developed countries: UN classification:835140827
Low & middle income:5767157445
Low income:796342000
Lower middle income:2518690865
Middle East & North Africa (all income levels):382803000
Middle East & North Africa (developing only):331263000
Middle income:4970815445
North America:343539600
OECD members:1236521688
Other small states:18293000
Pacific island small states:3345337
##### 省略好多执行结果
```

6.2.2 获取两个字母的国别码

编写实例文件 fan02.py，功能是获取两个字母的国别码，首先导入 pygal_maps_world.i18n 中的国家和地区信息，然后循环输出每个国家和地区的简写名字。具体实现代码如下所示：

```
from pygal_maps_world.i18n import COUNTRIES
for country_code in sorted(COUNTRIES.keys()):
    print(country_code, COUNTRIES[country_code])
```

在运行上述代码前需要通过如下命令安装 pygal_maps_world：

```
pip install pygal_maps_world
```

执行后会输出两个字母的国别码：

```
ad Andorra
ae United Arab Emirates
af Afghanistan
al Albania
am Armenia
ao Angola
aq Antarctica
ar Argentina
at Austria
au Australia
az Azerbaijan
ba Bosnia and Herzegovina
bd Bangladesh
be Belgium
bf Burkina Faso
```

```
bg Bulgaria
##### 省略好多执行结果
```

6.2.3　制作世界地图

编写实例文件 fan03.py，功能是绘制美洲的人口地图，首先使用“pip install”命令安装 pygal_maps_world，然后访问 pygal.maps.world 模块，通过 wm.add() 绘制指定国家和地区的地图。具体实现代码如下所示：

```
import pygal
wm = pygal.maps.world.World()
wm.title = '北美 中部 南美'
wm.add('北美', ['ca', 'mx', 'us'])
wm.add('中部', ['bz', 'cr', 'gt', 'hn', 'ni', 'pa', 'sv'])
wm.add('南美', ['ar', 'bo', 'br', 'cl', 'co', 'ec', 'gf',
            'gy', 'pe', 'py', 'sr', 'uy', 've'])
wm.render_to_file('americas.svg')
```

执行后会创建地图文件 americas.svg，打开后的效果如图 6-3 所示。

图 6-3　美洲人口地图

6.3　数据挖掘并分析处理日志文件

在现有软件的日志系统中，绝大多数使用 JSON 格式的文件保存日志信息，例如 Nginx 和 Logtail 等。通过 Python 语言可以分析 JSON 日志文件中的数据信息，并对信息进行归类和统计处理。在本节的内容中，将详细讲解使用 Python 挖掘并分析处理 JSON 日志数据的知识。

↑扫码看视频（本节视频课程时间：10 分 27 秒）

6.3.1　检查 JSON 日志的 Python 脚本

在日常开发的过程中，经常需要处理 JSON 格式的日志文件，在解析的过程中会经常发现问题，例如格式错误，缺少字段等。如果一旦发现就让开发人员去修改，改完再检查，一来一回耽误时间也比较被动，所以建议编写一个 Python 的脚本，可以自动检测日志中的问题。

（1）检查 JSON 格式

第一步是使用 Python 解析 JSON 字符串，例如嵌套的 JSON 数据在拼接时会不规范，比如多加了双引号，此时可以用如下所示的文件 log01.py 进行检查。

```
import json
line="JSON 串 "
try:
    data = json.loads(line.strip())   # line 是待检查的 json 串
except:
    print("json 格式有误，请检查 ===>", line)
```

（2）检查日期格式

有时 JSON 文件中的日期格式不规范，可以用如下所示的文件 log02.py 进行检查。

```
import time
line="JSON 串 "
try:
    dt = data['dt']
    time.strptime(dt, "%Y-%m-%d %H:%M:%S")
except:
    print(" 日期时间字段 (dt) 格式有误，请检查 ===>", line)
```

（3）禁用中文

有时不希望在 JSON 文件的一些字段中出现中文，可以用如下所示的文件 log03.py 检查是否有中文。

```
import re

zh_pattern = re.compile(u'[\u4e00-\u9fa5]+')
tmp = zh_pattern.search(data['z'])
if tmp:
    print("z 字段中，含有中文，请使用英文 ===>", line)
```

下面是一个完整的脚本文件 log04.py，功能是自动检测指定 JSON 日志文件中的数据信息，每一条数据信息由时间、字符、字符串和数字组成。文件 log04.py 的具体实现代码如下所示：

```
import sys
import re
import json
import time

def checklog():
    check = ['a', 'b', 'c']

    cnt = 0;
    json_error = 0;
    dt_error = 0;
    zh_error = 0;
    miss_error = 0

    for line in sys.stdin:
        if not line or not line.strip():
            continue
        line = "".join(i for i in line if ord(i) > 31)   # 去除特殊字符
        cnt += 1

        # json 格式
        try:
            data = json.loads(line.strip())
        except:
```

```
            print("json 格式有误，请检查 ===>", line)
            json_error += 1
            continue

        # dt 字段
        try:
            dt = data['dt']
            time.strptime(dt, "%Y-%m-%d %H:%M:%S")
        except:
            print(" 日期时间字段 (dt) 格式有误，请检查 ===>", line)
            dt_error += 1

        # 禁用中文字段
        zh_pattern = re.compile(u'[\u4e00-\u9fa5]+')
        tmp = zh_pattern.search(data['z'])
        if tmp:
            print("z 字段中含有中文，请使用英文 ===>", line)
            zh_error += 1

        # 其他必要字段

        tmp = ""
        for tag in check:
            if tag not in data or data[tag] is None or data[tag] == '':
                tmp += tag + ","
        if len(tmp) > 0:
            tmp = tmp[:-1]
            print(tmp, " 字段缺失或值为空，请检查 ===>", line)
            miss_error += 1

    print('=================== 完成 ===================')
    print(' 本次检查共 %d 条日志 , json 格式错误 %d 条，dt 字段错误 %d 条，z 字段错误或缺失 %d 条，
其他必要字段缺失 %d 条 ' % (cnt, json_error, dt_error, zh_error, miss_error))

if __name__ == '__main__':
    checklog()
```

假设存在如下所示的日志文件 t.json：

```
{"dt":"2018-11-02 11:11:11","z":"hello","a":1,"b":2,"c":3,"js":{"d":4}}
{"dt":"2018-11-02","z":"hello","a":1,"b":2,"c":3,"js":{"d":4}}
{"dt":"2018-11-02 11:11:11","z":" 中 ","a":1,"b":2,"c":3,"js":{"d":4}}
{"dt":"2018-11-02 11:11:11","z":"hello","a":1,"b":2,"c":3,"js":"{"d":4}"}
{"dt":"2018-11-02 11:11:11","z":"hello","a":1,"js":{"d":4}}
```

在 Linux 系统中，通过如下命令即可使用脚本文件 log04.py 检测上述日志文件 t.json：

```
cat t.json | python log04.py
```

例如下面是在笔者电脑中的检测结果：

```
日期时间字段 (dt) 格式有误，请检查 ===> {"dt":"2018-11-02","z":"hello","a":1,"b":2,"c"
:3,"js":{"d":4}}
z 字段中含有中文，请使用英文 ===> {"dt":"2018-11-02 11:11:11","z":" 中 ","a":1,"b":2,"
c":3,"js":{"d":4}}
json 格式有误，请检查 ===> {"dt":"2018-11-02 11:11:11","z":"hello","a":1,"b":2,"c":
3,"js":"{"d":4}"}
b,c 字段缺失或值为空，请检查 ===> {"dt":"2018-11-02 11:11:11","z":"hello","a":1,"js"
:{"d":4}}
=================== 完成 ===================
本次检查共 5 条日志 , json 格式错误 1 条，dt 字段错误 1 条，z 字段错误或缺失 1 条，其他必要字段缺失 1 条
```

6.3.2 将 MySQL 操作日志保存到数据库文件

本实例的功能是建立和 MySQL 数据库的连接，分析 MySQL 日志中的 JSON 数据，并将分析的日志信息提取保存到 MySQL 数据库中。

我们来看一下具体的操作步骤

（1）编写配置文件 config.py，功能是建立和指定 MySQL 数据库的连接，并设置要提取的日志文件目录和路径，然后重新生成日期格式的日志文件进行处理，例如 2019 年 1 月 2 日将分析前一天 2019 年 1 月 1 日的日志数据，系统重新生成的日志文件名是“nps_2019-01-01.log”。文件 config.py 的具体实现代码如下所示：

```
    # DATA_DIR
    data_dir = os.path.join(expanduser('~'), 'data')
    # 配置 log 文件路径
    log_root_path = os.path.join(data_dir, 'logs')
    # 配置 log 文件的名字
    log_names = 'nps_'+str((date.today() + timedelta(days=-1)).strftime("%Y-%m-%d")) + '.log'
    # 配置数据库信息
    db_config = {
        'user': 'root',
        'passwd': '66688888',
        'host': '126.0.0.1',
        'port': 3306,
        'db': 'log_analysis',
        'charset': 'utf8'
    }
    # 数据库类型配置
    db_type = 'mysql'
```

（2）编写日志分析文件 analysis.py，可以分为如下 5 个步骤：

① 建立和指定 MySQL 数据库的连接，对应实现代码如下所示：

```
    def connect_sql():
        # 创建连接
        try:
                conn = MySQLdb.connect(db_config['host'], db_config['user'], db_config['passwd'],
                                        db_config['db'], charset='utf8')
            print("MySQL 数据库连接成功！ ")
        except:
            print(" 数据库连接失败！ ")
            exit(1)
        return conn
```

② 分析重新生成的昨天的日志文件，将里面的 JSON 数据匹配出来。对应代码如下所示：

```
    def analysis():
        # 分析 log 日志文件，将 json 数据匹配出来
        result = []
        # 查找到文件 handle
        log_path = os.path.join(log_root_path, log_names)
        if not os.path.exists(log_path):
            print(u' 对不起，' + log_names + u' 文件不存在 ')
            os._exit(0)
        else:
            with open(log_path.decode('utf-8'), 'r') as txt:
                # txt.read().encode('utf-8')
                for line in txt:
                    pattern = re.compile(r'yuanli dumping status:(.+)')
                    goal = pattern.search(line)
```

```
                if goal is not None:
                    result.append(goal.groups()[0])
            return result
```

③ 将匹配出的数据添加到 MySQL 数据库中，对应代码如下所示：

```
    def storage():
        # 将匹配出来的数据存储进数据库
        if db_type == 'mysql':
            conn = connect_sql()
            cur = conn.cursor(cursorclass=MySQLdb.cursors.DictCursor)
            # 获取当前年月日，数据库中查找，如果没有，则添加
            yesterday = (date.today() + timedelta(days=-1)).strftime("%Y-%m-%d")
            cur.execute(
                "SELECT 'event','user','date_time' FROM log WHERE 'date_time' =
'%s'"%(yesterday))
            if len(cur.fetchall()) == 0:
              data = analysis()
              print(data)
              # os._exit(0)
              new_data = []
              for x in data:
                # 截取 yuanli dumping status: 字符
                # x = x[22:]
                x = x.decode('gbk')
                x = json.loads(x)
                # print x
                new_data.append((str(x['user']), str(x['event']), str(x['ts']),
                  str(x['app'] if x['app'] else None),
                  str(x['city'] if 'city'in x.keys() else None),
                  str(x['modType'] if 'modType'in x.keys() else None),
                  str(x['ip'] if 'ip'in x.keys() else None),
                  str(x['module'] if 'module'in x.keys() else None),
                  str(x['userName'] if 'userName'in x.keys() else None),
                  str(x['url'] if 'url'in x.keys() else None),
                  str(x['modId'] if 'modId'in x.keys() else None),
                  str(x['conditions'] if 'conditions'in x.keys() else None)
                  ))

                print(new_data)

                try:
                    sql = 'insert into log(user,event,date_time,app,city,modType,ip,
module,userName,url,modId,conditions) value(%s,%s,%s,%s,%s,%s,%s,%s,%s,%s,%s,%s)'
                    print(new_data)
                    cur.executemany(sql, new_data)
                    conn.commit()
                    print ('插入数据')
                    conn.close()
                    cur.close()
                except MySQLdb.Error as e:
                    conn.rollback()
                    print("执行MySQL: %s 时出错: %s" % (sql, e))
            else:
                print(u'当前日期的数据已经新增过数据库')
```

④ 统计当前日期和用户的操作次数，对应实现代码如下所示：

```
    def statistics():
        # 统计当前日期，用户操作时间的次数
        yesterday = (date.today() + timedelta(days=-1)).strftime("%Y-%m-%d")
        # nowTime = '2018-08-09'
        # 执行数据库操作
        conn = connect_sql()
        cur = conn.cursor(cursorclass=MySQLdb.cursors.DictCursor)
```

```
        cur.execute("SELECT COUNT('event') as num,'event','user' FROM log WHERE 'date_
time' = '%s' GROUP BY 'event','user'"%(yesterday))

        if len(cur.fetchall()) == 0:
            print(str(yesterday) + u'当前日期没有数据')
        else:
            for row in cur:
               print(str(row['user']) + "用户处理事件" + str(row['event']) + "共计" + str(
                    row['num']) + "次")

        conn.close()
        cur.close()
```

⑤ 创建 MySQL 数据库表的 SQL 文件 log.sql，执行后会在数据库“log_analysis”中创建表“log”，如图 6-4 所示。

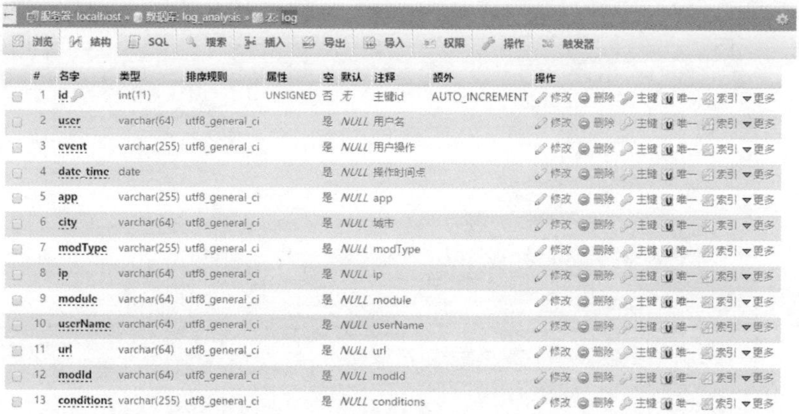

服务器: localhost » 数据库: log_analysis » 表: log

浏览 结构 SQL 搜索 插入 导出 导入 权限 操作 触发器

#	名字	类型	排序规则	属性	空	默认	注释	额外	操作
1	id	int(11)		UNSIGNED	否	无	主键id	AUTO_INCREMENT	修改 删除 主键 唯一 索引 更多
2	user	varchar(64)	utf8_general_ci		是	NULL	用户名		修改 删除 主键 唯一 索引 更多
3	event	varchar(255)	utf8_general_ci		是	NULL	用户操作		修改 删除 主键 唯一 索引 更多
4	date_time	date			是	NULL	操作时间点		修改 删除 主键 唯一 索引 更多
5	app	varchar(255)	utf8_general_ci		是	NULL	app		修改 删除 主键 唯一 索引 更多
6	city	varchar(64)	utf8_general_ci		是	NULL	城市		修改 删除 主键 唯一 索引 更多
7	modType	varchar(255)	utf8_general_ci		是	NULL	modType		修改 删除 主键 唯一 索引 更多
8	ip	varchar(64)	utf8_general_ci		是	NULL	ip		修改 删除 主键 唯一 索引 更多
9	module	varchar(64)	utf8_general_ci		是	NULL	module		修改 删除 主键 唯一 索引 更多
10	userName	varchar(64)	utf8_general_ci		是	NULL	userName		修改 删除 主键 唯一 索引 更多
11	url	varchar(64)	utf8_general_ci		是	NULL	url		修改 删除 主键 唯一 索引 更多
12	modId	varchar(64)	utf8_general_ci		是	NULL	modId		修改 删除 主键 唯一 索引 更多
13	conditions	varchar(255)	utf8_general_ci		是	NULL	conditions		修改 删除 主键 唯一 索引 更多

图 6-4　创建的 MySQL 数据库表

6.3.3　将日志中 JSON 数据保存为 CSV 格式

在下面的实例中，能够处理指定日期范围（包含起止日期）内的日志文件，将日志文件中的 JSON 数据保存为 CSV 格式的文件。

（1）编写配置文件 config.py，功能是设置要提取的日志文件目录和路径，然后设置要处理的文件的日期。文件 config.py 的具体实现代码如下所示：

```
# DATA_DIR
data_dir = os.path.join(expanduser('~'), 'data')

# 配置 log 文件路径
log_root_path = os.path.join(data_dir, 'logs')
# 配置 csv 文件路径
csv_root_path = os.path.join(data_dir, 'csv')
# 配置需要日期区间，包含起止日期
date_start = '2019-01-07'
date_end = '2019-01-10'
```

（2）编写日志分析文件 analysis_csv.py，可分为如下 3 个步骤：

① 定义方法 analysis()，功能是分析 log 日志文件，将里面的 JSON 数据匹配出来。具体实现代码如下所示：

```
def analysis():
    # 根据配置文件里的起止日期，查询出所有日期
    datestart = datetime.strptime(date_start, '%Y-%m-%d')
```

```
    dateend = datetime.strptime(date_end, '%Y-%m-%d')
    date_list = [datestart.strftime('%Y-%m-%d')]
    while datestart < dateend:
        datestart += timedelta(days=1)
        date_list.append(datestart.strftime('%Y-%m-%d'))

    result = []
    if date_list is not None:
        log_names = {}
        for key, value in enumerate(date_list):
            # 拼接日志文件名称
            log_names[key] = 'nps_' + str(value) + '.log'
            # 查找到文件 handle
            log_path = os.path.join(log_root_path, log_names[key])
            if not os.path.exists(log_path):
                print(u'对不起，' + data_dir+log_names[key] + u'文件不存在')
            else:
                with open(log_path, 'r') as txt:
                    # txt.read().encode('utf-8')
                    for line in txt:
                        pattern = re.compile(r'yuanli dumping status:(.+)')
                        goal = pattern.search(line)
                        if goal is not None:
                            result.append(goal.groups()[0])

    return result
```

② 编写函数 create_csv()，功能是根据提取的 JSON 数据创建指定格式的 CSV 文件，具体实现代码如下所示：

```
def create_csv():
    # 获取匹配到的数据
    data = analysis()
    # print data
    new_data = []
    for x in data:
        # 截取 yuanli dumping status: 字符
        # x = x[22:]
        x = x.decode('gbk')
        x = json.loads(x)
        # print x
        new_data.append([
            str(x['user']), str(x['event']), str(x['ts']),
            str(x['app'] if x['app'] else None),
            str(x['city'] if 'city' in x.keys() else None),
            str(x['modType'] if 'modType' in x.keys() else None),
            str(x['ip'] if 'ip' in x.keys() else None),
            str(x['module'] if 'module' in x.keys() else None),
            str(x['userName'] if 'userName' in x.keys() else None),
            str(x['url'] if 'url' in x.keys() else None),
            str(x['modId'] if 'modId' in x.keys() else None),
            str(x['conditions'] if 'conditions' in x.keys() else None),
            str(x['modAllCount'] if 'modAllCount' in x.keys() else None),
            str(x['mod1Count'] if 'mod1Count' in x.keys() else None),
            str(x['mod0Count'] if 'mod0Count' in x.keys() else None),
            str(x['modAUC'] if 'modAUC' in x.keys() else None),
            str(x['modShare'] if 'modShare' in x.keys() else None),
            str(x['forecastCount'] if 'forecastCount' in x.keys() else None),
            str(x['forecastDown'] if 'forecastDown' in x.keys() else None),
            str(x['dataPeriods'] if 'dataPeriods' in x.keys() else None),
            str(x['dataType'] if 'dataType' in x.keys() else None)
        ])
```

```
        t = datetime.now().strftime('%Y%m%d%H%M%S')
        file_name = 'new_analysis_'+str(t)+'.csv'
        csv_new = os.path.join(csv_root_path, file_name)
        print(u'我们创建了新的文件'+file_name + u'在' + csv_root_path + u'路径下，请注意查收')

        fileHeader = [u'用户名ID', u'操作类型', u'访问时间', u'应用名', u'所属地市',
                      u'模型类型', u'访问ip', u'所属模块名', u'用户姓名', u'访问url',
                      u'模型id', u'条件', u'更新模型时使用的样本数据总量',
                      u'更新模型时使用的正样本数据总量',
                      u'更新模型时使用的负样本数据总量',
                      u'模型生成后的auc', u'模型共享范围', u'预测前的数据量',
                      u'下载的数据量', u'数据期数', u'数据类型']

        csvFile = open(csv_new, "w")
        writer = csv.writer(csvFile)
        # 写入的内容都是以列表的形式传入函数
        writer.writerow(fileHeader)
        for item in new_data:
            writer.writerow(item)

        csvFile.close()
```

③ 执行后会分析指定日期范围内的日志文件，并将里面的 JSON 数据保存到指定的 CSV 文件中，如图 6-5 所示。

apple › data › csv

名称	修改日期	类型	大小
new_analysis_20190109145355.csv	2019/1/9 14:53	Microsoft Excel ...	1 KB
new_analysis_20190109145440.csv	2019/1/9 14:54	Microsoft Excel ...	1 KB
new_analysis_20190109145500.csv	2019/1/9 14:55	Microsoft Excel ...	1 KB
new_analysis_20190109145539.csv	2019/1/9 14:55	Microsoft Excel ...	1 KB
new_analysis_20190109145627.csv	2019/1/9 14:56	Microsoft Excel ...	1 KB
new_analysis_20190109145659.csv	2019/1/9 14:56	Microsoft Excel ...	1 KB
new_analysis_20190109145848.csv	2019/1/9 14:58	Microsoft Excel ...	1 KB

图 6-5　创建的 CSV 文件

6.4　实践案例：统计分析朋友圈的数据

在本节的内容中，将通过一个具体实例讲解如何将朋友圈信息导出为 JSON 文件的方法，并介绍使用 Python 语言统计分析这些 JSON 数据的过程。

↑扫码看视频（本节视频课程时间：2 分 34 秒）

6.4.1　将朋友圈数据导出到 JSON 文件

使用开源工具 WeChatMomentExport 导出微信朋友圈的数据，WeChatMomentExport 的源码地址和使用教程请参考 https://github.com/Chion82/WeChatMomentExport，如图 6-6 所示。

Deprecated. Try WeChatMomentStat-Android instead!

本项目已不再维护。请移步至无须依赖Xposed的WeChatMomentStat-Android

WeChatModuleExport is an Xposed module which helps you export WeChat moment data to JSON files.

本工具为Xposed模块，用于导出微信朋友圈数据。
支持导出朋友圈文字内容、图片、段视频、点赞、评论等数据。

Supported WeChat Versions（目前支持的微信版本）：

WeChat 6.3.13

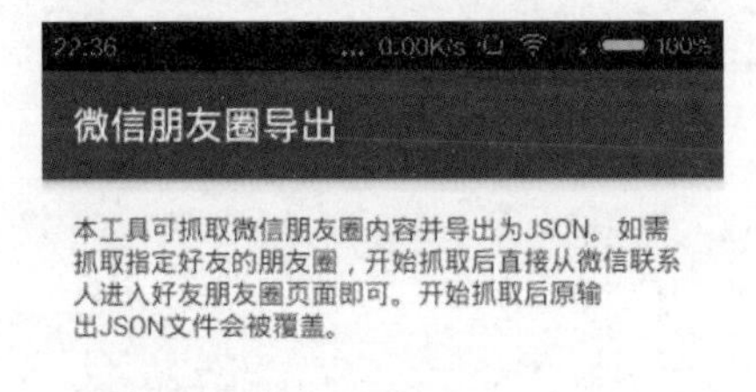

图 6-6　使用 WeChatMomentExport 导出数据

通过使用 WeChatMomentExport，可以导出如下所示的分类信息，每一个分类信息都被保存到对应的 JSON 文件中。

- 发朋友圈数量排名；
- 朋友圈点赞数排名；
- 被点赞数排名；
- 发评论数量排名；
- 朋友圈收到评论数量排名；
- 被无视概率排名（评论被回复数 / 写评论数，条件为写评论数≥ 15）；
- 发投票 / 问卷调查类广告数排名。

6.4.2　统计处理 JSON 文件中的朋友圈数据

使用 WeChatMomentExport 导出数据后，可以使用 Python 统计分析朋友圈的各类信息。实例文件 wechat_moment_stat.py 的功能是分别统计处理 7 个 JSON 文件中的数据，然后输出显示统计结果。文件 wechat_moment_stat.py 的具体实现流程可分为 10 个步骤。

（1）通过函数 get_user() 获取每个朋友圈用户的详细信息，具体实现代码如下所示。

```
def get_user(user_name):
    for user_info in result:
            if user_info['user'] == user_name:
                    return user_info
    user_info = {
            'user' : user_name,
            'moments' : [],
            'post_comments' : [],
            'replied_comments' : [],
            'received_comments' : [],
            'post_likes' : 0,
```

```
            'received_likes' : 0,
            'spam_counts' : 0,
        }
        result.append(user_info)
        return user_info
```

（2）编写函数 is_spam() 提取朋友圈留言中的信息，具体实现代码如下所示：

```
    def is_spam(moment_text):
        if ('投' in moment_text and '谢' in moment_text):
              return True
        if ('投票' in moment_text):
              return True
        if ('问卷' in moment_text):
              return True
        if ('填' in moment_text and '谢' in moment_text):
              return True
        return False
```

（3）编写函数 handle_moment() 处理留言信息，具体实现代码如下所示：

```
    def handle_moment(moment):
        user_info = get_user(moment['author'])
        user_info['moments'].append(moment)
        user_info['received_likes'] = user_info['received_likes'] +
len(moment['likes'])
        user_info['received_comments'].extend(moment['comments'])
        if (is_spam(moment['content'])):
              user_info['spam_counts'] = user_info['spam_counts'] + 1
        for comment_info in moment['comments']:
              comment_user = get_user(comment_info['author'])
              comment_user['post_comments'].append(comment_info)
              if (comment_info['to_user'] != ''):
                    replied_user = get_user(comment_info['to_user'])
                    replied_user['replied_comments'].append(comment_info)
        for like_info in moment['likes']:
              like_user = get_user(like_info)
              like_user['post_likes'] = like_user['post_likes'] + 1
```

（4）排序处理 7 类信息，具体实现代码如下所示：

```
    for moment_info in origin_data:
       handle_moment(moment_info)

    f = open('user_output.json', 'w')
    f.write(json.dumps(result))
    f.close()

    post_moment_rank = sorted(result, key=lambda user_info: len(user_
info['moments']), reverse=True)
    post_like_rank = sorted(result, key=lambda user_info: user_info['post_likes'],
reverse=True)
    received_like_rank = sorted(result, key=lambda user_info: user_info['received_
likes'], reverse=True)
    post_comment_rank = sorted(result, key=lambda user_info: len(user_info['post_
comments']), reverse=True)
    received_comment_rank = sorted(result, key=lambda user_info: len(user_
info['received_comments']), reverse=True)
    no_reply_rank = sorted(result, key=lambda user_info: ((float(len(user_
info['replied_comments']))/len(user_info['post_comments'])) if len(user_info['post_
comments'])>0 else 999))
    spam_rank = sorted(result, key=lambda user_info: user_info['spam_counts'],
reverse=True)
```

```
    f = open('post_moment_rank.json', 'w')
    f.write(json.dumps(post_moment_rank))
    f.close()
```

（5）打印输出发送朋友圈信息最多的前 5 位数据，具体实现代码如下所示：

```
    print('前5位发朋友圈最多：')
    temp_list = []
    for i in range(5):
        temp_list.append(post_moment_rank[i]['user'] + '(%d 条)' % len(post_moment_
rank[i]['moments']))
    print(', '.join(temp_list))

    f = open('post_like_rank.json', 'w')
    f.write(json.dumps(post_like_rank))
    f.close()
```

（6）打印输出前 5 位点赞狂魔的用户信息，具体实现代码如下所示：

```
    print('前5位点赞狂魔：')
    temp_list = []
    for i in range(5):
        temp_list.append(post_like_rank[i]['user'] + '(%d 赞)' % post_like_rank[i]
['post_likes'])
    print(', '.join(temp_list))

    f = open('received_like_rank.json', 'w')
    f.write(json.dumps(received_like_rank))
    f.close()
```

（7）打印输出前 5 位获得最多赞的用户信息，具体实现代码如下所示：

```
    print('前5位获得最多赞：')
    temp_list = []
    for i in range(5):
        temp_list.append(received_like_rank[i]['user'] + '(%d 赞)' % received_like_
rank[i]['received_likes'])
    print(', '.join(temp_list))

    f = open('post_comment_rank.json', 'w')
    f.write(json.dumps(post_comment_rank))
    f.close()
```

（8）打印输出前 5 位评论狂魔的用户信息，具体实现代码如下所示：

```
    print('前5位评论狂魔：')
    temp_list = []
    for i in range(5):
        temp_list.append(post_comment_rank[i]['user'] + '(%d  评 论)' % len(post_
comment_rank[i]['post_comments']))
    print(', '.join(temp_list))

    f = open('received_comment_rank.json', 'w')
    f.write(json.dumps(received_comment_rank))
    f.close()
```

（9）打印输出前 5 位朋友圈评论最多的用户信息，具体实现代码如下所示：

```
    print('前5位朋友圈评论最多：')
    temp_list = []
    for i in range(5):
        temp_list.append(received_comment_rank[i]['user'] + '(%d 评论)' % len(post_
comment_rank[i]['received_comments']))
    print(', '.join(temp_list))

    f = open('no_reply_rank.json', 'w')
    f.write(json.dumps(no_reply_rank))
```

```
    f.close()

    f = open('spam_rank.json', 'w')
    f.write(json.dumps(spam_rank))
    f.close()
```

（10）打印输出收到评论回复数 / 写评论数前 5 名且发出评论数≥ 15 用户的信息，且发出评论数≥ 15，具体实现代码如下：

```
    print('================================')
    print('前 5 名（收到评论回复数 / 写评论数 且发出评论数≥ 15）：')
    temp_list = []
    for user_info in no_reply_rank:
        if len(user_info['post_comments']) < 15:
                continue
        if (len(temp_list) > 5):
                break
        temp_list.append(user_info['user'] + ('(收到评论回复%d, 写评论%d)' % (len(user_
info['replied_comments']), len(user_info['post_comments']))))
    print(', '.join(temp_list))
```

执行后会打印输出朋友圈的统计结果，如图 6-1 所示。

```
    前 5 位发朋友圈最多：
    Joe (69 条), xxx (62 条), 嬴子夜。(58 条), xxx(46 条), psh(40 条)
    前 5 位点赞狂魔：
    Saruman(33 赞), xxx (28 赞), ChiaChia.Ý(27 赞), Max(27 赞), 郭含阳 (23 赞)
    前 5 位获得最多赞：
    陈思君 (38 赞), 404(32 赞), (29 赞), Justin Tan(27 赞), 杨宗炜 (26 赞)
    前 5 位评论狂魔：
    杨宗炜 (110 评论), 404(95 评论), Saruman(94 评论), MATTHEW °Д°(77 评论), Joe(57
评论)
    前 5 位朋友圈评论最多：
    杨宗炜 (209 评论), 404(130 评论), Joe(37 评论), MATTHEW °Д°(68 评论), Max(69 评论)
    ================================
    收到评论回复数 / 写评论数 且发出评论数≥ 15：0
    Joe(收到评论回复 13, 写评论 57), jjy(收到评论回复 4, 写评论 16), Saruman(收到评论回复 30,
写评论 94), 玺玺玺(收到评论回复 7, 写评论 21), ChiaChia.Ý(收到评论回复 6, 写评论 15), 郭含阳(收
到评论回复 14, 写评论 33)
```

6.5　实践案例：使用库 matplotlib 可视化 JSON 数据

在本节的实例中，将获取某个指定 JSON 文件中的数据，然后使用库 matplotlib 根据 JSON 文件中的数据绘制可视化图片，并将绘制的图片保存到本地电脑硬盘中。

↑扫码看视频（本节视频课程时间：3 分 02 秒）

6.5.1　准备 JSON 文件

新建文件夹“demo”，在里面创建 JSON 文件 demo.json，在里面保存了绘制图表曲线所需的两组数据。文件 demo.json 的代码如下所示：

```
    {
        "title": "A Simple Figure",
        "x_label": "X Axis",
        "y_label": "Y Axis",
        "legend_y_offset": -0.15,
```

```
        "x_lim": [0, 30],
        "y_lim": [],
        "x_scale": "linear",
        "y_scale": "linear",
        "x_log_base": 2,
        "y_log_base": 10,
        "x_ticks": [],
        "y_ticks": [],
        "x_ticks_labels": [],
        "y_ticks_labels": [],
        "x_ticks_minor": "False",
        "y_ticks_minor": "False",
        "grid_major": "True",
        "grid_minor": "False",
        "series": [
            {
                "name": "sin",
                "group": "Series A",
                "y_id": 0,
                "color": [0.1020, 0.6235, 0.4667, 1.0],
                "linestyle": "-",
                "marker": "o",
                "linewidth": 2.0,
                "marker_size": 5,
                "data": [
                    [0, 0.0],
                    [1, 0.20266793654820095],
                    [2, 0.39692414892492234],
                    [3, 0.57470604121617919],
                    [4, 0.7286347834693504],
                    [5, 0.85232156971961837],
                    [6, 0.94063278511248671],
                    [7, 0.98990307637212394],
                    [8, 0.99808748213471832],
                    [9, 0.96484630898376322],
                    [10, 0.89155923041100371],
                    [11, 0.78126802352626379],
                    [12, 0.63855032022660208],
                    [13, 0.46932961277720098],
                    [14, 0.28062939951435684],
                    [15, 0.080281674842813497],
                    [16, -0.12339813736217871],
                    [17, -0.32195631507261868],
                    [18, -0.50715170948451438],
                    [19, -0.67129779355193209],
                    [20, -0.80758169096833632],
                    [21, -0.91034694431078278],
                    [22, -0.97532828606704558],
                    [23, -0.99982866838408957],
                    [24, -0.98283120392563061],
                    [25, -0.92504137173820289],
                    [26, -0.82885773637304272],
                    [27, -0.69827239556539955],
                    [28, -0.53870528838615628],
                    [29, -0.35677924089893809],
                    [30, -0.16004508604325057]
                ]
            },
            {
                "name": "cos",
                "group": "Series B",
                "y_id": 0,
                "color": [1.0000, 0.1647, 0.5137, 1.0],
```

```
            "linestyle": "--",
            "marker": "s",
            "linewidth": 2.0,
            "marker_size": 6,
            "data": [
                    [0, 1.0],
                    [1, 0.97924752105649693],
                    [2, 0.91785141499058875],
                    [3, 0.81835992459896723],
                    [4, 0.68490244000045208],
                    [5, 0.52301810847301045],
                    [6, 0.33942593237925484],
                    [7, 0.14174589725634046],
                    [8, -0.061817295362854206],
                    [9, -0.26281476374132501],
                    [10, -0.4529041164186286],
                    [11, -0.62419570281712555],
                    [12, -0.76958007285694707],
                    [13, -0.883023054382162],
                    [14, -0.95981620122198996],
                    [15, -0.99677221705083296],
                    [16, -0.99235724398804326],
                    [17, -0.94675452530466453],
                    [18, -0.86185679991918307],
                    [19, -0.74118774434842594],
                    [20, -0.58975572266212262],
                    [21, -0.41384591454310704],
                    [22, -0.22075944916926946],
                    [23, -0.018510372154503518],
                    [24, 0.18450697707700792],
                    [25, 0.37986637199507933],
                    [26, 0.55945942914080526],
                    [27, 0.71583214624055413],
                    [28, 0.84249428025642303],
                    [29, 0.93418872465020553],
                    [30, 0.98710970536886555]
            ]
        }
    ]
}
```

在上述 JSON 文件中，首先设置了绘制图表的标题、x 轴标识、y 轴标识和图例信息，然后设置了两组 JSON 数据（分别表示两个方程曲线的 x 坐标和 y 坐标），最后将这两组数据作为绘制可视化图表的数据源。

6.5.2 数据可视化

新建程序文件 json2plot.py，功能是获取用户输入的参数提取 JSON 文件中的数据，然后使用库 matplotlib 和 seaborn 根据 JSON 数据绘制可视化图片，并将绘制的图片保存到指定的位置。文件 json2plot.py 的代码如下所示：

```
class Series:
    """ 类，用于方便地存储要绘制的数据。
    """

    def __init__(self, name, color, linewidth, linestyle, marker, marker_size, x, y):
        self.name = name
        self.color = color
        self.linewidth = linewidth
        self.linestyle = linestyle
```

```
        self.marker = marker
        self.marker_size = marker_size
        self.x = x
        self.y = y


class Fill:
    """ 用于方便存储填充数据的类
    """

    def __init__(self, color, x, y0, y1):
        self.color = color
        self.x = x
        self.y0 = y0
        self.y1 = y1


class PlotData:
    """ 类从 json 文件加载数据，并基于这些 JSON 数据绘制图形
    """

    # 包含 json 键的类 constnats
    TITLE_KEY = "title"
    X_LABEL_KEY = "x_label"
    Y_LABEL_KEY = "y_label"
    LEGEND_Y_OFFSET_KEY = "legend_y_offset"
    LEGEND_LOC_KEY = "legend_loc"
    X_LIM_KEY = "x_lim"
    Y_LIM_KEY = "y_lim"
    X_SCALE_KEY = "x_scale"
    Y_SCALE_KEY = "y_scale"
    X_LOG_BASE_KEY = "x_log_base"
    Y_LOG_BASE_KEY = "y_log_base"
    X_TICKS_KEY = "x_ticks"
    Y_TICKS_KEY = "y_ticks"
    X_TICKS_LABELS_KEY = "x_ticks_labels"
    Y_TICKS_LABELS_KEY = "y_ticks_labels"
    X_TICKS_MINOR_KEY = "x_ticks_minor"
    Y_TICKS_MINOR_KEY = "y_ticks_minor"
    GRID_MAJOR_KEY = "grid_major"
    GRID_MINOR_KEY = "grid_minor"
    SERIES_TAG = "series"
    SERIES_NAME_TAG = "name"
    SERIES_COLOR_TAG = "color"
    SERIES_LINEWIDTH_TAG = "linewidth"
    SERIES_LINESTYLE_TAG = "linestyle"
    SERIES_MARKER_TAG = "marker"
    SERIES_MARKER_SIZE_TAG = "marker_size"
    SERIES_DATA_TAG = "data"
    FILL_TAG = "fill"
    FILL_COLOR_TAG = "color"
    FILL_DATA_TAG = "data"

    FONTSIZE_TITLE_KEY = "fontsize_title"
    FONTSIZE_LABEL_KEY = "fontsize_label"
    FONTSIZE_LEGEND_KEY = "fontsize_legend"
    FONTWEIGHT_TITLE_KEY = "fontweight_title"
    LABELSIZE_TICK_KEY = "labelsize_tick"

    DEFUALT_LEGEND_Y_OFFSET = -0.15
    DEFUALT_LEGEND_LOC = "best"
    DEFAULT_LOG_BASE = 10
    SERIES_DEFAULT_LINEWIDTH = 2.0
```

```
    SERIES_DEFAULT_MARKER_SIZE = 4
    # DEFAULT_FIGSIZE = (8, 5.5)

    def __init__(self, path_to_file):
        self.title = None
        self.x_label = None
        self.y_label = None
        self.legend_y_offset = self.DEFUALT_LEGEND_Y_OFFSET
        self.legend_loc = self.DEFUALT_LEGEND_LOC
        self.x_lim = None
        self.y_lim = None
        self.x_scale = None
        self.y_scale = None
        self.x_log_base = self.DEFAULT_LOG_BASE
        self.y_log_base = self.DEFAULT_LOG_BASE
        self.x_ticks = None
        self.y_ticks = None
        self.x_ticks_labels = None
        self.y_ticks_labels = None
        self.x_ticks_minor = False
        self.y_ticks_minor = False
        self.grid_major = False
        self.grid_minor = False
        self.series_data = None
        self.fill_data = None

        self.fontsize_title = 16
        self.fontsize_label = 14
        self.fontsize_legend = 9
        self.fontweight_title = 600
        self.labelsize_tick = 13

        self.process_file(path_to_file)

    def __repr__(self):
          return "PlotData(\"{:}\", \"{:}\", \"{:}\", {:})".format(self.title,
self.x_label, self.y_label, self.series_data)
```

使用如下命令运行本实例：

```
python json2plot.py demo/demo.json -o demo/demo.png
```

在上述命令中，“demo/demo.json”表示JSON文件的位置，“demo/demo.png”表示绘制的是图片的保存位置。打开数据可视化图片demo.png后的效果如图6-7所示。

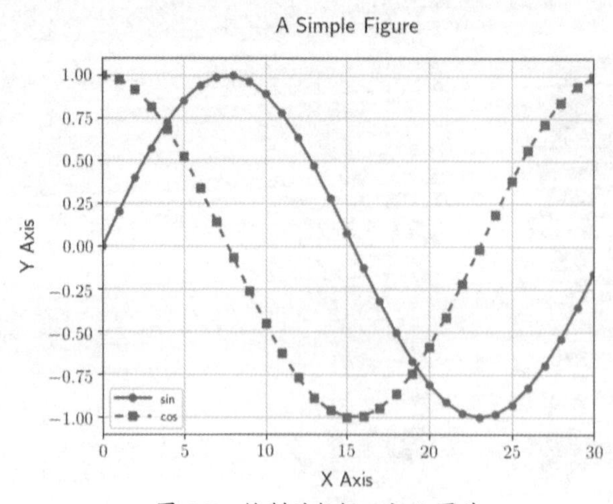

图 6-7　绘制的数据可视化图片

第 7 章

使用 NumPy 实现数据可视化处理

NumPy 是 Python 语言实现科学计算的一个库，提供了一个多维数组对象，各种派生对象（例如屏蔽的数组和矩阵）以及一系列用于数组快速操作的例程，包括数学、逻辑、形状操作、排序、选择、I/O、离散傅里叶变换、基本线性代数、基本统计操作和随机模拟等。在本章的内容中，将详细讲解在 Python 程序中使用库 NumPy 的知识。

7.1 NumPy 基础

本小节中我们将重点讲解 NumPy 库的安装、基本操作和通用的函数等基础知识；通过夯实这些内容，为本章后续实践内容的学习打下坚实的基础。

↑扫码看视频（本节视频课程时间：8 分 16 秒）

因为在标准的 Python 发行版中不会与 NumPy 库捆绑在一起，所以在使用库 NumPy 之前需要先安装 NumPy。最简单的安装方法是使用 Python 包安装程序 pip 实现，具体安装命令如下所示：

```
pip install numpy
```

在库 NumPy 中提供了一个 N 维数组类型 ndarray，用于描述相同类型的“元素”的集合，我们可以使用例如 N 个整数来对元素进行索引。在库 NumPy 中，所有的 ndarrays 都是同质的：每个元素占用相同大小的内存块，并且所有块都可以用完全相同的方式解释。如何解释数组中的每个元素由单独的数据类型对象指定，每个数组与其中一个对象相关联。除了基本类型之外（整数、浮点等），数据类型对象也可以表示数据结构。从数组中提取的元素（例如通过索引）由一个 Python 对象表示，该对象的类型为 NumPy 中内置的数组标量类型之一，数组标量允许简单地处理更为复杂的数据布置。

7.1.1 ndarray 操作

在库 NumPy 中，ndarray 是（通常大小固定的）一个多维容器，由相同类型和大小的元素组成。数组中的维度和元素数量由其 shape 定义，它是由 N 个正整数组成的元组，每个整数指定每个维度的大小。数组中元素的类型由单独的数据类型对象（dtype）指定，每个

ndarray 与其中一个对象相关联。

与 Python 中的其他容器对象一样，ndarray 的内容可以通过索引或切片（例如使用 N 个整数）以及 ndarray 的方法和属性来访问和修改数组。不同的 ndarrays 可以共享相同的数据，使得在一个 ndarray 中进行的改变在另一个中可见。也就是说，ndarray 可以是到另一个 ndarray 的“view”，并且其引用的数据由“base” ndarray 处理。ndarrays 还可以是由 Python strings 或实现 buffer 或 array 接口的对象拥有的内存的视图。

例如在下面的实例文件 001.py 中，创建了一个 2×3 的二维数组，并由 4 字节整数元素组成，具体代码如下：

```
import numpy as np
x = np.array([[1, 2, 3], [4, 5, 6]], np.int32)
print(type(x))
print(x.shape)
print(x.dtype)
```

执行后输出结果如下：

```
<class 'numpy.ndarray'>
(2, 3)
int32
```

在库 NumPy 中，数组可以使用类似 Python 容器的语法进行索引，并且切片可以生成数组的视图。例如在下面的实例文件 002.py 中演示了用 Python 语法进行索引和用切片生成列表这两种用法。

```
import numpy as np
x = np.array([[1, 2, 3], [4, 5, 6]], np.int32)
print(x[1, 2])
y = x[:,1]
print(y)
y[0] = 9                                         # 这也改变了 x 中的对应元素
print(y)
print(x)
```

执行后输出结果如下：

```
6
[2 5]
[9 5]
[[1 9 3]
[4 5 6]]
```

7.1.2 NumPy 中的通用函数

在库 NumPy 中内置了多个功能函数，通过这些函数可以实现矩阵操作。在接下来的内容中，将详细讲解这些内置函数的知识。

（1）字符串函数

在库 NumPy 中，通过使用如表 7-1 所示的函数，可以对 dtype 为 numpy.string_ 或 numpy.unicode_ 的数组执行向量化字符串操作，这些是基于 Python 内置库中的标准字符串函数。

表 7-1

函数名称	说明
add()	返回两个 str 或 Unicode 数组的逐个字符串连接
multiply()	返回按元素多重连接后的字符串
center()	返回给定字符串的副本，其中元素位于特定字符串的中央
capitalize()	返回给定字符串的副本，其中只有第一个字符串大写
title()	返回字符串或 Unicode 的按元素标题转换版本
lower()	返回一个数组，其元素转换为小写
upper()	返回一个数组，其元素转换为大写
split()	返回字符串中的单词列表，并使用分隔符来分隔
splitlines()	返回元素中的行列表，以换行符分隔
strip()	返回数组副本，其中元素移除了开头或者结尾处的特定字符
join()	返回一个字符串，它是序列中字符串的连接
replace()	返回字符串的副本，其中所有子字符串的出现位置都被新字符串取代
decode()	按元素调用 str.decode
encode()	按元素调用 str.encode

例如在下面的实例文件 003.py 中，演示了使用上述字符串函数的过程。例如使用函数 add() 将字符串“hello”和“ xyz”连接，使用函数 capitalize() 设置字符串“hello world”中的第一个字符大写等等。

```
import numpy as np
print('连接两个字符串：')
print(np.char.add(['hello'],[' xyz']) )
print('连接示例：')
print(np.char.add(['hello', 'hi'],[' abc', ' xyz']))

print(np.char.multiply('Hello ',3))

print(np.char.center('hello', 20,fillchar = '*'))

print(np.char.capitalize('hello world'))

print(np.char.title('hello how are you?'))

print(np.char.splitlines('hello\nhow are you?') )
print(np.char.splitlines('hello\rhow are you?'))

print(np.char.replace ('He is a good boy', 'is', 'was'))
```

执行后输出结果如下：

```
连接两个字符串：
['hello xyz']
连接示例：
['hello abc' 'hi xyz']
Hello Hello Hello
*******hello********
Hello world
Hello How Are You?
['hello', 'how are you?']
['hello', 'how are you?']
```

（2）算数运算函数

在库 NumPy 中包含大量实现各种数学运算功能的函数，例如三角函数、算术运算的函数和复数处理函数等。例如在下面的实例文件 004.py 中，演示了使用正弦、余弦和正切函数的过程。

```
import numpy as np
a = np.array([0,30,45,60,90])
print ('不同角度的正弦值：')
# 通过乘 pi/180 转化为弧度
print(np.sin(a*np.pi/180)  )
print ('数组中角度的余弦值：' )
print(np.cos(a*np.pi/180) )
print ('数组中角度的正切值：' )
print(np.tan(a*np.pi/180))
```

执行后会输出：

```
不同角度的正弦值：
[ 0.          0.5         0.70710678  0.8660254   1.        ]
数组中角度的余弦值：
[  1.00000000e+00   8.66025404e-01   7.07106781e-01   5.00000000e-01
   6.12323400e-17]
数组中角度的正切值：
[  0.00000000e+00   5.77350269e-01   1.00000000e+00   1.73205081e+00
   1.63312394e+16]
```

例如在下面的实例文件 005.py 中，演示了使用算数函数 add()、subtract()、multiply() 和 divide() 实现四则运算的过程。在使用这 4 个函数时，要求输入发数组必须具有相同的形状或符合数组广播规则。

```
import numpy as np
a = np.arange(9, dtype = np.float_).reshape(3,3)
print ('第一个数组：')
print(a )
print ('\n')
print ('第二个数组：' )
b = np.array([10,10,10])
print(b )
print ('\n' )
print ('两个数组相加：')
print(np.add(a,b))
print('\n')
print('两个数组相减：')
print(np.subtract(a,b)  )
print('\n' )
print('两个数组相乘：' )
print(np.multiply(a,b)  )
print('\n'  )
print('两个数组相除：' )
print(np.divide(a,b))
```

执行后输出结果如下：

```
第一个数组：
[[ 0.  1.  2.]
 [ 3.  4.  5.]
 [ 6.  7.  8.]]


第二个数组：
[10 10 10]


两个数组相加：
[[ 7.  11.  12.]
 [ 13.  14.  15.]
```

```
 [ 16.  17.  18.]]


两个数组相减：
[[-7.  -9.  -8.]
 [ -7.  -6.  -5.]
 [ -4.  -3.  -2.]]


两个数组相乘：
[[  0.   7.  20.]
 [ 30.  40.  50.]
 [ 60.  70.  80.]]


两个数组相除：
[[ 0.   0.1  0.2]
 [ 0.3  0.4  0.5]
 [ 0.6  0.7  0.8]]
```

（3）统计函数

在库 NumPy 中有很多有用的统计函数，用于从数组中给定的元素中查找最小元素、最大元素、百分标准差和方差等。例如在下面的实例文件 06.py 中，演示了使用算数函数从给定数组中的元素沿指定轴返回最小值和最大值的过程。

```
import numpy as np
a = np.array([[3,7,5],[8,4,3],[2,4,9]])
print('我们的数组是：')
print(a )
print('\n' )
print('调用 amin() 函数：' )
print(np.amin(a,1)  )
print('\n')
print('再次调用 amin() 函数：'  )
print(np.amin(a,0)  )
print('\n'  )
print('调用 amax() 函数：')
print(np.amax(a)  )
print('\n')
print('再次调用 amax() 函数：' )
print(np.amax(a, axis =  0))
```

执行后输出结果如下：

```
我们的数组是：
[[3 7 5]
[8 4 3]
[2 4 9]]

调用 amin() 函数：
[3 3 2]

再次调用 amin() 函数：
[2 4 3]

调用 amax() 函数：
9

再次调用 amax() 函数：
[8 7 9]
```

（4）排序、搜索和计数函数

在库 NumPy 中提供了各种排序相关功能，这些排序函数实现了不同的排序算法，每个排序算法的特征在于执行速度，最坏情况性能，所需的工作空间和算法的稳定性。表 7-2 中显示了三种排序算法的比较。

表 7-2

种类	速度	最坏情况	工作空间	稳定性
quicksort（快速排序）	1	$O(n\text{^}2)$	0	否
mergesort（归并排序）	2	$O(n*\log n)$	~n/2	是
heapsort（堆排序）	3	$O(n*\log n)$	0	否

例如在下面的实例文件 007.py 中，演示了使用算数函数 sort() 对数组中的元素实现快速排序的过程。

```
import numpy as np
a = np.array([[3,7],[9,1]])
print('我们的数组是：')
print(a )
print('\n' )
print('调用 sort() 函数：' )
print(np.sort(a)  )
print('\n' )
print('沿轴 0 排序：' )
print(np.sort(a, axis =  0)  )
print('\n')
# 在 sort 函数中排序字段
dt = np.dtype([('name',  'S10'),('age',  int)])
a = np.array([("raju",21),("anil",25),("ravi",  17),  ("amar",27)], dtype = dt)
print('我们的数组是：' )
print(a)
print('\n')
print('按 name 排序：')
print(np.sort(a, order =  'name'))
```

执行后输出结果如下：

```
我们的数组是：
[[3 7]
 [9 1]]


调用 sort() 函数：
[[3 7]
 [1 9]]


沿轴 0 排序：
[[3 1]
 [9 7]]


我们的数组是：
[(b'raju', 21) (b'anil', 25) (b'ravi', 17) (b'amar', 27)]


按 name 排序：
[(b'amar', 27) (b'anil', 25) (b'raju', 21) (b'ravi', 17)]
```

（5）矩阵库

在库 NumPy 中包含一个 Matrix 模块 numpy.matlib，此模块中的函数能够返回矩阵而不是返回 ndarray 对象。例如里面的 empty() 函数能够返回一个新的矩阵，而不会初始化元素。在下面的实例文件 008.py 中，演示了使用函数 empty() 返回一个矩阵的过程。

```
import numpy.matlib
import numpy as np
print(np.matlib.empty((2,2))  )
# 填充为随机数据
```

执行后输出结果如下：

```
[[  9.90263869e+067   8.01304531e+262]
 [  2.60799828e-310   9.48818959e+077]]
```

在模块 numpy.matlib 中还包含如下常用的内置函数：

- 函数 numpy.matlib.zeros()：创建一个以 0 填充的矩阵；
- 函数 numpy.matlib.ones()：创建一个以 1 填充的矩阵；
- 函数 numpy.matlib.eye()：返回一个矩阵，对角线元素为 1，其他位置为零。

7.2　当 NumPy 遇到 matplotlib

在库 NumPy 中可以使用 matplotlib 绘图库，这样可以将统计的数据以图形化的方式显示出来。在本节的内容中，将详细讲解联合使用 NumPy 和 matplotlib 的知识。

↑扫码看视频（本节视频课程时间：2 分 22 秒）

7.2.1　在 NumPy 中使用 matplotlib 绘制直线图

在下面的实例文件 014.py 中，演示了在 NumPy 中使用 matplotlib 绘制直线图的过程。

```
import numpy as np
from matplotlib import pyplot as plt

x = np.arange(1,11)
y =  2  * x +  5
plt.title("Matplotlib demo")
plt.xlabel("x axis caption")
plt.ylabel("y axis caption")
plt.plot(x,y)
plt.show()
```

在上述代码中，ndarray 对象 x 由 np.arange() 函数创建为 x 轴上的值。y 轴上的对应值存储在另一个数组对象 y 中。这些值使用 matplotlib 软件包的 pyplot 子模块的 plot() 函数绘制。最后绘制的图形由 show() 函数展示。执行效果如图 7-1 所示。

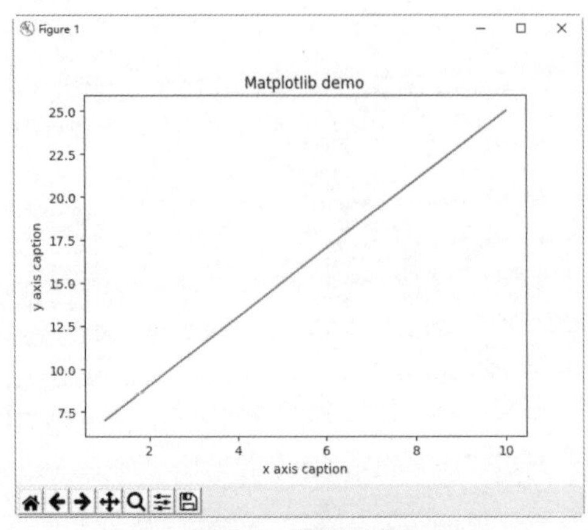

图 7-1　执行效果

7.2.2　在 NumPy 中使用 matplotlib 绘制正弦波图

在下面的实例文件 015.py 中，演示了在 NumPy 中使用 matplotlib 绘制正弦波图的过程。

```
import numpy as np
import matplotlib.pyplot as plt
# 计算正弦曲线上点的 x 和 y 坐标
x = np.arange(0,  3  * np.pi,  0.1)
y = np.sin(x)
plt.title("sine wave form")
# 使用 matplotlib 来绘制点
plt.plot(x, y)
plt.show()
```

执行效果如图 7-2 所示。

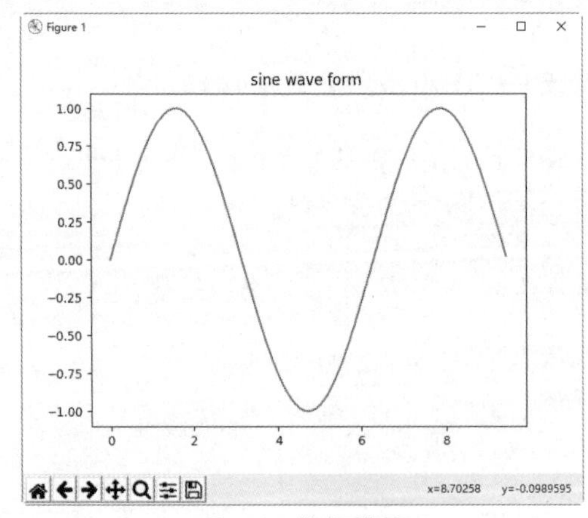

图 7-2　执行效果

7.2.3　在 NumPy 中使用 matplotlib 绘制直方图

在下面的实例文件 016.py 中，演示了在 NumPy 中使用 matplotlib 绘制直方图的过程。

```
a = np.array([22,87,5,43,56,73,55,54,11,20,51,5,79,31,27])
plt.hist(a, bins =  [0,20,40,60,80,100])
plt.title("histogram")
plt.show()
```

执行效果如图 7-3 所示。

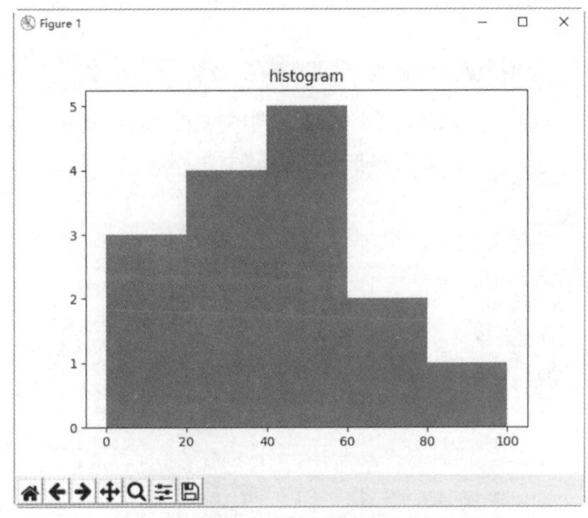

图 7-3　执行效果

7.3　实践案例：大数据分析国内主要城市的 PM2.5 状况

绿水青山就是金山银山，当前人们越来越重视环境的重要性。在本节的内容中，将通过具体实例讲解抓取指定网络中某个城市的历史环境数据的方法，并介绍将抓取的数据进行 PM2.5 大数据可视化分析的过程。

↑扫码看视频（本节视频课程时间：5 分 16 秒）

7.3.1　抓取某城市的历史天气数据

编写文件 123.py，抓取某个城市的历史天气数据，主要实现代码如下所示。

```
import time
import requests
from bs4 import BeautifulSoup

headers = {
    'User-Agent':'Mozilla/5.0 (Windows NT 6.1; WOW64) AppleWebKit/537.36 (KHTML, like Gecko) Chrome/63.0.3239.132 Safari/537.36'
}
for i in range(1, 13):
    time.sleep(5)
    # 把 1 转换为 01
    url = 'http://www. 网站域名 .com/aqi/tianjin-2017' + str("%02d" % i) + '.html'
    response = requests.get(url=url, headers=headers)
    soup = BeautifulSoup(response.text, 'html.parser')
    tr = soup.find_all('tr')
    # 去除标签栏
    for j in tr[1:]:
        td = j.find_all('td')
        Date = td[0].get_text().strip()
        Quality_grade = td[1].get_text().strip()
        AQI = td[2].get_text().strip()
        AQI_rank = td[3].get_text().strip()
        PM = td[4].get_text()
```

```
            with open('air_tianjin_2017.csv', 'a+', encoding='utf-8-sig') as f:
                f.write(Date + ',' + Quality_grade + ',' + AQI + ',' + AQI_rank + ','
+ PM + ' ' + '\n')
```

在上述代码中，首先在 headers 中设置 User-Agent，然后通过 url 设置要抓取的目标网页地址，然后通过 for 循环遍历每个目标 URL 中的指定标签信息，并将抓取的信息保存到 CSV 文件中。执行上述代码后，会抓取天津市 2017 年的天气状况，并将抓取到的数据保存到 CSV 文件 air_tianjin_2017.csv 中，如图 7-4 所示。

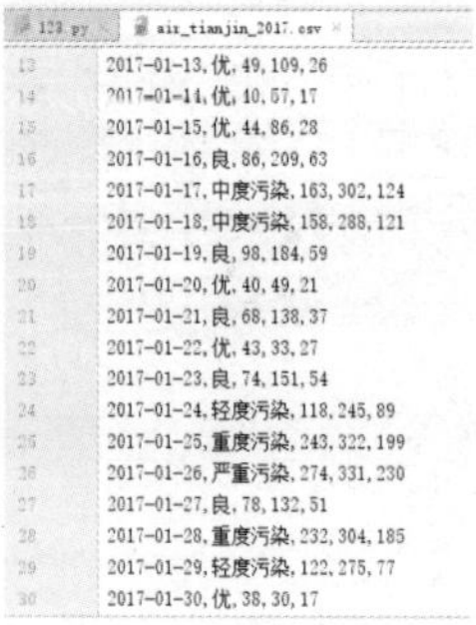

```
123.py    air_tianjin_2017.csv
13   2017-01-13,优,49,109,26
14   2017-01-14,优,40,57,17
15   2017-01-15,优,44,86,28
16   2017-01-16,良,86,209,63
17   2017-01-17,中度污染,163,302,124
18   2017-01-18,中度污染,158,288,121
19   2017-01-19,良,98,184,59
20   2017-01-20,优,40,49,21
21   2017-01-21,良,68,138,37
22   2017-01-22,优,43,33,27
23   2017-01-23,良,74,151,54
24   2017-01-24,轻度污染,118,245,89
25   2017-01-25,重度污染,243,322,199
26   2017-01-26,严重污染,274,331,230
27   2017-01-27,良,78,132,51
28   2017-01-28,重度污染,232,304,185
29   2017-01-29,轻度污染,122,275,77
30   2017-01-30,优,38,30,17

350  2017-12-15,良,89,255,58
351  2017-12-16,良,56,167,16
352  2017-12-17,良,68,185,43
353  2017-12-18,良,67,153,34
354  2017-12-19,良,52,83,27
355  2017-12-20,良,70,157,44
356  2017-12-21,良,89,197,65
357  2017-12-22,轻度污染,112,216,77
358  2017-12-23,良,94,127,68
359  2017-12-24,良,77,110,39
360  2017-12-25,优,49,42,27
361  2017-12-26,良,55,73,29
362  2017-12-27,良,84,131,59
363  2017-12-28,中度污染,152,257,115
364  2017-12-29,重度污染,209,281,159
365  2017-12-30,重度污染,218,307,170
366  2017-12-31,中度污染,193,307,153
```

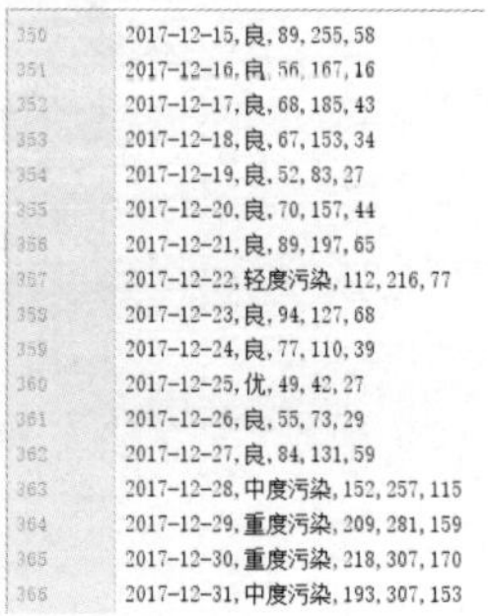

图 7-4　抓取到的天气数据（1 年的数据）

7.3.2　使用 numpy 对五个城市的 PM2.5 进行数据分析

我们可以根据项目需求来抓取不同城市的天气信息，在抓取不同的城市信息时，需要将文件 123.py 的目标 URL 修改为对应的城市即可。

（1）根据前面的爬虫原理，我们分别抓取北京、上海、广州、成都和沈阳这 5 座城市的天气数据，并且抓取的是 2010 年 1 月 11 日到 2015 年 12 月 31 日的详细数据，并且还抓取了每天整点的数据。每座城市的数据保存在一个 CSV 文件中，如图 7-5 所示。

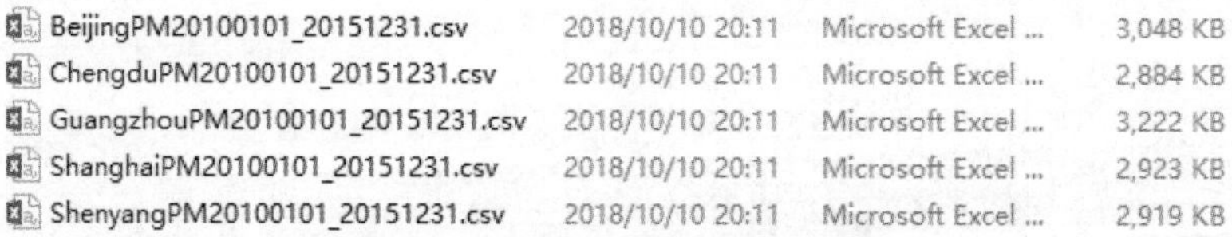

BeijingPM20100101_20151231.csv	2018/10/10 20:11	Microsoft Excel ...	3,048 KB
ChengduPM20100101_20151231.csv	2018/10/10 20:11	Microsoft Excel ...	2,884 KB
GuangzhouPM20100101_20151231.csv	2018/10/10 20:11	Microsoft Excel ...	3,222 KB
ShanghaiPM20100101_20151231.csv	2018/10/10 20:11	Microsoft Excel ...	2,923 KB
ShenyangPM20100101_20151231.csv	2018/10/10 20:11	Microsoft Excel ...	2,919 KB

图 7-5　5 座城市的详细天气数据

（2）编写文件 config.py，功能是设置要处理分析的 CSV 文件，设置分析处理后 CSV 文件的保存目录。文件 config.py 的主要实现代码如下所示：

```
import os

# 指定数据集路径
dataset_path = './data'

# 结果保存路径
output_path = './output'
if not os.path.exists(output_path):
    os.makedirs(output_path)

# 公共列
common_cols = ['year', 'month']

# 每个城市对应的文件名及所需分析的列名
```

```
# 以字典形式保存，如：{ 城市：( 文件名 , 列名 )}
data_config_dict = {'beijing': ('BeijingPM20100101_20151231.csv',
                              ['Dongsi', 'Dongsihuan', 'Nongzhanguan']),
                    'chengdu': ('ChengduPM20100101_20151231.csv',
                              ['Caotangsi', 'Shahepu']),
                    'guangzhou': ('GuangzhouPM20100101_20151231.csv',
                               ['City Station', '5th Middle School']),
                    'shanghai': ('ShanghaiPM20100101_20151231.csv',
                               ['Jingan', 'Xuhui']),
                    'shenyang': ('ShenyangPM20100101_20151231.csv',
                               ['Taiyuanjie', 'Xiaoheyan'])
                    }
```

（3）编写文件 pm2.5_demoes2.py，功能是分析北京、上海、广州、成都和沈阳这 5 座城市的天气数据，分析 5 座城市的污染状态和五城市每个区空气质量的月度差异。具体实现流程可分为 6 个步骤。

① 导入用到的模块，对应代码如下所示：

```
import csv
import os
import numpy as np
import config
```

② 通过函数 load_data() 加载要分析处理的 CSV 文件，首先提取文件中的每行数据，然后组合为一个列表放入数据列表中，最后将 data 转换为 ndarray。对应的实现代码如下所示：

```
def load_data(data_file, usecols):
    """
        参数:
            - data_file:    文件路径
            - usecols:       所使用的列
        返回:
            - data_arr:      数据的多维数组表示
    """
    data = []
    with open(data_file, 'r') as csvfile:
        data_reader = csv.DictReader(csvfile)
        print(data_reader)
        # === 数据处理 ===
        for row in data_reader:
            # 取出每行数据，组合为一个列表放入数据列表中
            row_data = []
            # 注意 csv 模块读入的数据全部为字符串类型
            for col in usecols:
                str_val = row[col]
                # 数据类型转换为 float，如果是 'NA'，则返回 nan
                row_data.append(float(str_val) if str_val != 'NA' else np.nan)
            # 如果行数据中不包含 nan 才保存该行记录
            if not any(np.isnan(row_data)):
                data.append(row_data)

    # 将 data 转换为 ndarray
    data_arr = np.array(data)
    return data_arr
```

③ 通过 get_polluted_perc() 获取污染占比的小时数，对应实现代码如下所示：

```
def get_polluted_perc(data_arr):
    """
            重度污染 (heavy)      PM2.5 > 150
            中度污染 (medium)     75 < PM2.5 <= 150
            轻度污染 (light)      35 < PM2.5 <= 75
            优良空气 (good)       PM2.5 <= 35
```

```
        参数：
            - data_arr：数据的多维数组表示
        返回：
            - polluted_perc_list：污染小时数百分比列表
    """
    # 将每个区的 PM 值平均后作为该城市小时的 PM 值
    # 按行取平均值
    hour_val = np.mean(data_arr[:, 2:], axis=1)
    # 总小时数
    n_hours = hour_val.shape[0]
    # 重度污染小时数
    n_heavy_hours = hour_val[hour_val > 150].shape[0]
    # 中度污染小时数
    n_medium_hours = hour_val[(hour_val > 75) & (hour_val <= 150)].shape[0]
    # 轻度污染小时数
    n_light_hours = hour_val[(hour_val > 35) & (hour_val <= 75)].shape[0]
    # 优良空气小时数
    n_good_hours = hour_val[hour_val <= 35].shape[0]
    polluted_perc_list = [n_heavy_hours / n_hours, n_medium_hours / n_hours,
                          n_light_hours / n_hours, n_good_hours / n_hours]
    return polluted_perc_list
```

④ 通过函数 get_avg_pm_per_month() 获取每个区每月的平均 PM 值，对应实现代码如下所示：

```
def get_avg_pm_per_month(data_arr):
    """
        参数：
            - data_arr：数据的多维数组表示
        返回：
            - results_arr：  多维数组结果
    """
    results = []
   # 获取年份
    years = np.unique(data_arr[:, 0])
    for year in years:
        # 获取当前年份数据
        year_data_arr = data_arr[data_arr[:, 0] == year]
        # 获取数据的月份
        month_list = np.unique(year_data_arr[:, 1])

        for month in month_list:
            # 获取月份的所有数据
            month_data_arr = year_data_arr[year_data_arr[:, 1] == month]
            # 计算当前月份 PM 的均值
            mean_vals = np.mean(month_data_arr[:, 2:], axis=0).tolist()

            # 格式化字符串
            row_data = ['{:.0f}-{:02.0f}'.format(year, month)] + mean_vals
            results.append(row_data)
    results_arr = np.array(results)
    return results_arr
```

⑤ 通过函数 save_stats_to_csv() 将统计结果保存至 csv 文件中，对应实现代码如下所示：

```
def save_stats_to_csv(results_arr, save_file, headers):
    """
       参数：
           - results_arr:          多维数组结果
           - save_file:            文件保存路径
           - headers:              csv 表头
    """
    with open(save_file, 'w', newline='') as csvfile:
        writer = csv.writer(csvfile)
```

```
        writer.writerow(headers)
        for row in results_arr.tolist():
            writer.writerow(row)
```

⑥ 在主函数 main() 中实现详细的数据分析，并打印输出分析结果。对应实现代码如下所示：

```
def main():
    """
        主函数
    """
    polluted_state_list = []

    for city_name, (filename, cols) in config.data_config_dict.items():
        # === 数据获取 + 数据处理 ===
        data_file = os.path.join(config.dataset_path, filename)
        usecols = config.common_cols + ['PM_' + col for col in cols]
        data_arr = load_data(data_file, usecols)

        print('{} 共有 {} 行有效数据 '.format(city_name, data_arr.shape[0]))
        # 预览前 10 行数据
        print('{} 的前 10 行数据: '.format(city_name))
        print(data_arr[:10])

        # === 数据分析 ===
        # 五城市污染状态，统计污染小时数的占比
        polluted_perc_list = get_polluted_perc(data_arr)
        polluted_state_list.append([city_name] + polluted_perc_list)
        print('{} 的污染小时数百分比 {}'.format(city_name, polluted_perc_list))

        # 五城市每个区空气质量的月度差异，分析计算每个月，每个区的平均 PM 值
        results_arr = get_avg_pm_per_month(data_arr)
        print('{} 的每月平均 PM 值预览: '.format(city_name))
        print(results_arr[:10])

        # === 结果展示 ===
        # 保存月度统计结果至 csv 文件
        save_filename = city_name + '_month_stats.csv'
        save_file = os.path.join(config.output_path, save_filename)
        save_stats_to_csv(results_arr, save_file, headers=['month'] + cols)
        print(' 月度统计结果已保存至 {}'.format(save_file))
        print()
    # 污染状态结果保存
    save_file = os.path.join(config.output_path, 'polluted_percentage.csv')
    with open(save_file, 'w', newline='') as csvfile:
        writer = csv.writer(csvfile)
        writer.writerow(['city', 'heavy', 'medium', 'light', 'good'])
        for row in polluted_state_list:
            writer.writerow(row)
    print(' 污染状态结果已保存至 {}'.format(save_file))
if __name__ == '__main__':
    main()
```

执行后输出分析结果如下：

```
beijing 共有 19613 行有效数据
beijing 的前 10 行数据:
[[2013.    3.  117.  166.  140.]
 [2013.    3.  131.  165.  152.]
 [2013.    3.  141.  173.  128.]
 [2013.    3.  169.  182.    3.]
 [2013.    3.  169.  169.    3.]
 [2013.    3.  174.  183.  163.]
```

```
 [2013.    3.  194.  195.  192.]
 [2013.    3.  208.  212.  203.]
 [2013.    3.  213.  207.  195.]
 [2013.    3.  203.  198.  185.]]
beijing的污染小时数百分比[0.1723346759802172, 0.26956610411461784, 0.24611227247233977,
0.3119869474328252]
beijing 的每月平均 PM 值预览:
[['2013-03' '117.99354838709678' '128.4725806451613' '116.1774193548387']
 ['2013-04' '64.298937784522' '63.165402124430955' '56.88770864946889']
 ['2013-05' '91.35816618911174' '101.55014326647564' '77.11174785100286']
 ['2013-06' '17.01160092807424' '119.17169373549883'
  '108.27146171693735']
 ['2013-07' '72.19110378912686' '85.35090609555189' '74.67051070840198']
 ['2013-08' '63.986301369863014' '69.77168949771689' '64.64687975646879']
 ['2013-09' '83.79607250755286' '82.89577039274924' '80.97129909365559']
 ['2013-10' '102.78525641025641' '101.52403846153847' '94.6923076923077']
 ['2013-11' '83.16338028169014' '84.2338028169014' '83.55211267605634']
 ['2013-12' '87.7453505007153' '92.02718168812589' '89.99570815450643']]
月度统计结果已保存至 ./output\beijing_month_stats.csv

<csv.DictReader object at 0x0000028579113E10>
chengdu 共有 23816 行有效数据
chengdu 的前 10 行数据:
[[2.013e+03 1.000e+00 1.210e+02 1.380e+02]
 [2.013e+03 1.000e+00 1.340e+02 1.590e+02]
 [2.013e+03 1.000e+00 2.030e+02 1.620e+02]
 [2.013e+03 1.000e+00 2.170e+02 1.570e+02]
 [2.013e+03 1.000e+00 2.200e+02 1.700e+02]
 [2.013e+03 1.000e+00 2.140e+02 2.250e+02]
 [2.013e+03 1.000e+00 2.090e+02 2.440e+02]
 [2.013e+03 1.000e+00 2.280e+02 2.420e+02]
 [2.013e+03 1.000e+00 2.190e+02 2.770e+02]
 [2.013e+03 1.000e+00 2.250e+02 2.810e+02]]
chengdu 的污染小时数百分比[0.10971615720524018, 0.2613789049378569, 0.394902586496473,
0.23400235136042996]
chengdu 的每月平均 PM 值预览:
[['2013-01' '170.09582689335394' '189.5625965996909']
 ['2013-02' '126.59324758842443' '118.9807073954984']
 ['2013-03' '141.24685534591194' '139.7059748427673']
 ['2013-04' '102.12990196078431' '94.19607843137256']
 ['2013-05' '77.12660944206009' '66.92703862660944']
 ['2013-06' '52.236486486486484' '47.11711711711712']
 ['2013-07' '50.69642857142857' '40.565934065934066']
 ['2013-08' '66.55602240896359' '56.627450980392155']
 ['2013-09' '60.584' '58.364']
 ['2013-10' '100.51994301994301' '99.68518518518519']]
月度统计结果已保存至 ./output\chengdu_month_stats.csv

<csv.DictReader object at 0x0000028578EFA358>
guangzhou 共有 20074 行有效数据
guangzhou 的前 10 行数据:
[[2.013e+03 1.000e+00 8.300e+01 7.800e+01]
 [2.013e+03 1.000e+00 9.500e+01 7.000e+01]
 [2.013e+03 1.000e+00 5.500e+01 6.600e+01]
 [2.013e+03 1.000e+00 6.000e+01 6.900e+01]
 [2.013e+03 1.000e+00 4.100e+01 5.100e+01]
########### 后面省略好多分析结果
```

并且执行后还会生成包含每座城市的月度 PM2.5 统计数据的 CSV 文件，如图 7-6 所示。

beijing_month_stats.csv
chengdu_month_stats.csv
guangzhou_month_stats.csv
polluted_percentage.csv
shanghai_month_stats.csv
shenyang_month_stats.csv

	A	B	C	D	E
1	month	Dongsi	Dongsihua	Nongzhanguan	
2	Mar-13	117.9935	128.4726	116.1774	
3	Apr-13	64.29894	63.1654	56.88771	
4	May-13	91.35817	101.5501	77.11175	
5	Jun-13	110.0116	119.1717	108.2715	
6	Jul-13	72.1911	85.35091	74.67051	
7	Aug-13	63.9863	69.77169	64.64688	
8	Sep-13	83.79607	82.89577	80.9713	
9	Oct-13	102.7853	101.524	94.69231	
10	Nov-13	83.16338	84.2338	83.55211	
11	Dec-13	87.74535	92.02718	89.99571	
12	Jan-14	106.0099	103.33	107.7109	
13	Feb-14	143.0712	138.8689	144.0307	
14	Mar-14	98.71291	100.6003	99.53846	
15	Apr-14	89.04089	95.57922	86.31857	
16	May-14	63.06165	58.78195	63.28271	
17	Jun-14	57.82436	63.53541	56.9221	
18	Jul-14	100.7775	95.694	86.22987	
19	Aug-14	69.81568	70.18019	61.01926	
20	Sep-14	68.49786	72.26819	64.70613	
21	Oct-14	112.7324	145.4609	128.3574	
22	Nov-14	77.9721	119.2599	105.6535	
23	Dec-14	52.0389	78.89625	70.83285	
24	Jan-15	103.1328	123.2836	109.2731	
25	Feb-15	112.3959	112.264	104.9272	
26	Mar-15	95.56158	103.5337	95.66716	
27	Apr-15	74.36215	83.15498	73.81892	
28	May-15	60.36438	66.66667	57.85621	

图 7-6　包含月度 PM2.5 统计数据的 CSV 文件

7.3.3　使用 Numpy 和 matplotli 绘制 PM2.5 数据统计图

接下来将以 7.3.2 小节中源码为基础，使用 Numpy 和 matplotli 绘制 5 个城市的 PM2.5 数据统计图。实例文件 main.py 的具体实现代码如下所示：

```
matplotlib.rcParams['font.sans-serif'] = ['SimHei']

def load_data(data_file,usecols):
    """
    :param data_file: 读取的文件的路径
    :param usecols: 需要取的列
    :return: data_arr:  数据的多维数组，形式 [[2013    3  117  166  140]]
    """
    data = []
    # === 读取数据 ===
    with open(data_file,'r') as csvfile:
        data_reader = csv.DictReader(csvfile)

        # === 数据处理 ===
        for row in data_reader:
            # 取出每行数据，组合为一个列表放入数据列表中
            row_data = []
            # 注意 csv 模块读入的数据全部为字符串类型
            for col in usecols:
                str_val = row[col]
                # 数据类型转换为 float，如果是 'NA'，则返回 nan
                row_data.append(float(str_val) if str_val !='NA' else np.nan)
            # 如果数据中不包含 nan 才保存该行记录
            if not any(np.isnan(row_data)):
                data.append(row_data)
    # 将 data 转换为 ndarray
    data_arr = np.array(data).astype(int)
    return data_arr

def get_polluted_perc(data_arr):
    """
    获取污染占比小时数
    规则：
```

```
                重度污染 (heavy)      PM2.5 > 150
                中度污染 (medium)     75 < PM2.5 <= 150
                轻度污染 (light)      35 < PM2.5 <= 75
                优良空气 (good)       PM2.5 <= 35
        :param data_arr:  数据的多维数组，形式 [[2013    3  117  166  140]]
        :return: polluted_perc_list: 污染小时数百分比列表
        """
        # 将每个区的 PM 值平均后作为该城市小时的 PM 值
        # 按行取平均值
        hour_val = np.mean(data_arr[:,2:-1],axis=1)
        # us_hour_val = data_arr[:,-1]
        # 总小时数
        n_hours = hour_val.shape[0]
        # usn_hours = us_hour_val.shape[0]
        # 重度污染小时数
        n_heavy_hours = hour_val[hour_val > 150].shape[0]
        # usn_heavy_hours = us_hour_val[us_hour_val > 150].shape[0]
        # print(usn_heavy_hours / usn_hours)
        # 中度污染小时数
        n_medium_hours = hour_val[(hour_val > 75)&(hour_val <= 150)].shape[0]
        # 轻度污染小时数
        n_light_hours = hour_val[(hour_val > 35) & (hour_val <= 75)].shape[0]
        # 优良空气小时数
        n_good_hours = hour_val[hour_val <= 35].shape[0]
         polluted_prc_list = [n_heavy_hours / n_hours, n_medium_hours/ n_hours, n_
light_hours / n_hours, n_good_hours / n_hours]
        return polluted_prc_list


    def get_us_polluted_perc(data_arr):
        """
        获取污染占比小时数
        规则:
                重度污染 (heavy)      PM2.5 > 150
                中度污染 (medium)     75 < PM2.5 <= 150
                轻度污染 (light)      35 < PM2.5 <= 75
                优良空气 (good)       PM2.5 <= 35
        :param data_arr: 数据的多维数组，形式 [[2013    3  117  166  140]]
        :return: us_polluted_perc_list: 污染小时数百分比列表
        """
        # 取到 us 给出的数据
        us_hours = data_arr[:,-1]

        # 取到总小时数
        usn_hours = us_hours.shape[0]

        # 重度污染小时数
        usn_heavy_hours = us_hours[us_hours > 150].shape[0]

        # 中度污染小时数
        usn_medium_hours = us_hours[(us_hours > 75) & (us_hours <= 150)].shape[0]

        # 轻度污染小时数
        usn_light_hours = us_hours[(us_hours > 35) & (us_hours <= 75)].shape[0]

        # 优良空气小时数
        usn_good_hours = us_hours[us_hours <= 35].shape[0]
         us_polluted_perc_list = [usn_heavy_hours / usn_hours, usn_medium_hours /
usn_hours, usn_light_hours / usn_hours, usn_good_hours / usn_hours]
        return us_polluted_perc_list

    def get_avg_pm_per_month(data_arr):
        """
```

```
        获取每个区每月的平均 PM 值
        :param data_arr: 数据的数组表示
        :return: results_arr: 多维数组结果
        """
        results = []
        # 获取年份
        years = np.unique(data_arr[:, 0])
        for year in years:
            # 获取当前年份数据
            year_data_arr = data_arr[data_arr[:,0] == year]
            # 获取数据的月份
            month_list = np.unique(year_data_arr[:, 1])

            for month in month_list:
                # 获取月份的所有数据
                month_data_arr = year_data_arr[year_data_arr[:, 1] == month]
                # 计算当前月份 PM 的均值，并转换为列表
                mean_vals = np.mean(month_data_arr[:, 2:-1],axis=0).tolist()
                # 格式化字符串
                row_data = ['{:.0f}-{:02.0f}'.format(year, month)] + mean_vals
                results.append(row_data)
        results_arr = np.array(results)
        return results_arr

    def draw_hist(polluted_perc_list,us_polluted_perc_list,city_name):
        plt.figure()
        labels = ['重度污染','中度污染','轻度污染','优良空气']
        explode = [0.05,0,0,0]
        colors = ['#FF0000', 'orange', '#0099CC', 'lime']
        plt.subplot(121)
         plt.pie(polluted_perc_list,explode=explode,colors=colors,labels=labels,auto
pct='%1.1f%%',shadow=True,labeldistance=1.1,pctdistance=0.6)
        plt.title(city_name + '污染情况，数据来自中国')
        plt.axis('equal')
        plt.subplot(122)
        plt.pie(us_polluted_perc_list, explode=explode, colors=colors,labels=labels,
autopct='%1.1f%%', shadow=True, labeldistance=1.1,pctdistance=0.6)
        plt.title(city_name + '污染情况，数据来自国外')
        plt.axis('equal')
        plt.legend()
        plt.show()


    def main():
        """
            主函数
        """
        for city_name,(filename, cols) in config.data_config_dict.items():
            #   数据获取 + 数据处理
            data_file = os.path.join(config.dataset_path, filename)
            usecols = config.common_cols + [col for col in cols]
            data_arr = load_data(data_file, usecols)
            # # 五城市的污染状态，统计污染小时数的占比
            polluted_perc_list = get_polluted_perc(data_arr)
            us_polluted_perc_list = get_us_polluted_perc(data_arr)
            draw_hist(polluted_perc_list, us_polluted_perc_list, city_name)
    if __name__ == '__main__':
        main()
```

执行后会显示北京、上海、广州、成都和沈阳这 5 座城市的 PM2.5 数据统计图，例如沈阳市的环境状况统计图如图 7-7 所示。

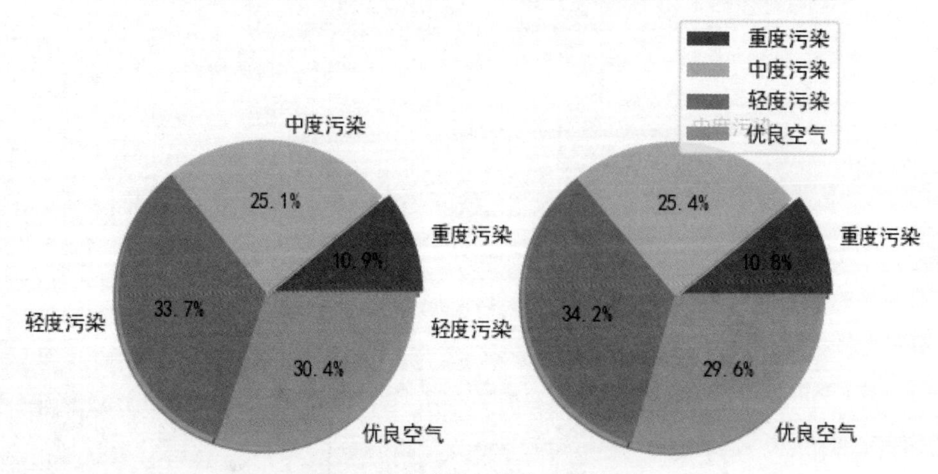

图 7-7　沈阳市的环境状况统计图

广州市的环境状况统计图如图 7-8 所示。

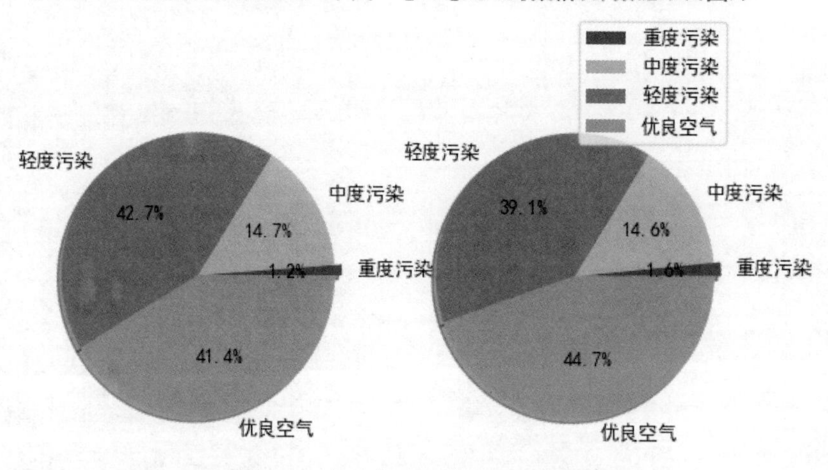

图 7-8　广州市的环境状况统计图

第 8 章

使用库 pygal 实现数据可视化

在 Python 程序中，可以使用库 pygal 生成 SVG 格式的图形文件，从而实现数据的可视化功能。SVG 是一种常见的数据分析文件格式，是一种矢量图格式。我们可以使用浏览器打开 SVG 文件，可以方便与之交互。在本章的内容中，将详细讲解使用库 pygal 实现数据可视化的知识，为读者步入本书后面知识的学习打下基础。

8.1 pygal 的基本操作

在本节的内容中，将首先通过 10 个具体实例的实现过程，讲解使用 pygal 绘制数据分析图表的知识；在这些实例中蕴含着 pygal 库的基本操作，读者可体会掌握。

↑扫码看视频（本节视频课程时间：6 分 31 秒）

8.1.1 使用库 pygal 绘制条形图

使用 pygal 绘制条形图的方法十分简单，只需调用库 pygal 中的 Bar() 方法即可。例如在下面的实例文件 tiao01.py 中，绘制了 2002~2013 年网页浏览器的使用变化数据的条形图。

```
import pygal

line_chart = pygal.Bar()
line_chart.title = '网页浏览器的使用变化 (in %)'
line_chart.x_labels = map(str, range(2002, 2013))
line_chart.add('Firefox', [None, None, 0, 16.6, 25, 31, 36.4, 45.5, 46.3, 42.8, 37.1])
line_chart.add('Chrome',  [None, None, None, None, None, None, 0, 3.9, 10.8, 23.8, 35.3])
line_chart.add('IE',      [85.8, 84.6, 84.7, 74.5, 66, 58.6, 54.7, 44.8, 36.2, 26.6, 20.1])
line_chart.add('Others',  [14.2, 15.4, 15.3,  8.9, 9, 10.4,  8.9,  5.8,  6.7,  6.8, 7.5])
line_chart.render_to_file('bar_chart.svg')
```

执行后会创建生成条形图文件 bar_chart.svg，打开后的效果如图 8-1 所示。

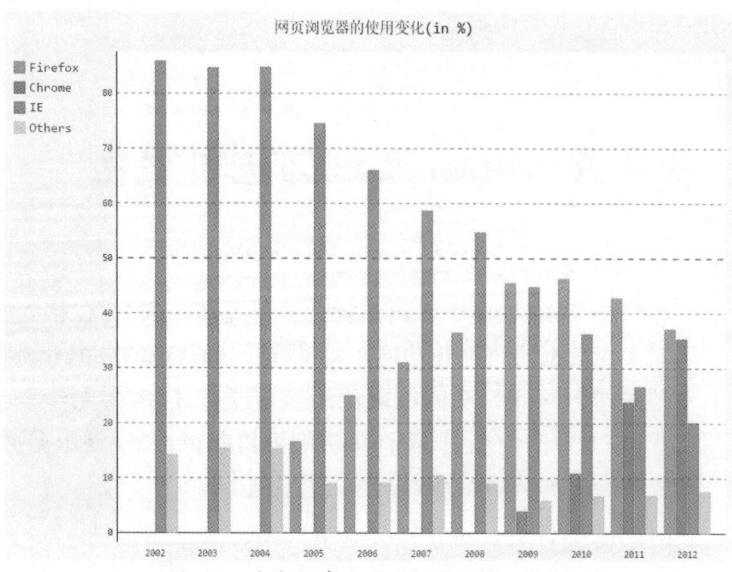

图 8-1　生成的条形图文件 bar_chart.svg

8.1.2　使用库 pygal 绘制直方图

使用 pygal 绘制直方图的方法十分简单，只需调用库 pygal 中的 Histogram() 方法即可。例如在下面的实例文件 tiao02.py 中，使用 Histogram() 方法分别绘制了宽直方图和窄直方图。

```
import pygal
hist = pygal.Histogram()
hist.add('Wide bars', [(5, 0, 10), (4, 5, 13), (2, 0, 15)])
hist.add('Narrow bars',  [(10, 1, 2), (12, 4, 4.5), (8, 11, 13)])
hist.render_to_file('bar_chart.svg')
```

执行后会创建生成直方图文件 bar_chart.svg，打开后的效果如图 8-2 所示。

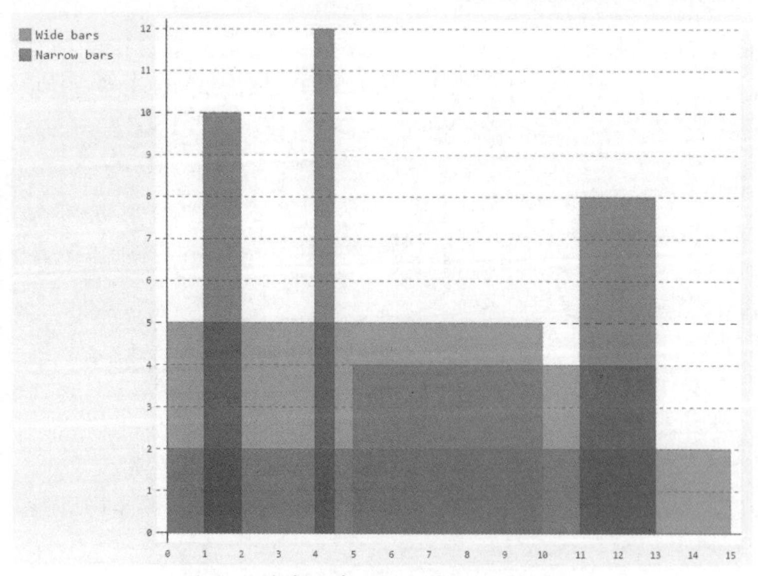

图 8-2　生成的直方图文件 bar_chart.svg

8.1.3 使用库 pygal 绘制 XY 线图

XY 线是将各个点用直线连接起来的折线图，在绘制时需提供一个横纵坐标元组作为元素的列表。使用 pygal 绘制 XY 线图的方法十分简单，只需调用库 pygal 中的 XY() 方法即可。例如在下面的实例文件 tiao03.py 中，演示了使用 XY() 方法绘制两条 XY 余弦曲线图的过程。

```
import pygal
from math import cos

xy_chart = pygal.XY()
xy_chart.title = 'XY 余弦曲线图'
xy_chart.add('x = cos(y)', [(cos(x / 10.), x / 10.) for x in range(-50, 50, 5)])
xy_chart.add('y = cos(x)', [(x / 10., cos(x / 10.)) for x in range(-50, 50, 5)])
xy_chart.add('x = 1',  [(1, -5), (1, 5)])
xy_chart.add('x = -1', [(-1, -5), (-1, 5)])
xy_chart.add('y = 1',  [(-5, 1), (5, 1)])
xy_chart.add('y = -1', [(-5, -1), (5, -1)])
xy_chart.render_to_file('bar_chart.svg')
```

执行后会创建生成 XY 余弦曲线图文件 bar_chart.svg，打开后的效果如图 8-3 所示。

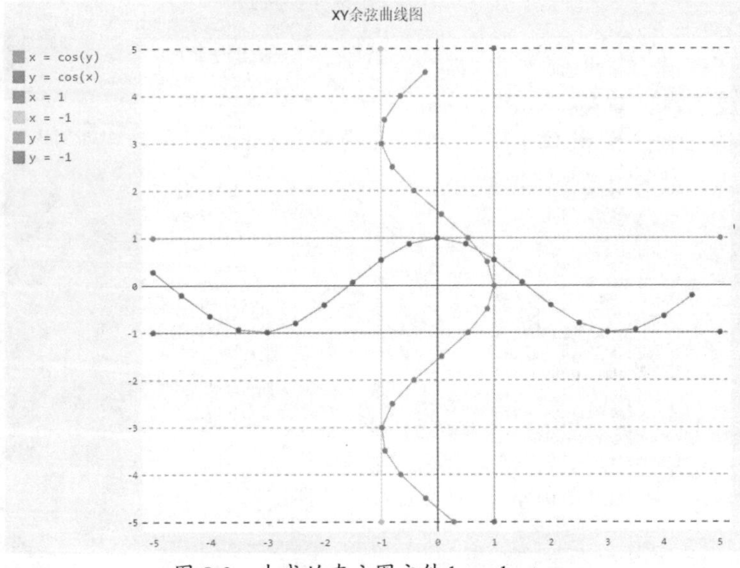

图 8-3　生成的直方图文件 bar_chart.svg

8.1.4 使用库 pygal 绘制饼状图

使用 pygal 绘制饼状图的方法十分简单，只需调用库 pygal 中的 Pie() 方法即可。

1. 基本饼状图

例如在下面的实例文件 tiao04.py 中演示了使用 Pie() 方法绘制 2012 年度浏览器使用数据饼状图的过程。

```
import pygal
pie_chart = pygal.Pie()
pie_chart.title = '2012 年主流网页浏览器的使用率 (in %)'
pie_chart.add('IE', 18.4)
pie_chart.add('Firefox', 36.6)
pie_chart.add('Chrome', 36.3)
pie_chart.add('Safari', 4.5)
```

```
pie_chart.add('Opera', 2.3)
pie_chart.render_to_file('bar_chart.svg')
```

执行后会创建生成饼状图文件 bar_chart.svg，打开后的效果如图 8-4 所示。

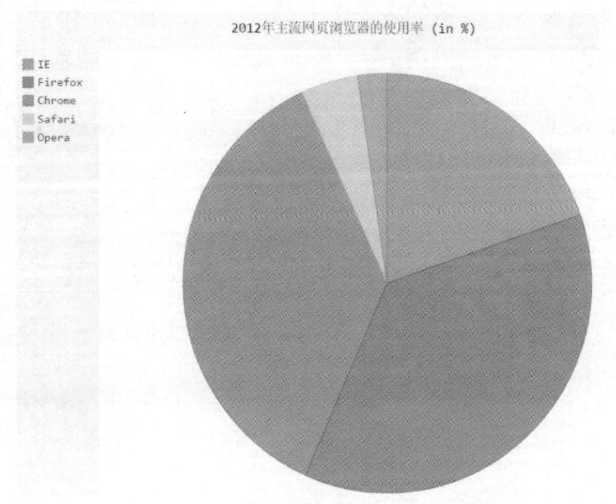

图 8-4　生成的饼状图文件 bar_chart.svg

2. 多系列饼状图

请看下面的实例文件 pie09.py，功能是使用库 pygal 绘制多系列饼状图，可视化展示 2012 年度浏览器产品的市场份额数据。

```
import pygal

pie_chart = pygal.Pie()
pie_chart.title = 'Browser usage by version in February 2012 (in %)'
pie_chart.add('IE', [5.7, 10.2, 2.6, 1])
pie_chart.add('Firefox', [.6, 16.8, 7.4, 2.2, 1.2, 1, 1, 1.1, 4.3, 1])
pie_chart.add('Chrome', [.3, .9, 17.1, 15.3, .6, .5, 1.6])
pie_chart.add('Safari', [4.4, .1])
pie_chart.add('Opera', [.1, 1.6, .1, .5])
pie_chart.render_to_file('bar_chart.svg')
```

执行后会生成文件 bar_chart.svg，效果如图 8-5 所示。

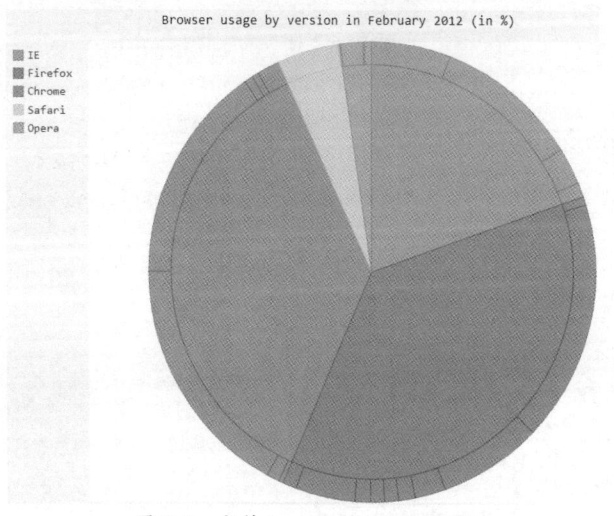

图 8-5　文件 bar_chart.svg 的效果

3. 圈状饼状图

请看下面的实例文件 pie10.py，功能是使用库 pygal 绘制圈状饼状图，可视化展示某年度浏览器产品的市场份额数据。

```
import pygal

pie_chart = pygal.Pie(inner_radius=.4)
pie_chart.title = 'Browser usage in February 2012 (in %)'
pie_chart.add('IE', 18.4)
pie_chart.add('Firefox', 36.6)
pie_chart.add('Chrome', 36.3)
pie_chart.add('Safari', 4.5)
pie_chart.add('Opera', 2.3)
pie_chart.render_to_file('bar_chart.svg')
```

执行后会生成文件 bar_chart.svg，效果如图 8-6 所示。

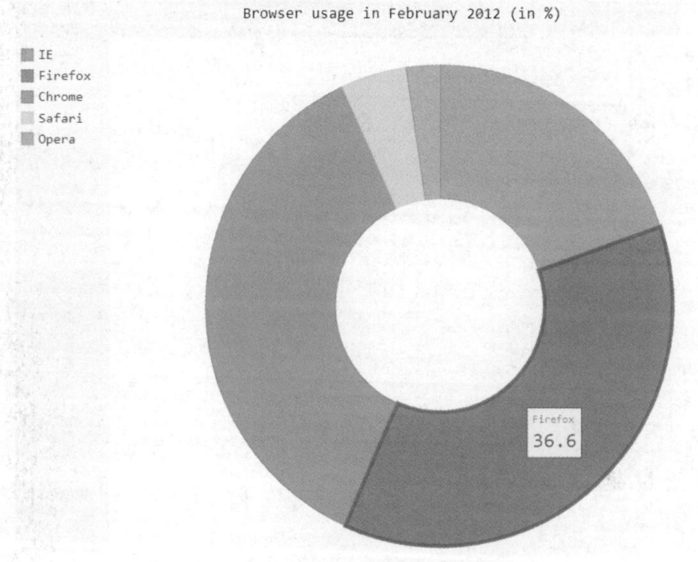

图 8-6　文件 bar_chart.svg 的效果

4. 环状饼状图

请看下面的实例文件 pie11.py，功能是使用库 pygal 绘制环状饼状图，可视化展示 2012 年度浏览器产品的市场份额数据。

```
import pygal

pie_chart = pygal.Pie(inner_radius=.75)
pie_chart.title = 'Browser usage in February 2012 (in %)'
pie_chart.add('IE', 18.4)
pie_chart.add('Firefox', 36.6)
pie_chart.add('Chrome', 36.3)
pie_chart.add('Safari', 4.5)
pie_chart.add('Opera', 2.3)
pie_chart.render_to_file('bar_chart.svg')
```

执行后会生成文件 bar_chart.svg，效果如图 8-7 所示。

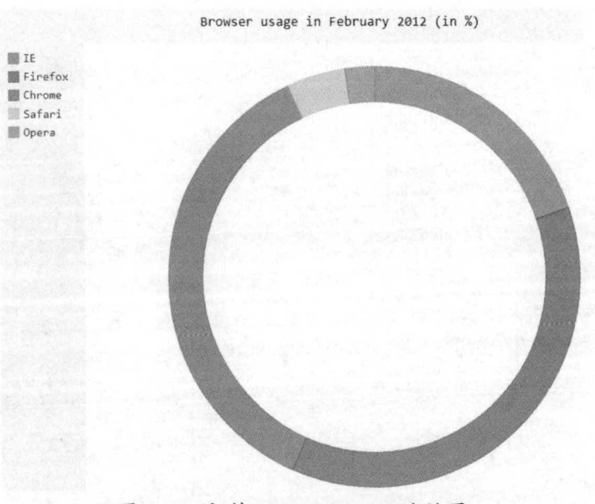

图 8-7　文件 bar_chart.svg 的效果

5. 半饼状图

请看下面的实例文件 pie12.py，功能是使用库 pygal 绘制半饼状图，可视化展示某年度浏览器产品的市场份额数据。

```
import pygal

pie_chart = pygal.Pie(half_pie=True)
pie_chart.title = 'Browser usage in February 2012 (in %)'
pie_chart.add('IE', 18.4)
pie_chart.add('Firefox', 36.6)
pie_chart.add('Chrome', 36.3)
pie_chart.add('Safari', 4.5)
pie_chart.add('Opera', 2.3)
pie_chart.render_to_file('bar_chart.svg')
```

执行后会生成文件 bar_chart.svg，效果如图 8-8 所示。

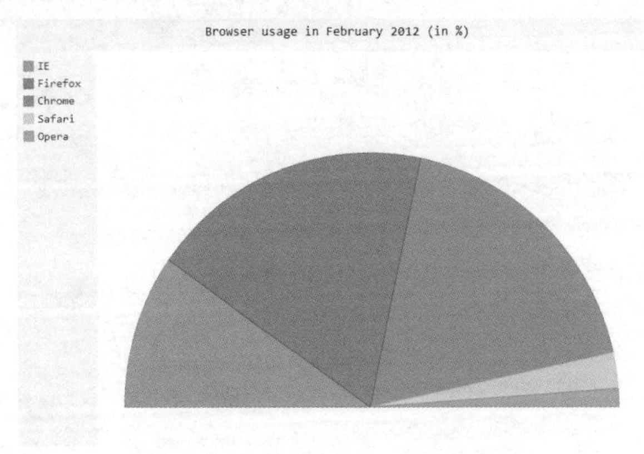

图 8-8　文件 bar_chart.svg 的效果

8.1.5　使用库 pygal 绘制雷达图

使用 pygal 绘制雷达图的方法十分简单，只需调用库 pygal 中的 Radar() 方法即可。例如在下面的实例文件 tiao05.py 中，演示了使用 Radar() 法绘制主流浏览器 V8 基准测试雷达图的过程。

```
import pygal

radar_chart = pygal.Radar()
radar_chart.title = 'V8 基准测试结果 '
radar_chart.x_labels = ['Richards', 'DeltaBlue', 'Crypto', 'RayTrace',
'EarleyBoyer', 'RegExp', 'Splay', 'NavierStokes']
radar_chart.add('Chrome', [6395, 8212, 7520, 7218, 12464, 1660, 2123, 8607])
radar_chart.add('Firefox', [7473, 8099, 11700, 2651, 6361, 1044, 3797, 9450])
radar_chart.add('Opera', [3472, 2933, 4203, 5229, 5810, 1828, 9013, 4669])
radar_chart.add('IE', [43, 41, 59, 79, 144, 136, 34, 102])
radar_chart.render_to_file('bar_chart.svg')
```

执行后会创建生成雷达图文件 bar_chart.svg，打开后的效果如图 8-9 所示。

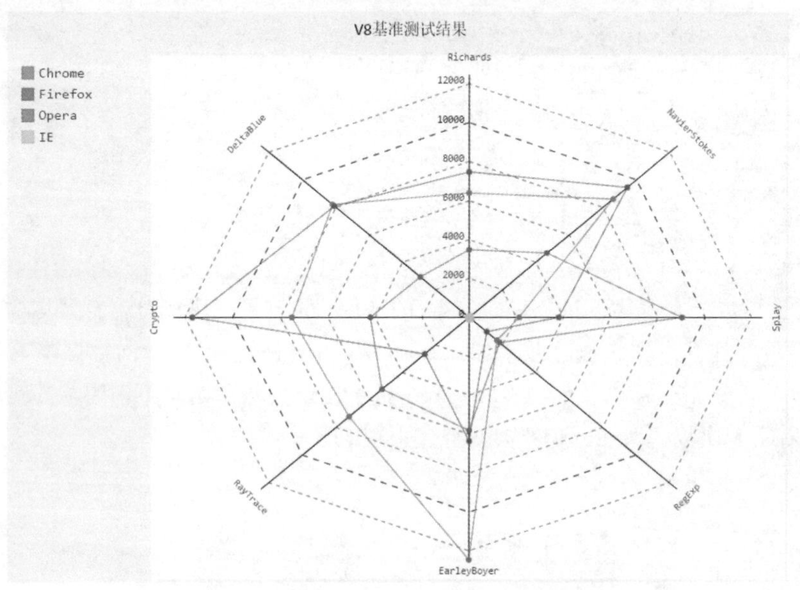

图 8-9　生成的雷达图文件 bar_chart.svg

8.1.6　使用库 pygal 模拟掷骰子并可视化点数概率

在下面的实例文件 01.py 中，演示了使用 pygal 库实现模拟掷骰子功能，然后通过柱形图可视化点数出现概率的过程。首先定义骰子类 Die，然后使用函数 range() 模拟掷骰子 1 000 次，然后统计每个骰子点数的出现次数，最后在柱形图中显示统计结果。

文件 01.py 的具体实现代码如下所示：

```
import random

class Die:
    """
    一个骰子类
    """
    def __init__(self, num_sides=6):
        self.num_sides = num_sides

    def roll(self):
        return random.randint(1, self.num_sides)

import pygal

die = Die()
```

```
result_list = []
# 掷 1000 次
for roll_num in range(1000):
    result = die.roll()
    result_list.append(result)

frequencies = []
# 范围 1~6，统计每个数字出现的次数
for value in range(1, die.num_sides + 1):
    frequency = result_list.count(value)
    frequencies.append(frequency)

# 条形图
hist = pygal.Bar()
hist.title = 'Results of rolling one D6 1000 times'
# x 轴坐标
hist.x_labels = [1, 2, 3, 4, 5, 6]
# x、y 轴的描述
hist.x_title = 'Result'
hist.y_title = 'Frequency of Result'
# 添加数据， 第一个参数是数据的标题
hist.add('D6', frequencies)
# 保存到本地，格式必须是 svg
hist.render_to_file('die_visual.svg')
```

执行后会生成一个名为“die_visual.svg”的文件，我们可以用浏览器打开这个 SVG 文件，打开后会显示统计柱形图。执行效果如图 8-10 所示。如果将鼠标指向数据，可以看到显示了标题“D6”，x 轴坐标以及 y 轴坐标。六个数字出现的频次是差不多的，其实理论上概率是 1/6，随着实验次数的增加，趋势会变得越来越明显。

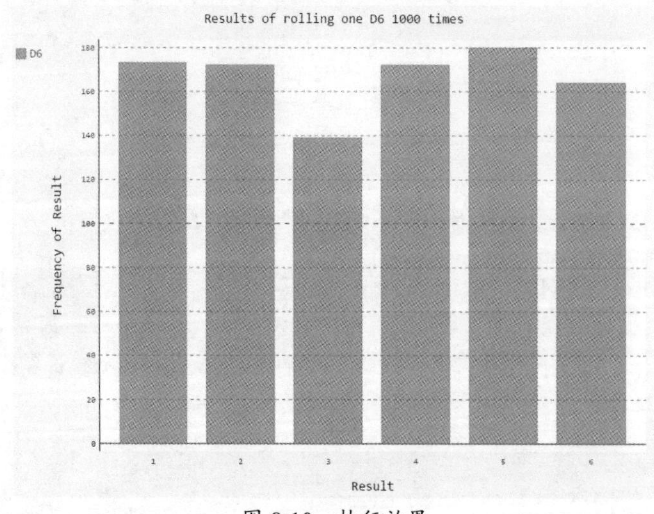

图 8-10　执行效果

我们可以对上面的实例进行升级，例如同时掷两个骰子，可以通过下面的实例文件 02.py 实现。在具体实现时，首先定义骰子类 Die，然后使用函数 range() 模拟掷两个骰子 5 000 次，然后统计每次掷两个骰子点数的最大数的次数，最后在柱形图中显示统计结果。文件 02.py 的具体实现代码如下所示：

```
class Die:
    """
    一个骰子类
```

```
    """
    def __init__(self, num_sides=6):
        self.num_sides = num_sides

    def roll(self):
        return random.randint(1, self.num_sides)
die_1 = Die()
die_2 = Die()

result_list = []
for roll_num in range(5000):
    # 两个骰子的点数和
    result = die_1.roll() + die_2.roll()
    result_list.append(result)

frequencies = []
# 能掷出的最大数
max_result = die_1.num_sides + die_2.num_sides

for value in range(2, max_result + 1):
    frequency = result_list.count(value)
    frequencies.append(frequency)

# 可视化
hist = pygal.Bar()
hist.title = 'Results of rolling two D6 dice 5000 times'
hist.x_labels = [x for x in range(2, max_result + 1)]
hist.x_title = 'Result'
hist.y_title = 'Frequency of Result'
# 添加数据
hist.add('two D6', frequencies)
# 格式必须是svg
hist.render_to_file('2_die_visual.svg')
```

执行后会生成一个名为“2_die_visual.svg”的文件，我们可以用浏览器打开这个 SVG 文件，打开后会显示统计柱形图。执行效果如图 8-11 所示。由此可以看出，两个骰子之和为 7 的次数最多，和为 2 的次数最少。因为能掷出 2 的只有一种情况（1, 1）；而掷出 7 的情况有（1, 6），（2, 5），（3, 4），（4, 3），（5, 2），（6, 1）共 6 种情况，其余数字的情况都没有 7 得多，故掷得 7 的概率最大。

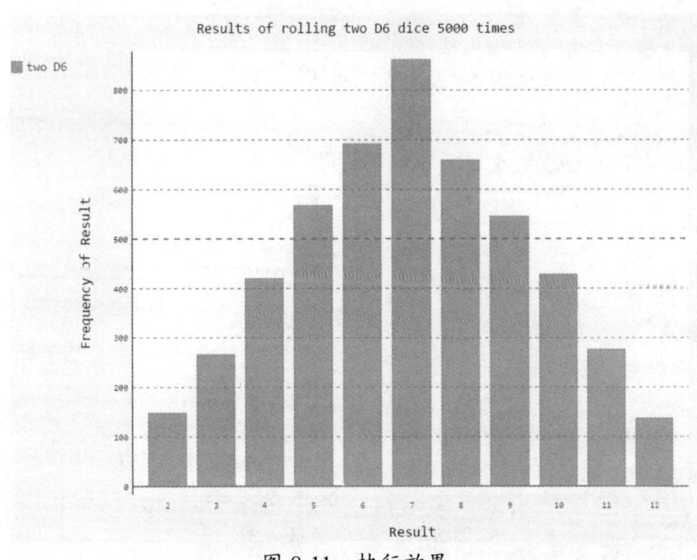

图 8-11　执行效果

8.1.7 使用库 pygal 绘制散点图

在 Python 程序中，可以使用库 pygal 实现数据的可视化操作功能，生成 SVG 格式的图形文件。SVG 是一种常见的数据分析文件格式，是一种矢量图格式，全称是 Scalable Vector Graphics，被翻译为可缩放矢量图形。我们可以使用浏览器打开 SVG 文件，可以方便与之交互。对于需要在尺寸不同的屏幕上显示的图表，SVG 会变得很有用，可以自动缩放，自适应观看者的屏幕。例如在下面的实例文件 dian.py 中演示了使用库 pygal 绘制指定样式散点图的过程。

```
import random
import pygal

scatter_plot_chart = pygal.XY(stroke=False)
scatter_plot_chart.title = "散点图"
list_a = []
list_b = []
list_c = []
for i in range(20):
    a = random.uniform(0,5)
    b = random.uniform(0,5)
    list_a.append((a,b))
for j in range(20):
    a = random.uniform(0,5)
    b = random.uniform(0,5)
    list_b.append((a,b))
for s in range(20):
    a = random.uniform(0,5)
    b = random.uniform(0,5)
    list_c.append((a,b))
scatter_plot_chart.add('A', list_a)
scatter_plot_chart.add('B', list_b)
scatter_plot_chart.add('C', list_c)
scatter_plot_chart.render_to_file('xy_scatter_plot.svg')
```

执行后会生成文件 xy_scatter_plot.svg，用浏览器打开这个文件后会看到绘制的散点图，将鼠标放在某个散点时会悬浮显示这个点的值。如图 8-12 所示。

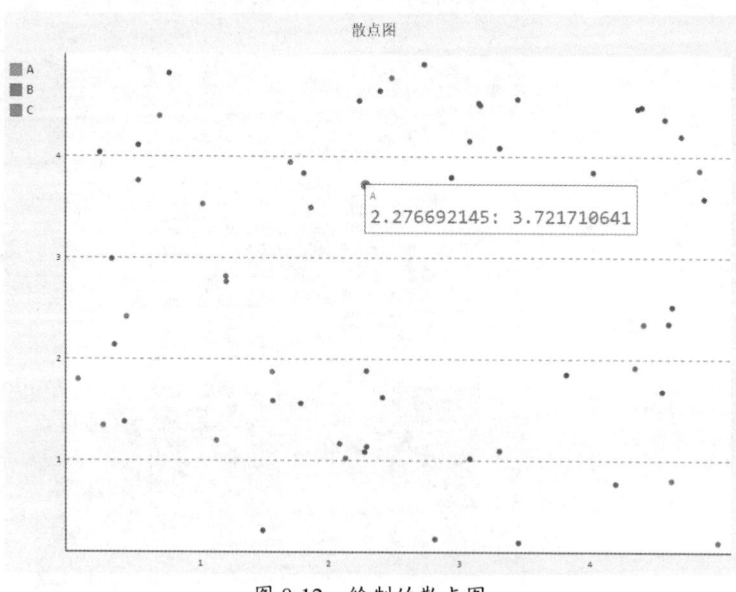

图 8-12　绘制的散点图

8.1.8　使用库 pygal 绘制水平样式的浏览器市场占有率变化折线图

请看下面的实例文件 zhe11.py，功能是使用库 pygal 绘制在某时间段内浏览器产品的市场占有率变化折线图，注意：这个折线图是水平样式的。

```
    import pygal

    line_chart = pygal.HorizontalLine()
    line_chart.title = 'Browser usage evolution (in %)'
    line_chart.x_labels = map(str, range(2002, 2013))
    line_chart.add('Firefox', [None, None,      0, 16.6,    25,    31, 36.4, 45.5,
46.3, 42.8, 37.1])
    line_chart.add('Chrome',   [None, None, None, None, None, None,      0,   3.9,
10.8, 23.8, 35.3])
    line_chart.add('IE',         [85.8, 84.6, 84.7, 74.5,    66, 58.6, 54.7, 44.8,
36.2, 26.6, 20.1])
    line_chart.add('Others',   [14.2, 15.4, 15.3,  8.9,     9, 10.4,  8.9,  5.8,
6.7,  6.8,  7.5])
    line_chart.range = [0, 100]
    line_chart.render_to_file('bar_chart.svg')
```

执行效果如图 8-13 所示。

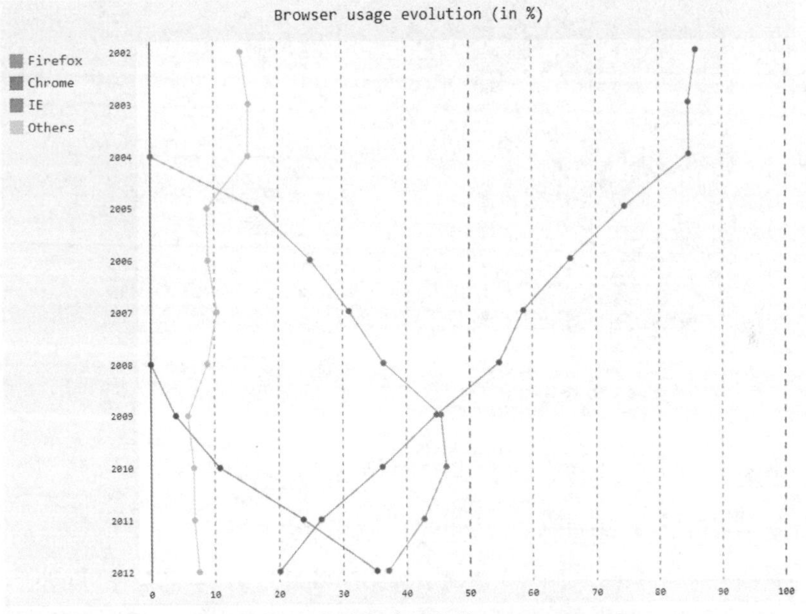

图 8-13　生成的折线图文件 bar_chart.svg

8.1.9　使用库 pygal 绘制叠加折线图

请看下面的实例文件 zhe12.py，功能是使用 pygal 绘制浏览器市场占有率的叠加折线图。

```
    import pygal

    line_chart = pygal.StackedLine(fill=True)
    line_chart.title = 'Browser usage evolution (in %)'
    line_chart.x_labels = map(str, range(2002, 2013))
    line_chart.add('Firefox', [None, None, 0, 16.6,    25,    31, 36.4, 45.5, 46.3,
42.8, 37.1])
    line_chart.add('Chrome',   [None, None, None, None, None, None,      0,   3.9,
10.8, 23.8, 35.3])
```

```
    line_chart.add('IE',        [85.8, 84.6, 84.7, 74.5,    66, 58.6, 54.7, 44.8,
36.2, 26.6, 20.1])
    line_chart.add('Others',   [14.2, 15.4, 15.3,  8.9,     9, 10.4,  8.9,  5.8,
6.7,  6.8,  7.5])
    line_chart.render_to_file('bar_chart.svg')
```

执行效果如图 8-14 所示。

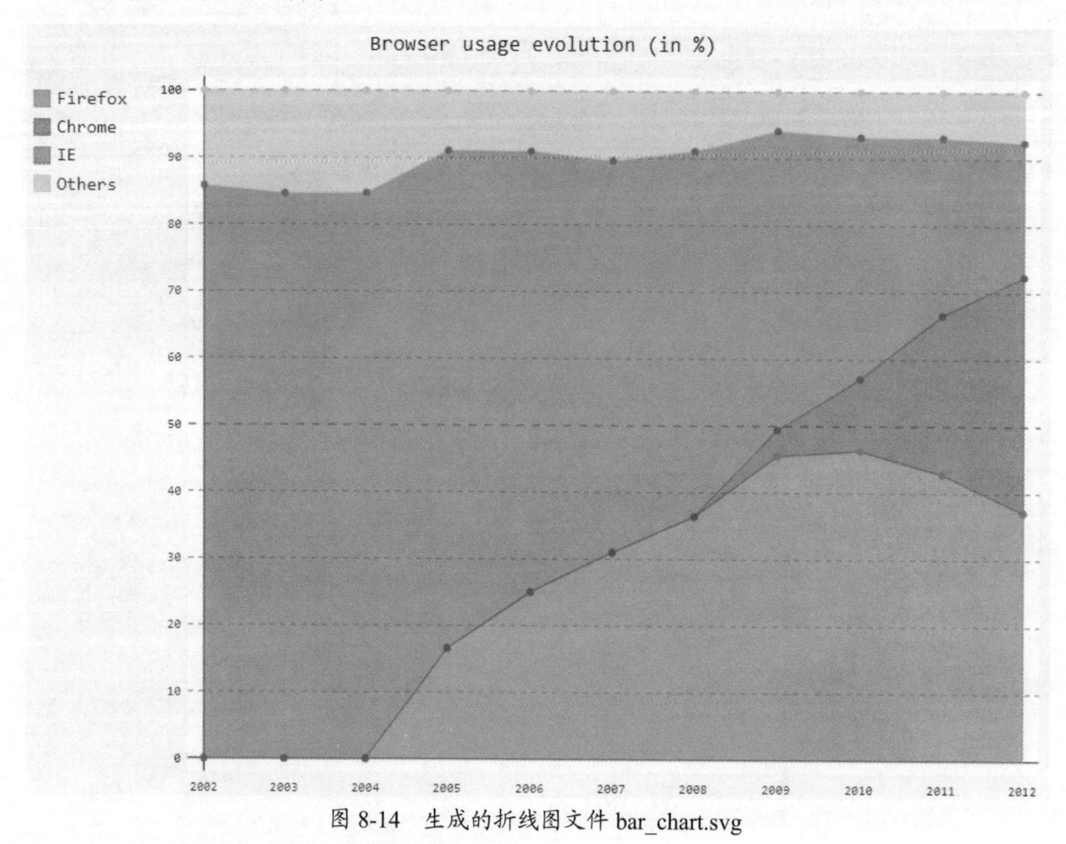

图 8-14　生成的折线图文件 bar_chart.svg

8.1.10　使用库 pygal 绘制某网站用户访问量折线图

在使用库 pygal 绘制折线图时，可以设置坐标轴标签的显示样式，例如我们可以在 x 轴标签显示“年 - 月 - 日”信息。请看下面的实例文件 zhe13.py，功能是使用 pygal 绘制某时间段内某网站用户访问量折线图，x 轴标签显示的是日期信息。

```
from datetime import datetime
import pygal

date_chart = pygal.Line(x_label_rotation=20)
date_chart.x_labels = map(lambda d: d.strftime('%Y-%m-%d'), [
 datetime(2013, 1, 2),
 datetime(2013, 1, 12),
 datetime(2013, 2, 2),
 datetime(2013, 2, 22)])
date_chart.add("Visits", [300, 412, 823, 672])
date_chart.render_to_file('bar_chart.svg')
```

执行效果如图 8-15 所示。

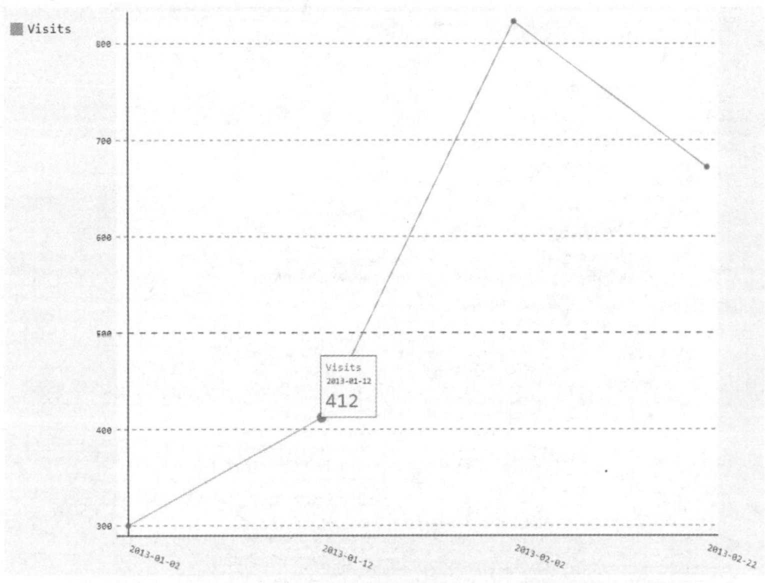

图 8-15　生成的折线图文件 bar_chart.svg

8.2　实践案例：可视化分析最受欢迎的 Python 库

在现实应用中，我们经常需要分析网络数据。例如，我们想可视化分析 Github 网站中最受欢迎的 Python 库，要求以 stars 进行排序，应该如何实现呢。在本节的内容中，将详细讲解实现这一功能的方法。

↑扫码看视频（本节视频课程时间：3 分 49 秒）

8.2.1　统计前 30 名最受欢迎的 Python 库

对于广大开发者来说，××× 网站是大家心中的殿堂，在里面有无数个开源程序供开发者学习和使用。为了便于给开发者了解 ××× 网站中的每一个项目的基本信息，××× 网站官方提供了一个 JSON 网页，在里面存储了按照某个标准排列的项目信息。例如，通过如下网址可以查看关键字是“python”、按照“stars”从高到低排列的项目信息。如图 8-16 所示。

```
https://api.XXX 网站 .com/search/repositories?q=language:python
```

对上述 JSON 数据中，在 items 里面保存了前 30 名 stars 最多的 Python 项目信息。其中 name 表示库名称，owner 下的 login 是库的拥有者，html_url 表示该库的网址（注意：owner 下也有个 html_url，但那个是用户的 ××× 网站网址，我们要定位到该用户的具体库，所以不要用 owner 下的 html_url），stargazers_count 表示所得的 stars 数目。

另外，total_count 表示 Python 语言的仓库的总数。incomplete_results 表示响应的值是否完整，一般来说是 false，表示响应的数据完整。

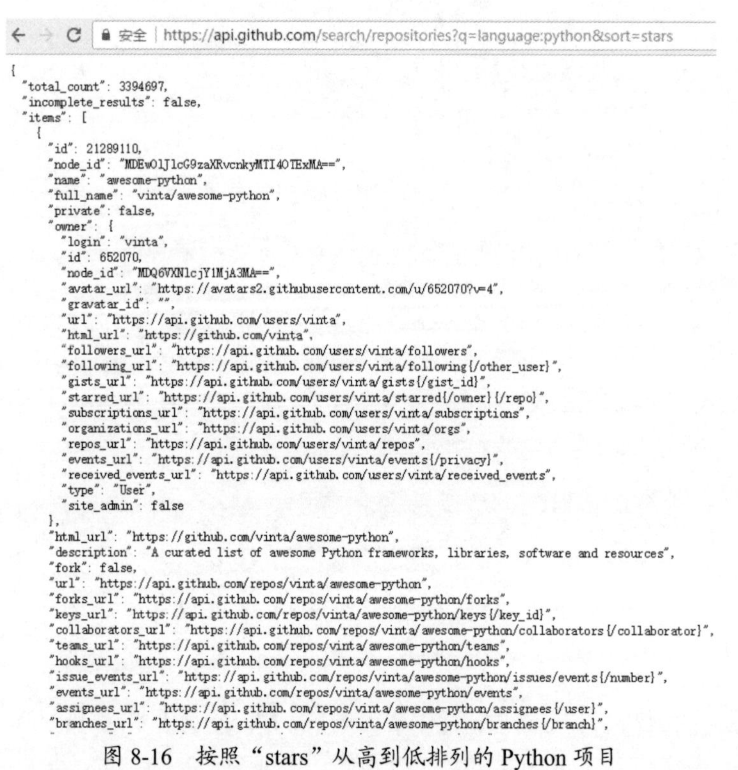

图 8-16　按照“stars”从高到低排列的 Python 项目

在下面的实例文件 github01.py 中，演示了使用 requests 获取 ××× 网站中前 30 名最受欢迎的 Python 库信息的方法。

```
import requests

url = 'https://api.××× 网站 .com/search/repositories?q=language:python&sort=stars'
response = requests.get(url)
# 200 为响应成功
print(response.status_code, '响应成功！ ')
response_dict = response.json()

total_repo = response_dict['total_count']
repo_list = response_dict['items']
print('总仓库数：', total_repo)
print('top', len(repo_list))
for repo_dict in repo_list:
    print('\n名字：', repo_dict['name'])
    print('作者： ', repo_dict['owner']['login'])
    print('Stars: ', repo_dict['stargazers_count'])
    print('网址：', repo_dict['html_url'])
    print('简介：', repo_dict['description'])
```

执行后会提取 JSON 数据中的信息，输出显示 ××× 网站中前 30 名最受欢迎的 Python 库信息如下：

```
200 响应成功！
总仓库数：  3394688
top 30

名字：  awesome-python
作者：   vinta
Stars:  60032
网址：  https://××× 网站 .com/vinta/awesome-python
```

```
    简介 :  A curated list of awesome Python frameworks, libraries, software and resources

    名字 :  system-design-primer
    作者:   donnemartin
    Stars:  54886
    网址 :  https://××× 网站 .com/donnemartin/system-design-primer
    简介:  Learn how to design large-scale systems. Prep for the system design interview.
Includes Anki flashcards.

    名字 :  models
    作者:   tensorflow
    Stars:  47172
    网址 :  https://××× 网站 .com/tensorflow/models
    简介 :  Models and examples built with TensorFlow

    名字 :  public-apis
    作者:   toddmotto
    Stars:  46373
    网址 :  https://××× 网站 .com/toddmotto/public-apis
    简介 :  A collective list of free APIs for use in software and web development.
    ######## 在后面省略其余的结果
```

8.2.2　使用库 pygal 实现数据可视化

虽然通过实例文件 ××× 网站 01.py 可以提取 JSON 页面中的数据，但是数据还不够直观，接下来编写实例文件 github02.py，将从 ××× 网站的总仓库中提取最受欢迎的 Python 库（前 30 名），并绘制统计直方图。文件 github02.py 的具体实现代码如下所示：

```
import requests

import pygal
from pygal.style import LightColorizedStyle, LightenStyle

url = 'https://api.××× 网站 .com/search/repositories?q=language:python&sort=stars'
response = requests.get(url)
# 200 为响应成功
print(response.status_code, ' 响应成功！ ')
response_dict = response.json()

total_repo = response_dict['total_count']
repo_list = response_dict['items']
print(' 总仓库数 : ', total_repo)
print('top', len(repo_list))

names, plot_dicts = [], []
for repo_dict in repo_list:
    names.append(repo_dict['name'])
    # 加上 str 强转，否则会遇到 'NoneType' object is not subscriptable 错误
    plot_dict = {
        'value' : repo_dict['stargazers_count'],
        # 有些描述很长，选择前一部分
        'label' : str(repo_dict['description'])[:200]+'...',
        'xlink' : repo_dict['html_url']
    }
    plot_dicts.append(plot_dict)

# 改变默认主题颜色，偏蓝色
my_style = LightenStyle('#333366', base_style=LightColorizedStyle)
# 配置
my_config = pygal.Config()
# x 轴的文字旋转 45 度
my_config.x_label_rotation = -45
```

```
# 隐藏左上角的图例
my_config.show_legend = False
# 标题字体大小
my_config.title_font_size = 30
# 副标签，包括 x 轴和 y 轴大部分
my_config.label_font_size = 20
# 主标签是 y 轴某数倍数，相当于一个特殊的刻度，让关键数据点更醒目
my_config.major_label_font_size = 24
# 限制字符为 15 个，超出的以 ... 显示
my_config.truncate_label = 15
# 不显示 y 参考虚线
my_config.show_y_guides = False
# 图表宽度
my_config.width = 1000

# 第一个参数可以传配置
chart = pygal.Bar(my_config, style=my_style)
chart.title = 'XXX 网站最受欢迎的 Python 库（前 30 名）'
# x 轴的数据
chart.x_labels = names
# 加入 y 轴的数据，无须 title 设置为空，注意这里传入的字典，
# 其中的键 --value 也就是 y 轴的坐标值了
chart.add('', plot_dicts)
chart.render_to_file('30_stars_python_repo.svg')
```

执行后会创建生成数据统计直方图文件 30_stars_python_repo.svg，并输出如下所示的提取信息：

```
200 响应成功！
总仓库数： 3394860
top 30
```

数据统计直方图文件 30_stars_python_repo.svg 的效果如图 8-17 所示。

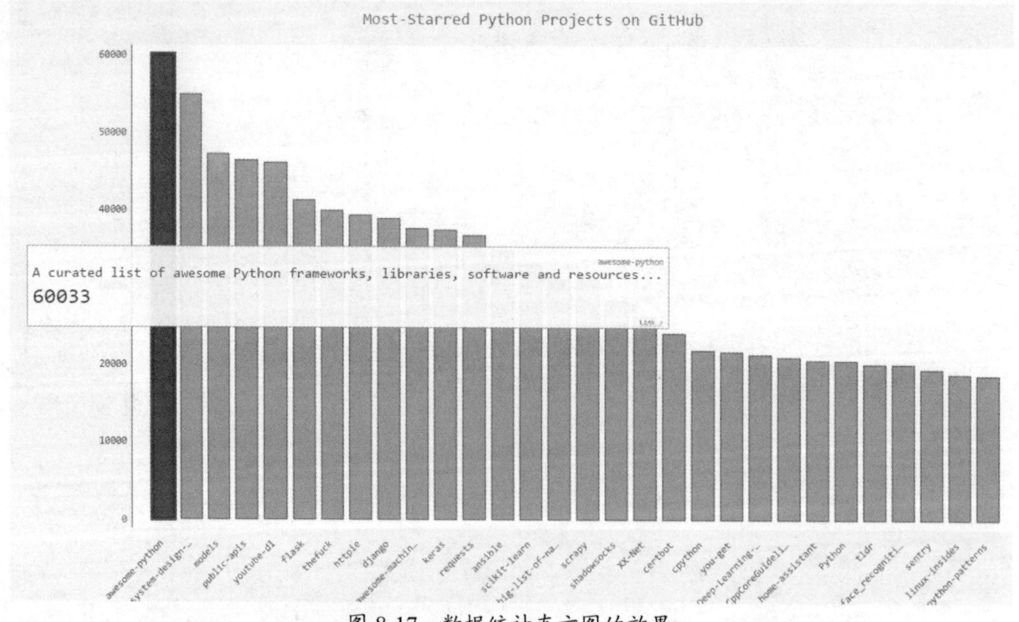

图 8-17 数据统计直方图的效果

8.3　实践案例：可视化分析 SQLite 数据

在本节的内容中，将使用 Flask+pygal+SQLite3 实现数据可视化分析功能。将需要分析的数据保存在 SQLite3 数据库中，然后在 Flask Web 网页中使用 pygal 绘制出对应的统计图。

↑扫码看视频（本节视频课程时间：2 分 31 秒）

8.3.1　创建数据库

首先使用 PyCharm 创建一个 Flask Web 项目，然后通过文件 models.py 设计 SQLite 数据库的结构，主要实现代码如下所示：

```
from dbconnect import db

# 许可证申请数量
class Appinfo(db.Model):
    __tablename__ = 'appinfo'
    # 注意这句，网上有些实例上并没有
    # 必须设置主键
    id = db.Column(db.Integer, primary_key=True)
    year = db.Column(db.String(20))
    month = db.Column(db.String(20))
    cnt = db.Column(db.String(20))

    def __init__(self, year, month, cnt):
        self.year = year
        self.month = month
        self.cnt = cnt

    def __str__(self):
        return self.year + ":" + self.month + ":" + self.cnt

    def __repr__(self):
        return self.year + ":" + self.month + ":" + self.cnt

    def save(self):
        db.session.add(self)
        db.session.commit()
```

在上述代码中，通过变量 id、year、year 和 cnt 设置 4 个数据库字段的数据类型，然后根据变量 id、year、year 和 cnt 的值在数据库表 appinfo 中添加数据，如图 8-18 所示。

8.3.2　绘制统计图

在本项目中，使用 Web 程序展示绘制的可视化数据。因为本项目的规模不是很大，所以使用轻量级的 Flask 框架实现 Web 部分。

（1）编写 Flask Web 启动文件 pygal_test.py，首先建立 URL 路径导航指向模板文件 index.htm，然后提取数据库中的数据，并使用 pygal 绘制出统计图表。文件 pygal_test.py 的主要实现代码如下所示：

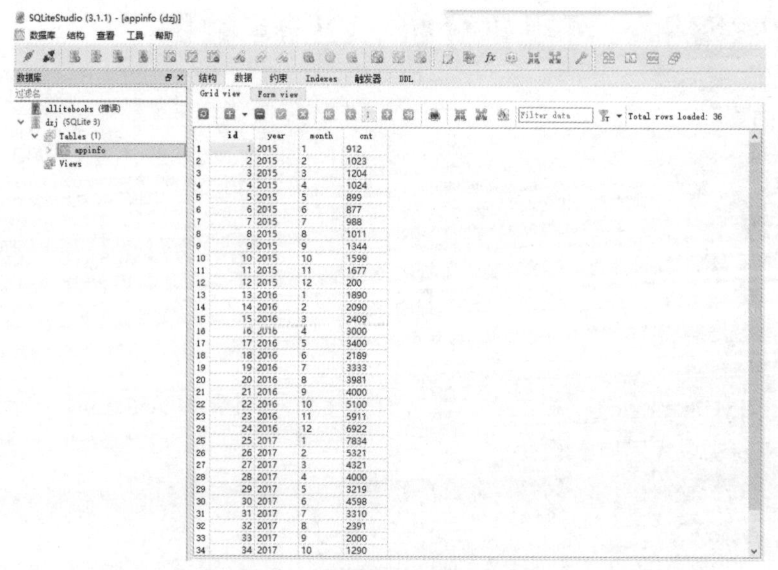

	id	year	month	cnt
1	1	2015	1	912
2	2	2015	2	1023
3	3	2015	3	1204
4	4	2015	4	1024
5	5	2015	5	899
6	6	2015	6	877
7	7	2015	7	988
8	8	2015	8	1011
9	9	2015	9	1344
10	10	2015	10	1599
11	11	2015	11	1677
12	12	2015	12	200
13	13	2016	1	1890
14	14	2016	2	2090
15	15	2016	3	2409
16	16	2016	4	3000
17	17	2016	5	3400
18	18	2016	6	2189
19	19	2016	7	3333
20	20	2016	8	3981
21	21	2016	9	4000
22	22	2016	10	5100
23	23	2016	11	5911
24	24	2016	12	6922
25	25	2017	1	7834
26	26	2017	2	5321
27	27	2017	3	4321
28	28	2017	4	4000
29	29	2017	5	3219
30	30	2017	6	4598
31	31	2017	7	3310
32	32	2017	8	2391
33	33	2017	9	2000
34	34	2017	10	1290

图 8-18　数据库 dzj.db 中的数据

```
app = Flask(__name__)
dbpath = app.root_path
# 注意斜线的方向
app.config['SQLALCHEMY_DATABASE_URI'] = r'sqlite:///' + dbpath + '/dzj.db'
app.config['SQLALCHEMY_TRACK_MODIFICATIONS'] = True
print(app.config['SQLALCHEMY_DATABASE_URI'])

db.init_app(app)

@app.route('/')
def APPLYTBLINFO():
    db.create_all()
    # 在第一次调用时执行就可以
    appinfos = Appinfo.query.all()
    # 选择年份
    list_year = []
    # 选择月份
    list_month = []
    # 月份对应的数字
    map_cnt = {}
    for info in appinfos:
        if info.year not in list_year:
            list_year.append(info.year)
            map_cnt[info.year] = [int(info.cnt)]
        else:
            map_cnt[info.year].append(int(info.cnt))
        if info.month not in list_month:
            list_month.append(info.month)
    line_chart = pygal.Line()
    line_chart.title = '信息'
    line_chart.x_labels = map(str, list_month)
    for year in list_year:
        line_chart.add(str(year) + "年", map_cnt[year])
    return render_template('index.html', chart=line_chart)


if __name__ == '__main__':
    app.run(debug=True)
```

（2）编写 Flask 部分的模板文件 index.html，具体实现代码如下所示：

```
    <body style="width: 1000px;margin: auto">
    <div  id="container">
        <div id="header" style="background: burlywood;height: 50px;">
                <h2 style="font-size: 30px;  position: absolute; margin-top:
10px;margin-left: 300px;
            text-align:center;">数据走势图分析</h2>
        </div>
          <div  id="leftbar"  style="width: 200px;height: 600px;background:
cadetblue;float: left">
            <h2 style="margin-left: 20px">数据图总览</h2><br/>
            <table>
                <tr>
                    <td>
                       <a name="appinfo"  style="margin-left: 20px;">数量分析图</a><br>
                    </td>
                </tr>
            </table>
        </div>
        <div id="chart" style="width: 800px;float: left">
           <embed type="image/svg+xml" src= {{ chart.render_data_uri()|safe }} />
        </div>
    </div>
    </body>
```

执行 Flask Web 项目，在浏览器中输入 http://127.0.0.1:5000/ 后会显示绘制的统计图，执行效果如图 8-19 所示。

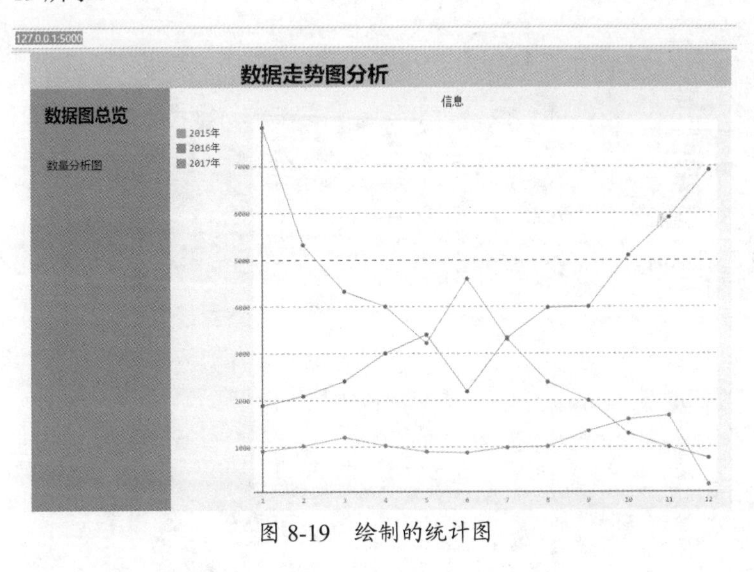

图 8-19　绘制的统计图

8.4　实践案例：可视化租房信息

在本节的内容中，将开发一个网络爬虫程序，使用第三方爬虫框架 Scrapy 抓取某租房信息网中某个城市的租房信息，然后将租房信息保存到 JSON 文件中，最后使用 Pygal 可视化展示租房信息。

↑扫码看视频（本节视频课程时间：4 分 40 秒）

8.4.1 网络爬虫

为了提高开发效率，本项目将使用专业的第三方爬虫框架 Scrapy 实现。具体实现流程如下所示：

（1）通过如下命令创建一个 Scrapy 项目：

```
scrapy startproject house_position
```

（2）通过文件 settings.py 设置 Scrapy 项目的配置信息，包括默认的请求头信息和使用 Pipeline，主要实现代码如下所示：

```
BOT_NAME = 'XuzhouSpider'

SPIDER_MODULES = ['XuzhouSpider.spiders']
NEWSPIDER_MODULE = 'XuzhouSpider.spiders'
# 配置默认的请求头
DEFAULT_REQUEST_HEADERS = {
      "User-Agent" : "Mozilla/5.0 (Windows NT 6.1; Win64; x64; rv:61.0) 
Gecko/20100101 Firefox/61.0",
     'Accept': 'text/html,application/xhtml+xml,application/xml;q=0.9,*/*;q=0.8'
}
# 配置使用 Pipeline
ITEM_PIPELINES = {
    'XuzhouSpider.pipelines.XuzhouspiderPipeline': 300,
}
```

（3）在文件 items.py 设置需要提取的爬虫信息的数据结构，主要实现代码如下所示：

```
import scrapy
class XuzhouspiderItem(scrapy.Item):
    # define the fields for your item here like:
    # name = scrapy.Field()
    #楼盘名称
    name=scrapy.Field()
    #位置
    address=scrapy.Field()
    #价格
    price=scrapy.Field()
    #描述
    size=scrapy.Field()
```

（4）在文件 pipelines.py 中进一步处理在 items 中提取的数据，将抓取到的信息写入本地 JSON 文件 Xuzhouhouse.json 中。主要实现代码如下所示：

```
import json

class XuzhouspiderPipeline(object):
    def __init__(self):
        self.json_file=open("Xuzhouhouse.json","wb+")
        self.json_file.write('[\n'.encode("utf-8"))
    def close_spider(self,spider):
        print("---------closefile----------")
        self.json_file.seek(-2,1)
        self.json_file.write('\n]'.encode("utf-8"))
        self.json_file.close()
    def process_item(self, item, spider):
        text=json.dumps(dict(item),ensure_ascii=False) + ",\n"
        self.json_file.write(text.encode("utf-8"))
```

（5）在文件 house_position.py 中实现具体的爬虫功能，首先设置要爬取的目标 URL，然后使用 xpath 提取要抓取信息的页面元素。主要实现代码如下所示：

```
import scrapy
```

```
    from XuzhouSpider.items import XuzhouspiderItem
    class HousePositionSpider(scrapy.Spider):
        name = 'house_position'
        allowed_domains = ['xuzhou.ganji.com']
        start_urls = ['http://xuzhou.ganji.com/zufang/']

        def parse(self, response):
                for house in response.xpath('//div[@class="f-list-item ershoufang-list"]'):
                item=XuzhouspiderItem()
                    item['name']=house.xpath('./dl/dd[@class="dd-item title"]/a/text()').extract()
                    item['size']=house.xpath('./dl/dd[@class="dd-item size"]/span/text()').extract()
                    item['address']=house.xpath('./dl/dd[@class="dd-item address"]/span/a/span/text()').extract()
                    item['price']=house.xpath('./dl/dd[@class="dd-item info"]/div/span[@class="num"]/text()').extract()
                yield item
            new_links = response.xpath('//div[@class="pageBox"]/a[@class="next"]/@href').extract()
            if new_links and len(new_links)>0:
                new_link = new_links[0]
                yield scrapy.Request(new_link,callback=self.parse)
```

（6）输入下面的命令执行爬虫程序，爬虫界面的截图效果如图 8-20 所示。

```
    scrapy crawl house_position
```

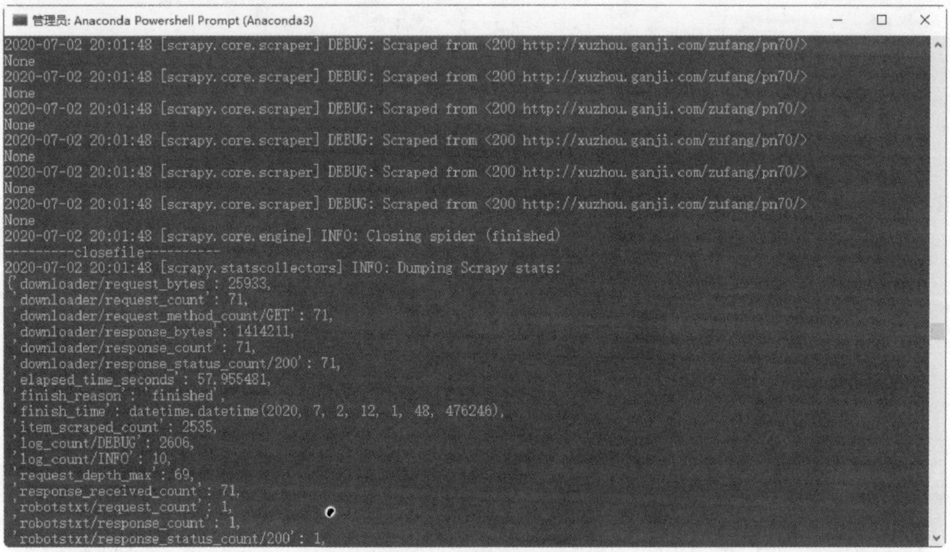

```
管理员: Anaconda Powershell Prompt (Anaconda3)
2020-07-02 20:01:48 [scrapy.core.scraper] DEBUG: Scraped from <200 http://xuzhou.ganji.com/zufang/pn70/>
None
2020-07-02 20:01:48 [scrapy.core.scraper] DEBUG: Scraped from <200 http://xuzhou.ganji.com/zufang/pn70/>
None
2020-07-02 20:01:48 [scrapy.core.scraper] DEBUG: Scraped from <200 http://xuzhou.ganji.com/zufang/pn70/>
None
2020-07-02 20:01:48 [scrapy.core.scraper] DEBUG: Scraped from <200 http://xuzhou.ganji.com/zufang/pn70/>
None
2020-07-02 20:01:48 [scrapy.core.scraper] DEBUG: Scraped from <200 http://xuzhou.ganji.com/zufang/pn70/>
None
2020-07-02 20:01:48 [scrapy.core.scraper] DEBUG: Scraped from <200 http://xuzhou.ganji.com/zufang/pn70/>
None
2020-07-02 20:01:48 [scrapy.core.engine] INFO: Closing spider (finished)
---------closefile---------
2020-07-02 20:01:48 [scrapy.statscollectors] INFO: Dumping Scrapy stats:
{'downloader/request_bytes': 25933,
 'downloader/request_count': 71,
 'downloader/request_method_count/GET': 71,
 'downloader/response_bytes': 1414211,
 'downloader/response_count': 71,
 'downloader/response_status_count/200': 71,
 'elapsed_time_seconds': 57.955481,
 'finish_reason': 'finished',
 'finish_time': datetime.datetime(2020, 7, 2, 12, 1, 48, 476246),
 'item_scraped_count': 2535,
 'log_count/DEBUG': 2606,
 'log_count/INFO': 10,
 'request_depth_max': 69,
 'response_received_count': 71,
 'robotstxt/request_count': 1,
 'robotstxt/response_count': 1,
 'robotstxt/response_status_count/200': 1,
```

图 8-20　爬虫界面截图

8.4.2 实现数据可视化

编写文件 pygal_xuzhouhouse.py，功能是提取在本地 JSON 文件 Xuzhouhouse.json 中保存的爬虫数据，使用库 pygal 绘制可视化统计图。

（1）设置打开文件 Xuzhouhouse.json，加载文件里面的 JSON 数据。代码如下所示：

```
    filename='Xuzhouhouse.json'
    with open(filename,'r',True,'utf-8') as f:
        house_list=json.load(f)
```

```
adict0 = {'1室1厅1卫':0,'2室1厅1卫':0,'2室2厅1卫':0,'3室2厅1卫':0,'其他':0}
adict1 = {'小于50平方米':0,'50~75平方米':0,'75~100平方米':0,'大于100平方米':0,'其他':0}
adict2 = {'南向':0,'南北向':0,'北向':0,'其他':0}
adict3 = {'毛坯':0,'简单装修':0,'中等装修':0,'精装修':0,'豪华装修':0,'其他':0}
adict4 = {'小于1000':0,'1000-1500':0,'1500-2000':0,'大于2000':0,'其他':0}
```

（2）绘制可视化房间数统计图，代码如下所示：

```
#房间数
for house_dict in house_list:
    str_size0=""
    try:
        for size0 in house_dict['size'][0]:
            str_size0+=size0
        if str_size0 == '1室1厅1卫':
            adict0['1室1厅1卫']+=1
        elif str_size0 == '2室1厅1卫':
            adict0['2室1厅1卫']+=1
        elif str_size0 == '2室2厅1卫':
            adict0['2室2厅1卫']+=1
        elif str_size0 =='3室2厅1卫':
            adict0['3室2厅1卫']+=1
        else:
            adict0['其他']+=1
    except:
        print('装修有误')
```

（3）绘制可视化房屋面积大小的数据统计图，代码如下所示：

```
pie0 = pygal.Pie()
for k in adict0.keys():
    pie0.add(k , adict0[k])
pie0.title="二手房屋房间数"
pie0.legend_at_bottom = True
pie0.render_to_file('房间数.svg')

#平方占比
for house_dict in house_list:
    str_size1=""
    try:
        for size1 in house_dict['size'][1][:-1]:
            if size1 == '.':
                break
            str_size1+=size1
        int_size1 = int(str_size1)
        if int_size1 < 50:
            adict1['小于50平方米']+=1
        elif 50 <= int_size1 < 75:
            adict1['50~75平方米']+=1
        elif 75 <= int_size1 < 100:
            adict1['75~100平方米']+=1
        elif 100 <= int_size1 :
            adict1['大于100平方米']+=1
        else:
            adict1['其他']+=1
    except:
        print("平方米有误")

pie1 = pygal.Pie()
for k in adict1.keys():
    pie1.add(k , adict1[k])
pie1.title="二手房屋平方米"
pie1.legend_at_bottom = True
pie1.render_to_file('房屋平方米.svg')
```

（4）绘制可视化房屋朝向统计图，代码如下所示：

```
# 朝向占比
for house_dict in house_list:
    str_size2=""
    try:
        for size2 in house_dict['size'][2]:
            str_size2+=size2
        if str_size2 == '南向':
            adict2['南向']+=1
        elif str_size2 == '南北向':
            adict2['南北向']+=1
        elif str_size2 ==' 北向':
            adict2['北向']+=1
        else:
            adict2['其他']+=1
    except:
        print('朝向有误')

pie2 = pygal.Pie()
for k in adict2.keys():
    pie2.add(k , adict2[k])
pie2.title="二手房屋朝向"
pie2.legend_at_bottom = True
pie2.render_to_file('房屋朝向.svg')
```

（5）绘制可视化房屋装修数据统计图，代码如下所示：

```
# 装修占比
for house_dict in house_list:
    str_size3=""
    try:
        for size3 in house_dict['size'][3]:
            str_size3+=size3
        if str_size3 == '毛坯':
            adict3['毛坯']+=1
        elif str_size3 == '简单装修':
            adict3['简单装修']+=1
        elif str_size3 == '中等装修':
            adict3['中等装修']+=1
        elif str_size3 ==' 精装修':
            adict3['精装修']+=1
        elif str_size3 ==' 豪华装修':
            adict3['豪华装修']+=1
        else:
            adict3['其他']+=1
    except:
        print('装修有误')

pie3 = pygal.Pie()
for k in adict3.keys():
    pie3.add(k , adict3[k])
pie3.title="二手房屋装修"
pie3.legend_at_bottom = True
pie3.render_to_file('房屋装修.svg')
```

（6）绘制可视化房租价格数据统计图，代码如下所示：

```
# 价格占比
for house_dict in house_list:
    int_price=0
    try:
        for price in house_dict['price']:
            int_price=int(price)
            if int_price < 1000:
```

```
                adict4[' 小于 1000']+=1
            elif 1000 <= int_price < 1500:
                adict4['1000-1500']+=1
            elif 1500 <= int_price < 2000:
                adict4['1500-2000']+=1
            elif 2000 <= int_price :
                adict4[' 大于 2000']+=1
            else:
                adict4[' 其他 ']+=1
    except:
        print(" 价格有误 ")

pie4 = pygal.Pie()
for k in adict4.keys():
    pie4.add(k , adict4[k])
pie4.title=" 二手房屋价格 "
pie4.legend_at_bottom = True
pie4.render_to_file(' 房屋价格 .svg')
```

执行后会分别绘制对应的可视化数据统计图，如图 8-21 所示。

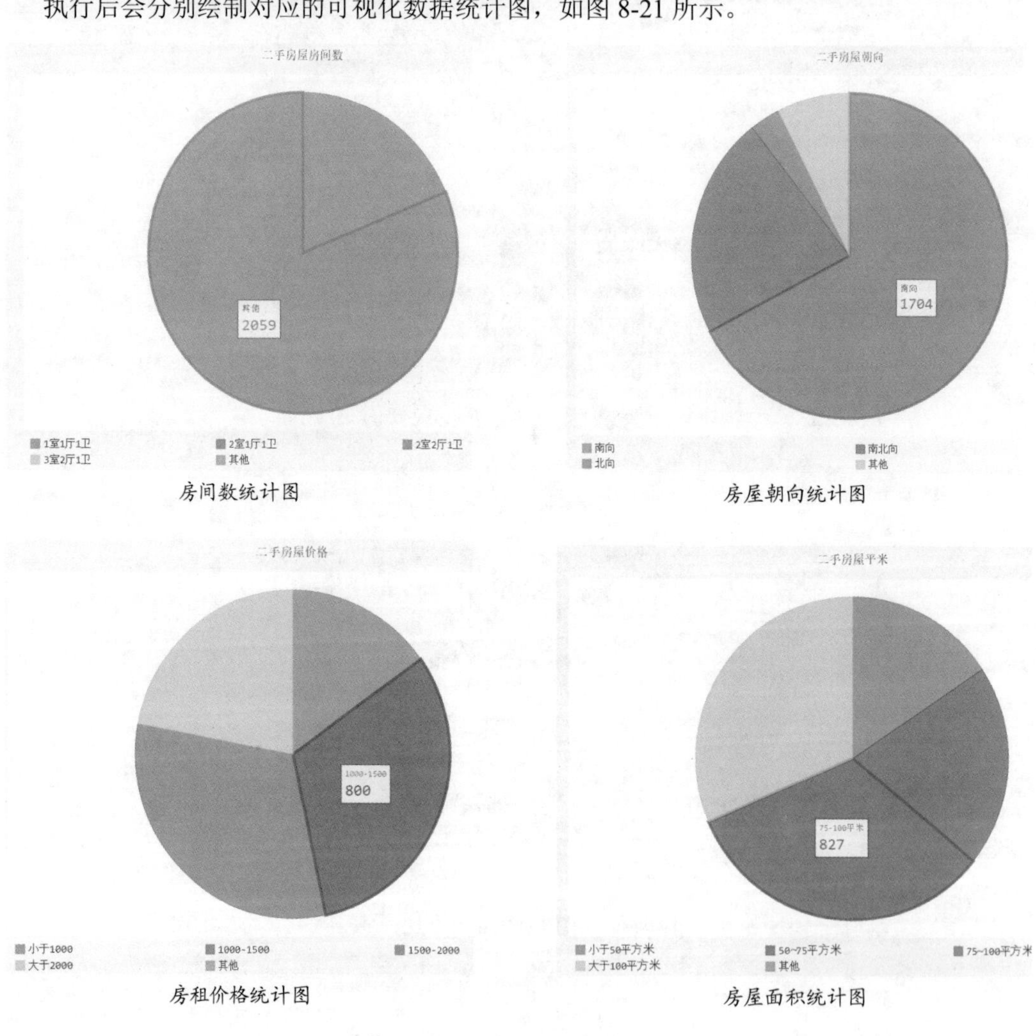

房间数统计图　房屋朝向统计图

房租价格统计图　房屋面积统计图

图 8-21　执行效果

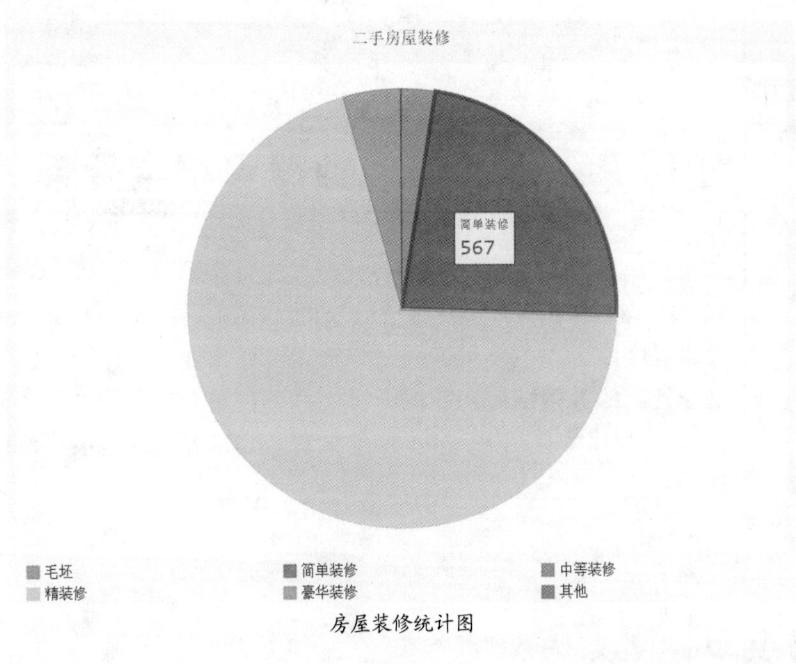

房屋装修统计图

图 8-21　执行效果（续）

第 9 章

使用 Pandas 实现数据可视化处理

Pandas（Python Data Analysis Library）是 Python 语言的一个扩展程序库，它主要用于数据分析和可视化处理工作，提供了高性能、易于使用的数据结构和数据分析工具。在本章的内容中，将详细讲解在 Python 程序中使用 Pandas 库的知识，为读者步入本书后面知识的学习打下基础。

9.1 安装 Pandas 库

在计算机系统中，我们无须安装即可使用 pandas，这时需要使用 Wakari 免费服务，可以在云中提供托管的 IPython Notebook 服务。开发者只需创建一个账户，即可在几分钟内通过 IPython Notebook 在浏览器中访问并使用 pandas。

↑扫码看视频（本节视频课程时间：1 分 11 秒）

对于大多数开发者来说，还是建议使用如下所示的命令来安装 pandas：

```
pip install pandas
```

然后可以通过如下所示的实例文件 001.py 来测试是否安装成功并成功运行。

```
import pandas as pd
print(pd.test())
```

因机器配置差异，执行效果会有所区别，在笔者的电脑中执行后会输出：

```
running: pytest --skip-slow --skip-network --skip-db C:\ProgramData\Anaconda3\lib\site-packages\pandas
============================== test session starts ==============================
platform win32 -- Python 3.7.6, pytest-5.3.5, py-1.8.1, pluggy-0.13.1
rootdir: D:\tiedao\data-daima\9\9-1
plugins: hypothesis-5.5.4, arraydiff-0.3, astropy-header-0.1.2, doctestplus-0.5.0, openfiles-0.4.0, remotedata-0.3.2
### 为节省本书篇幅，后面会省略好多执行效果
```

注意：为了节省本书的篇幅，在书中将不再详细讲解 pandas API 的语法知识，这方面知识请读者阅读 pandas 的官方文档，具体地址是：

http://pandas.pydata.org/pandas-docs/stable/generated/pandas.read_csv.html

9.2　从 CSV 文件读取数据

在库 pandas 中，可以使用方法 read_csv() 读取 CSV 文件中的数据。在默认情况下，read_csv() 方法会假设 CSV 文件中的字段是用逗号进行分隔的。假设存在一个名为“bikes.csv”的 CSV 文件，在里面保存了蒙特利尔的一些骑行数据，在里面保存了每天在蒙特利尔 7 条不同的路道路上有多少人骑自行车。在本节的内容中，将详细讲解从上述 CSV 文件中读取数据的知识。

↑扫码看视频（本节视频课程时间：6 分 0 秒）

9.2.1　读取显示 CSV 文件中的前 3 条数据

在下面的实例文件 002.py 中，读取并显示了文件 bikes.csv 中的前 3 条数据。

```
import pandas as pd
broken_df = pd.read_csv('bikes.csv')
print(broken_df[:3])
```

执行后会输出：

```
    Date;Berri 1;Brébeuf (données non disponibles);Côte-Sainte-
Catherine;Maisonneuve 1;Maisonneuve 2;du Parc;Pierre-Dupuy;Rachel1;St-Urbain
(données non disponibles)
  0                 01/01/2012;35;;0;38;51;26;10;16;
  1                 02/01/2012;83;;1;68;153;53;6;43;
  2                 03/01/2012;135;;2;104;248;89;3;58;
```

读者会发现上述执行效果显得比较凌乱，此时可以利用方法 read_csv() 中的参数选项进行设置。例如在下面的实例文件 003.py 中，使用规整的格式读取并显示了文件 bikes.csv 中的前 3 条数据。

```
import pandas as pd
fixed_df = pd.read_csv('bikes.csv', sep=';', encoding='latin1', parse_
dates=['Date'], dayfirst=True, index_col='Date')
print(fixed_df[:3])
```

执行后会输出：

```
            Berri 1  BrÃ©beuf (donnÃ©es non disponibles)  \
Date
2010-01-01       35                                   NaN
2010-01-02       83                                   NaN
2010-01-03      135                                   NaN

            CÃ´te-Sainte-Catherine  Maisonneuve 1  Maisonneuve 2  du Parc  \
Date
2010-01-01                       0             38             51       26
2010-01-02                       1             68            153       53
2010-01-03                       2            104            248       89

            Pierre-Dupuy  Rachel1  St-Urbain (donnÃ©es non disponibles)
Date
2010-01-01            10       16                                    NaN
2010-01-02             6       43                                    NaN
2010-01-03             3       58                                    NaN
```

9.2.2　读取显示 CSV 文件中指定列的数据

在读取 CSV 文件时，得到的是一种由行和列组成的数据帧，我们可以列出在帧中相同方式的元素。例如在下面的实例文件 004.py 中，读取并显示了文件 bikes.csv 中的“Berri 1”列的数据。

```
    import pandas as pd
    fixed_df = pd.read_csv('bikes.csv', sep=';', encoding='latin1', parse_
dates=['Date'], dayfirst=True, index_col='Date')
    print(fixed_df['Berri 1'])
```

执行后会输出：

```
Date
2010-01-01       35
2010-01-02       83
2010 01-03      135
……省略部分行数
2010-10-23    4177
2010-10-24    3744
2010-10-25    3735
2010-10-26    4290
2010-10-27    1857
2010-10-28    1310
2010-10-29    2919
2010-10-30    2887
2010-10-31    2634
2010-9-01    2405
2010-9-02    1582
2010-9-03     844
2010-9-04     966
2010-9-05    2247
Name: Berri 1, Length: 310, dtype: int64
```

9.2.3 可视化显示某一列的数据

为了使我们的应用程序更加美观，在下面的实例文件 005.py 中加入了 matplotlib 功能，以统计图表的方式展示了文件 bikes.csv 中的“Berri 1”列的数据。

```
    import pandas as pd
    import matplotlib.pyplot as plt
    plt.rcParams['figure.figsize'] = (15, 5)
    fixed_df = pd.read_csv('bikes.csv', sep=';', encoding='latin1', parse_
dates=['Date'], dayfirst=True, index_col='Date')
    fixed_df['Berri 1'].plot()
    plt.show()
```

执行后会显示每个月的骑行数据统计图，执行效果如图 9-1 所示。

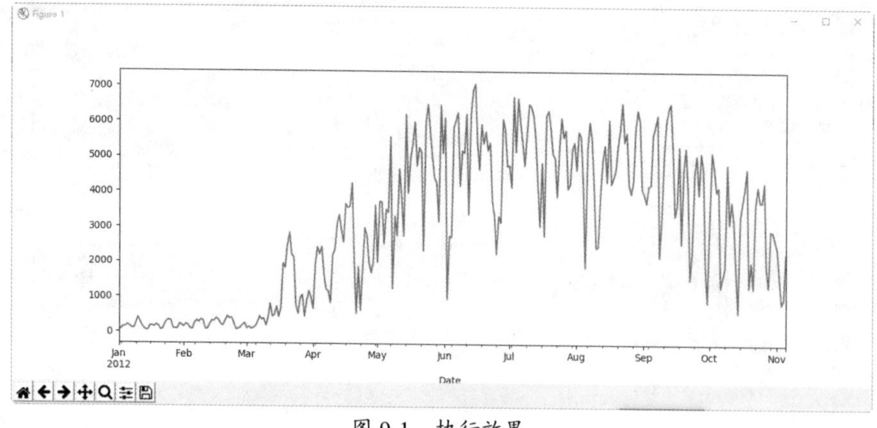

图 9-1　执行效果

9.2.4　可视化指定的骑行数据

请看下面的实例文件 006.py，功能是处理一个更大的数据集文件 39-service-requests.csv，打印输出这个文件中的数据信息。文件 39-service-requests.csv 是 311 的服务请求从纽约开放数据的一个子集，完整文件有 52MB，在书中我们只是截取了一小部分，完整文件可以从网络中获取。

```
import pandas as pd
complaints = pd.read_csv('39-service-requests.csv')
print(complaints)
```

执行后会显示读取文件 39-service-requests.csv 后的结果，并在最后统计数据数目。执行后会输出：

```
          Unique Key                    Created Date                          Closed Date Ag
ency                                        Agency Name                     Complaint Type
Descriptor                  Location Type   Incident Zip                Incident Address
Street Name    Cross Street 1                    Cross Street 2 Intersection Street
1 Intersection Street 2  Address Type                 City  Landmark Facility Type
Status               Due Date Resolution Action Updated Date          Community Board
Borough  X Coordinate (State Plane)  Y Coordinate (State Plane)  Park Facility Name
Park Borough            School Name School Number School Region  School Code School
Phone Number                                 School Address  School City School
State     School Zip School Not Found  School or Citywide Complaint  Vehicle Type
Taxi Company Borough Taxi Pick Up Location  Bridge Highway Name  Bridge Highway
Direction   Road Ramp  Bridge Highway Segment  Garage Lot Name  Ferry Direction
Ferry Terminal Name  Latitude  Longitude                                  Location
    0          26589651   10/31/2013 02:08:41  AM                                   NaN
NYPD             New York City Police Department    Noise - Street/Sidewalk
Loud Talking                      Street/Sidewalk         11432.0              90-03 169
STREET              169 STREET            90 AVENUE                          91 AVENUE
NaN                    NaN        ADDRESS                  JAMAICA         NaN     Precinct
Assigned  10/31/2013 10:08:41 AM          10/31/2013 02:35:17 AM                         12
QUEENS           QUEENS                          1042027.0                         197389.0
Unspecified           QUEENS          Unspecified   Unspecified   Unspecified  Unspecified
Unspecified                                                 Unspecified  Unspecified
Unspecified  Unspecified                     N                          NaN           NaN
NaN                   NaN                 NaN                          NaN           NaN
NaN                  NaN                NaN                   NaN  40.708275 -73.791604
(40.70827532593202, -73.79160395779721)
    1         26593698   10/31/2013 02:01:04 AM                  NaN    NYPD          New
York City Police Department          Illegal Parking                          Commercial
Overnight Parking                  Street/Sidewalk          11378.0                     58
AVENUE                 58 AVENUE          58 PLACE                             59 STREET
NaN                   NaN       BLOCKFACE             MASPETH         NaN       Precinct
Open   10/31/2013 10:01:04 AM                          NaN                     05 QUEENS
QUEENS                     1009349.0                     201984.0             Unspecified
QUEENS             Unspecified    Unspecified    Unspecified   Unspecified        Unspec
ified                                           Unspecified   Unspecified   Unspecified
Unspecified                         N                              NaN                NaN
NaN                  NaN                NaN                        NaN                NaN
NaN                 NaN                NaN                   NaN  40.721041 -73.909453
(40.721040535628305, -73.90945306791765)
    2          26594139    10/31/2013  02:00:24  AM   10/31/2013  02:40:32  AM
NYPD             New York City Police Department              Noise - Commercial
Loud Music/Party              Club/Bar/Restaurant          10032.0                    4060
BROADWAY              BROADWAY   WEST 171 STREET                      WEST 172 STREET
NaN                   NaN       ADDRESS                  NEW YORK       NaN       Precinct
Closed   10/31/2013 10:00:24  AM          10/31/2013  02:39:42  AM                        12
MANHATTAN      MANHATTAN                       1001088.0                         246531.0
Unspecified         MANHATTAN          Unspecified     Unspecified      Unspecified
Unspecified         Unspecified                                            Unspecified
```

```
Unspecified  Unspecified  Unspecified                  N                   NaN
NaN                    NaN                    NaN                          NaN
NaN          NaN                  NaN              NaN                     NaN
NaN  40.843330 -73.939144   (40.84332975466513, -73.93914371913482)
   3        26595721   10/31/2013  01:56:23  AM   10/31/2013  02:21:48  AM
NYPD           New  York  City  Police  Department          Noise  -  Vehicle
Car/Truck  Horn                 Street/Sidewalk        10023.0          WEST  72
STREET       WEST  72  STREET    COLUMBUS  AVENUE               AMSTERDAM  AVENUE
NaN                NaN     BLOCKFACE             NEW  YORK        NaN     Precinct
Closed    10/31/2013  09:56:23  AM         10/31/2013  02:21:10  AM           07
MANHATTAN       MANHATTAN                      989730.0                  222727.0
Unspecified        MANHATTAN              Unspecified    Unspecified   Unspecified
Unspecified          Unspecified                                     Unspecified
Unspecified  Unspecified  Unspecified                  N                   NaN
NaN                    NaN                    NaN                          NaN
NaN          NaN                  NaN              NaN                     NaN
NaN  40.778009 -73.980213    (40.7780087446372, -73.98021349023975)
   ……省略部分执行结果
   [263 rows x 52 columns]
```

在下面的实例文件 007.py 中，首先输出显示了文件 39-service-requests.csv 中“Complaint Type”列的信息，其次输出了文件 39-service-requests.csv 中的前 5 行信息，再次输出了文件 39-service-requests.csv 中前 5 行“Complaint Type”列的信息，再其次输出了文件 39-service-requests.csv 中“Complaint Type”和“Borough”这两列的信息，最后输出了文件 39-service-requests.csv 中“Complaint Type”和“Borough”这两列的前 10 行信息。

```
import pandas as pd
complaints = pd.read_csv('39-service-requests.csv')
print(complaints['Complaint Type'])
print(complaints[:5])
print(complaints[:5]['Complaint Type'])
print(complaints[['Complaint Type', 'Borough']])
print(complaints[['Complaint Type', 'Borough']][:10])
```

执行后会输出：

```
//下面首先输出"Complaint Type"列的信息
0       Noise - Street/Sidewalk
1               Illegal Parking
2            Noise - Commercial
3               Noise - Vehicle
4                        Rodent
5            Noise - Commercial
6              Blocked Driveway
7            Noise - Commercial
8            Noise - Commercial
9            Noise - Commercial
10     Noise - House of Worship
11           Noise - Commercial
12              Illegal Parking
13              Noise - Vehicle
14                       Rodent
15     Noise - House of Worship
16      Noise - Street/Sidewalk
17              Illegal Parking
18       Street Light Condition
19           Noise - Commercial
20     Noise - House of Worship
21           Noise - Commercial
22              Noise - Vehicle
23           Noise - Commercial
24             Blocked Driveway
```

```
25       Noise - Street/Sidewalk
26        Street Light Condition
27          Harboring Bees/Wasps
28       Noise - Street/Sidewalk
29        Street Light Condition
                   ...
233           Noise - Commercial
234               Taxi Complaint
235          Sanitation Condition
236      Noise - Street/Sidewalk
237           Consumer Complaint
238     Traffic Signal Condition
239       DOF Literature Request
240      Litter Basket / Request
241             Blocked Driveway
242      Violation of Park Rules
243       Collection Truck Noise
244               Taxi Complaint
245               Taxi Complaint
246       DOF Literature Request
247      Noise - Street/Sidewalk
248              Illegal Parking
249              Illegal Parking
250             Blocked Driveway
251      Maintenance or Facility
252           Noise - Commercial
253              Illegal Parking
254                        Noise
255                       Rodent
256              Illegal Parking
257                        Noise
258       Street Light Condition
259                 Noise - Park
260             Blocked Driveway
261              Illegal Parking
262           Noise - Commercial
Name: Complaint Type, Length: 263, dtype: object
//下面输出前5列信息
   Unique Key           Created Date             Closed Date Agency  \
0    26589651  10/31/2013 02:08:41 AM                    NaN   NYPD
1    26593698  10/31/2013 02:01:04 AM                    NaN   NYPD
2    26594139  10/31/2013 02:00:24 AM  10/31/2013 02:40:32 AM   NYPD
3    26595721  10/31/2013 01:56:23 AM  10/31/2013 02:21:48 AM   NYPD
4    26590930  10/31/2013 01:53:44 AM                    NaN  DOHMH

                             Agency Name           Complaint Type  \
0        New York City Police Department  Noise - Street/Sidewalk
1        New York City Police Department          Illegal Parking
2        New York City Police Department       Noise - Commercial
3        New York City Police Department          Noise - Vehicle
4  Department of Health and Mental Hygiene                  Rodent

                    Descriptor        Location Type  Incident Zip  \
0                 Loud Talking      Street/Sidewalk       11432.0
1  Commercial Overnight Parking      Street/Sidewalk       11378.0
2             Loud Music/Party  Club/Bar/Restaurant       10032.0
3               Car/Truck Horn      Street/Sidewalk       10023.0
4  Condition Attracting Rodents           Vacant Lot       10027.0

   Incident Address                     ...                  \
0  90-03 169 STREET                     ...
1         58 AVENUE                     ...
2      4060 BROADWAY                    ...
```

```
3    WEST 72 STREET                           ...
4   WEST 124 STREET                           ...

  Bridge Highway Name Bridge Highway Direction Road Ramp  \
0                 NaN                      NaN       NaN
1                 NaN                      NaN       NaN
2                 NaN                      NaN       NaN
3                 NaN                      NaN       NaN
4                 NaN                      NaN       NaN

  Bridge Highway Segment Garage Lot Name Ferry Direction Ferry Terminal Name  \
0                    NaN             NaN             NaN                 NaN
1                    NaN             NaN             NaN                 NaN
2                    NaN             NaN             NaN                 NaN
3                    NaN             NaN             NaN                 NaN
4                    NaN             NaN             NaN                 NaN

    Latitude  Longitude                                  Location
0  40.708275 -73.791604   (40.70827532593202, -73.79160395779721)
1  40.721041 -73.909453  (40.721040535628305, -73.90945306791765)
2  40.843330 -73.939144   (40.84332975466513, -73.93914371913482)
3  40.778009 -73.980213    (40.7780087446372, -73.98021349023975)
4  40.807691 -73.947387   (40.80769092704951, -73.94738703491433)

[5 rows x 52 columns]
//下面输出前5行"Complaint Type"列的信息
[5 rows x 52 columns]
0    Noise - Street/Sidewalk
1            Illegal Parking
2         Noise - Commercial
3            Noise - Vehicle
4                     Rodent
….省略部分
259             Noise - Park       BROOKLYN
260         Blocked Driveway         QUEENS
261          Illegal Parking       BROOKLYN
262       Noise - Commercial      MANHATTAN
[263 rows x 2 columns]
//下面输出"Complaint Type"和"Borough"这两列的信息
              Complaint Type        Borough
0    Noise - Street/Sidewalk         QUEENS
1            Illegal Parking         QUEENS
2         Noise - Commercial      MANHATTAN
3            Noise - Vehicle      MANHATTAN
4                     Rodent      MANHATTAN
5         Noise - Commercial         QUEENS
//下面输出"Complaint Type"和"Borough"这两列的前10行信息
           Complaint Type    Borough
0  Noise - Street/Sidewalk     QUEENS
1          Illegal Parking     QUEENS
2       Noise - Commercial  MANHATTAN
3          Noise - Vehicle  MANHATTAN
4                   Rodent  MANHATTAN
5       Noise - Commercial     QUEENS
6         Blocked Driveway     QUEENS
7       Noise - Commercial     QUEENS
8       Noise - Commercial  MANHATTAN
9       Noise - Commercial   BROOKLYN
```

在下面的实例文件008.py中，首先输出显示了文件39-service-requests.csv中“Complaint Type”列中数值前10名的信息，然后在图表中统计显示这前10名的信息。

```
    import pandas as pd
    import matplotlib.pyplot as plt

    pd.set_option('display.width', 5000)
    pd.set_option('display.max_columns', 60)

    plt.rcParams['figure.figsize'] = (10, 6)

    complaints = pd.read_csv('39-service-requests.csv')
    complaint_counts = complaints['Complaint Type'].value_counts()
    print(complaint_counts[:10])# 打印输出 "Complaint Type" 列中数值前 10 名的信息
    complaint_counts[:10].plot(kind='bar')# 绘制 "Complaint  Type" 列中数值前 10 名的图表
信息
    plt.show()
```

执行后会在控制台中输出显示“Complaint Type”列中数值前 10 名的信息：

```
    Noise - Commercial          51
    Noise                       27
    Noise - Street/Sidewalk     22
    Blocked Driveway            21
    Illegal Parking             18
    Taxi Complaint              13
    Traffic Signal Condition    10
    Rodent                      10
    Water System                 9
    Noise - Vehicle              7
    Name: Complaint Type, dtype: int64
```

并且会在 matplotlib 图表中统计列“Complaint Type”中数值前 10 名的信息，如图 9-2 所示。

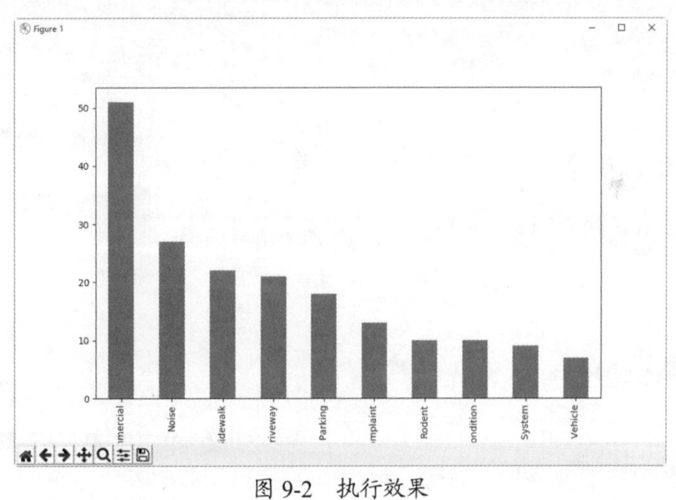

图 9-2　执行效果

9.3　可视化和日期相关的操作

在进行数据统计分析时，时间通常是一个重要的因素。在本节的内容中，将详细讲解和日期相关的操作知识，为读者步入本书后面知识的学习打下基础。

↑扫码看视频（本节视频课程时间：5 分 18 秒）

9.3.1 可视化统计每个月的骑行数据

在下面的实例文件 009.py 中，可以使用 matplotlib 统计出文件 bikes.csv 中每个月的骑行数据信息。

```
import pandas as pd
import matplotlib.pyplot as plt

plt.rcParams['figure.figsize'] = (10, 8)
plt.rcParams['font.family'] = 'sans-serif'

pd.set_option('display.width', 5000)
pd.set_option('display.max_columns', 60)

bikes = pd.read_csv('bikes.csv', sep=';', encoding='latin1', parse_
dates=['Date'], dayfirst=True, index_col='Date')
bikes['Berri 1'].plot()
plt.show()
```

执行后的效果如图 9-3 所示。

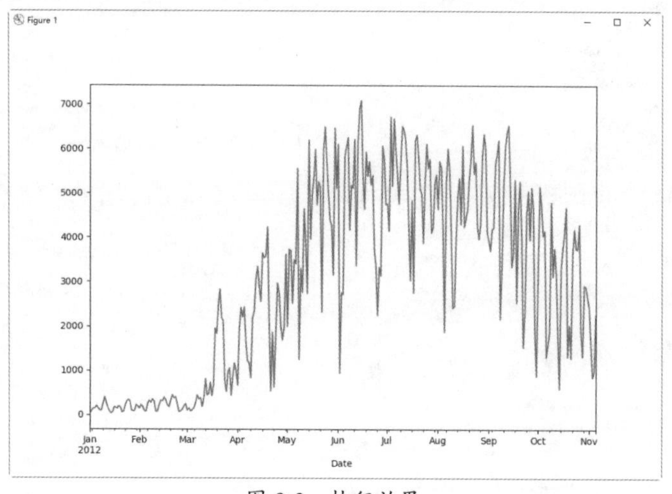

图 9-3　执行效果

9.3.2 统计某街道的前 5 天骑行数据

在下面的实例文件 010.py 中，首先输出显示文件 bikes.csv 中“Berri 1”街道前 5 天的骑行数据信息，然后使用“print(berri_bikes.index)”输出星期几的时间。

```
import pandas as pd

bikes = pd.read_csv('bikes.csv', sep=';', encoding='latin1', parse_
dates=['Date'], dayfirst=True, index_col='Date')
berri_bikes = bikes[['Berri 1']].copy()
print(berri_bikes[:5])
print(berri_bikes.index)
```

执行后会输出：

```
            Berri 1
Date
2010-01-01       35
2010-01-02       83
2010-01-03      135
```

```
    2010-01-04       144
    2010-01-05       197
    DatetimeIndex(['2010-01-01', '2010-01-02', '2010-01-03', '2010-01-04',
                   '2010-01-05', '2010-01-06', '2010-01-07', '2010-01-08',
                   '2010-01-09', '2010-01-10',
                   ...
                   '2010-10-27', '2010-10-28', '2010-10-29', '2010-10-30',
                   '2010-10-31', '2010-9-01', '2010-9-02', '2010-9-03',
                   '2010-9-04', '2010-9-05'],
                  dtype='datetime64[ns]', name='Date', length=310, freq=None)
```

由上述执行效果可知，只输出显示了 310 天的统计数据。其实 Pandas 有一系列非常好的时间序列功能，所以如果我们想得到每一行的月份，可以通过如下所示的文件 09.py 实现。

```
    import pandas as pd

    bikes = pd.read_csv('bikes.csv', sep=';', encoding='latin1', parse_
dates=['Date'], dayfirst=True, index_col='Date')
    berri_bikes = bikes[['Berri 1']].copy()
    print(berri_bikes.index.day)
    print(berri_bikes.index.weekday)
```

执行后会输出：

```
    Int64Index([ 1,  2,  3,  4,  5,  6,  7,  8,  9, 10,
                ...
                27, 28, 29, 30, 31,  1,  2,  3,  4,  5],
               dtype='int64', name='Date', length=310)
    Int64Index([6, 0, 1, 2, 3, 4, 5, 6, 0, 1,
                ...
                5, 6, 0, 1, 2, 3, 4, 5, 6, 0],
               dtype='int64', name='Date', length=310)
```

在上述输出结果中，0 表示星期一。我们可以使用库 pandas 灵活获取某一天是星期几，例如下面的实例文件 012.py。

```
    import pandas as pd

    bikes = pd.read_csv('bikes.csv', sep=';', encoding='latin1', parse_
dates=['Date'], dayfirst=True, index_col='Date')
    berri_bikes = bikes[['Berri 1']].copy()
    berri_bikes.loc[:,'weekday'] = berri_bikes.index.weekday
    print(berri_bikes[:5])
```

执行后会输出：

```
                Berri 1  weekday
    Date
    2010-01-01       35        6
    2010-01-02       83        0
    2010-01-03      135        1
    2010-01-04      144        2
    2010-01-05      197        3
```

9.3.3 统计周一到周日每天的数据

在现实应用中，我们当然也可以统计周一到周日每天的数据。例如在下面的实例文件 013.py 中，首先统计了周一到周日每天的数据，然后用更加通俗、易懂的星期几的英文名统计了周一到周日每天的骑行数据。

```
    import pandas as pd
    bikes = pd.read_csv('bikes.csv', sep=';', encoding='latin1', parse_
dates=['Date'], dayfirst=True, index_col='Date')
```

```
    berri_bikes = bikes[['Berri 1']].copy()
    berri_bikes.loc[:,'weekday'] = berri_bikes.index.weekday

    weekday_counts = berri_bikes.groupby('weekday').aggregate(sum)
    print(weekday_counts)

    weekday_counts.index = ['Monday', 'Tuesday', 'Wednesday', 'Thursday', 'Friday',
'Saturday', 'Sunday']
    print(weekday_counts)
```

执行后会输出：

```
              Berri 1
    weekday
    0          134298
    1          135305
    2          152972
    3          160131
    4          141771
    5          101578
    6           99310
                Berri 1
    Monday       134298
    Tuesday      135305
    Wednesday    152972
    Thursday     160131
    Friday       141771
    Saturday     101578
    Sunday        99310
```

9.3.4 可视化周一到周日每天的骑行数据

为了使统计数据显得更加直观，可以在程序中使用 matplotlib 技术。例如在下面的实例文件 014.py 中，使用 matplotlib 图表统计了周一到周日每天的骑行数据。

```
    import pandas as pd
    import matplotlib.pyplot as plt
    plt.rcParams['figure.figsize'] = (15, 5)
    bikes = pd.read_csv('bikes.csv',
                        sep=';', encoding='latin1',
                        parse_dates=['Date'], dayfirst=True,
                        index_col='Date')
    # 添加标识
    berri_bikes = bikes[['Berri 1']].copy()
    berri_bikes.loc[:,'weekday'] = berri_bikes.index.weekday

    # 开始统计
    weekday_counts = berri_bikes.groupby('weekday').aggregate(sum)
    weekday_counts.index = ['Monday', 'Tuesday', 'Wednesday', 'Thursday', 'Friday',
'Saturday', 'Sunday']
    weekday_counts.plot(kind='bar')

    plt.show()
```

执行效果如图 9-4 所示。

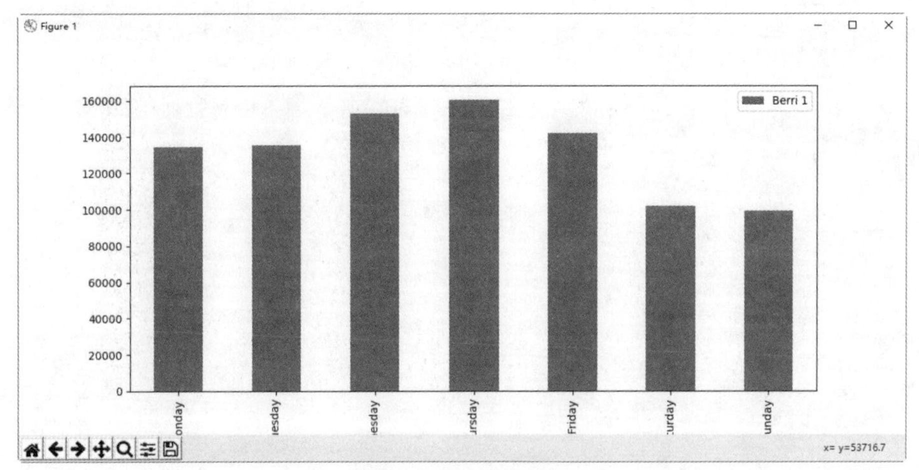

图 9-4　执行效果

9.3.5　可视化天气信息

在下面的实例文件 015.py 中，通过借助于素材文件 weather_2012.csv，使用 matplotlib 统计了加拿大 2012 年的全年天气数据信息。

```
import pandas as pd
import matplotlib.pyplot as plt
import numpy as np

plt.rcParams['figure.figsize'] = (15, 3)
plt.rcParams['font.family'] = 'sans-serif'
weather_2012_final = pd.read_csv('weather_2012.csv', index_col='Date/Time')
weather_2012_final['Temp (C)'].plot(figsize=(15, 6))
plt.show()
```

执行效果如图 9-5 所示。

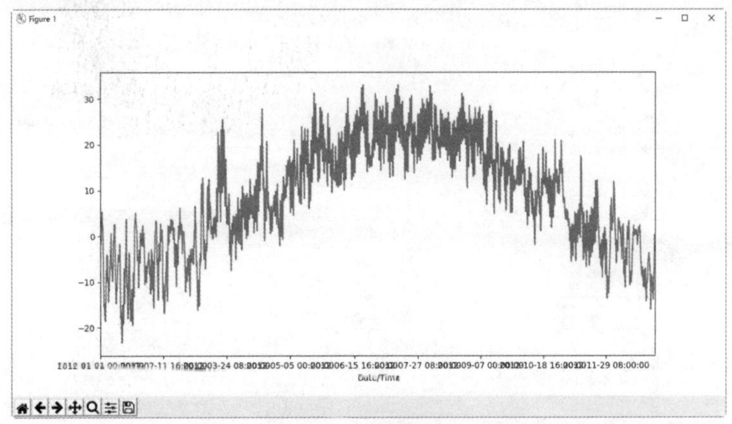

图 9-5　执行效果

只需通过下面简单的实例文件 016.py，即可打印输出文件 weather_2012.csv 中的全部天气信息。

```
import pandas as pd

weather_2012_final = pd.read_csv('weather_2012.csv', index_col='Date/Time')
```

```
print(weather_2012_final)
```

执行后会输出：

```
                     Temp (C)  Dew Point Temp (C)  Rel Hum (%)  \
Date/Time
2010-01-01 00:00:00      -1.8                -3.9           86
2010-01-01 01:00:00      -1.8                -3.7           87
2010-01-01 02:00:00      -1.8                -3.4           89
2010-01-01 03:00:00      -1.5                -3.2           88
2010-01-01 04:00:00      -1.5                -3.3           88
2010-01-01 05:00:00      -1.4                -3.3           87
2010-01-01 06:00:00      -1.5                -3.1           89
2010-01-01 07:00:00      -1.4                -3.6           85
2010-01-01 08:00:00      -1.4                -3.6           85
#在此省略好多输出结果
2010-10-31 19:00:00                 Snow
2010-10-31 20:00:00                 Snow
2010-10-31 21:00:00                 Snow
2010-10-31 22:00:00                 Snow
2010-10-31 23:00:00                 Snow

[8784 rows x 7 columns]
```

9.4 分析服务器日志数据

现在存在一个名为“log_data.csv”的文件，在里面保存了服务器的日志信息。一共有 8 157 277 条数据，这已经超过了 Excel 的最大行数。在本节的内容中，将详细讲解使用 pandas 分析这个服务器日志数据的过程。

↑扫码看视频（本节视频课程时间：3 分 56 秒）

9.4.1 分析统计每个 enrollment_id 事件的总数

在文件 log_data.csv 中，enrollment_id 由 username 和 course_id 构成，表示每个学生上课的学号；username 表示学生的姓名；course_id 表示课程号；time 表示这条日志数据发生的时间；source 表示发生日志数据产生源（server 或者 browser）；event 表示发生日志事件（有 7 种）；object 表示其他事件。

编写文件 question1.py，功能是统计文件 log_data.csv 中的每个 enrollment_id 事件总数，具体实现代码如下所示：

```
import pandas as pd
#导入数据
df=pd.read_csv('log_data.csv',dtype={'time': str},nrows=1000)
#将1000行的记录保存为csv格式文件
# df.to_csv('log_data.csv')
group1 = df.groupby('enrollment_id')
#按照enrollment_id来统计分组数量
d = group1.size()
print(d)
print(d*2)#有两列活动，所以就*2
```

执行后会输出：

```
enrollment_id
1    314
```

```
3    288
4     99
5    299
dtype: int64
enrollment_id
1    628
3    576
4    198
5    598
dtype: int64
```

9.4.2　统计每种事件的个数和占用比率

编写文件 question2.py，统计每种事件的个数，以及与总事件个数的比值。

文件 question2.py 的具体实现代码如下所示：

```
import pandas as pd
import math
#导入数据，并只用第一行
df=pd.read_csv('log_data.csv',dtype={'time': str},nrows=1000)
#按照enrollment_id来分组
group1 = df.groupby('enrollment_id').size()
#按照event，enrollment_id来统计分组数量
group2 = df.groupby(['enrollment_id','event']).size()
#按照enrollment_id来统计分组数量
print(group2)
print(group2/(group1*2))
```

执行后会输出：

```
enrollment_id  event
1              access        107
               nagivate       25
               page_close     66
               problem        87
               video          29
3              access         79
               discussion     26
               nagivate       14
               page_close     22
               problem       138
               video           9
4              access         64
               nagivate       15
               page_close     10
               problem         6
               video           4
5              access        112
               discussion     17
               nagivate       15
               page_close     48
               problem        73
               video          34
dtype: int64
enrollment_id  event
1              access        0.170382
               nagivate      0.039809
               page_close    0.105096
               problem       0.138535
               video         0.046178
3              access        0.137153
               discussion    0.045139
               nagivate      0.024306
```

```
                page_close    0.038194
                problem       0.239583
                video         0.015625
4               access        0.323232
                nagivate      0.075758
                page_close    0.050505
                problem       0.030303
                video         0.020202
5               access        0.187291
                discussion    0.028428
                nagivate      0.025084
                page_close    0.080268
                problem       0.122074
                video         0.056856
dtype: float64
```

9.5 实践案例：使用库 pandas 提取数据并构建 Neo4j 知识图谱

在本节将通过一个具体实例的实现过程，详细讲解使用 pandas 提取 Excel 数据的方法，并介绍了将提取出的数据保存到 Neo4j 数据库并构建知识图谱的过程。

↑扫码看视频（本节视频课程时间：4 分 52 秒）

9.5.1 准备工作

在本实例的 Excel 文件 Invoice_data_Demo.xls 中保存了多张发票的信息，包含每张发票的名称、机器编号、发票代码、发票号码、开票日期、校验码等信息。如图 9-6 所示。

	A	B	C	D	E	F	G	H	I	J	K	L	M	N	O
1	发票名称	机器编号	发票代码	发票号码	开票日期	校验码	购买方名称	购买方纳税	购买方地址	购买方开户	销售方名称	销售方纳税	销售方地址	销售方开户行及账号	项目名称
2	山东增值税	49909964	03700170	00476941	2018年03月	08385227	武汉市车城	914201007	武汉经济技	中国农业银	山东高速集	91370100	山东省济南	威海市商业银行济南舜耕支	*经营租赁
3	湖北增值税	49909966	04200170	13208805	2018年03月	103507112	武汉市车城	914201007	武汉经济技	中国农业银	湖北阿深南	914200007	湖北省大冶	工行黄石市杭州路支行180	*经营租赁
4	湖北增值税	49909892	04200170	27908343	2018年03月	081106411	武汉市车城	914201007	武汉经济技	中国农业银	湖北楚天鄂	91420200	黄石磁湖科	工行大冶城中支行1803021	*经营租赁
5	湖北增值税	499098892	04200170	47502662	2018年03月	087953274	武汉市车城	914201007	武汉经济技	中国农业银	湖北中交嘉	91421200	湖北咸宁市	农行咸宁市温泉支行17680	*经营租赁
6	湖北增值税	499098892	04200170	47153800	2018年03月	118085544	武汉市车城	914201007	武汉经济技	中国农业银	湖北中交咸	91421200	咸宁市咸安	中行咸宁分行57165753467	*经营租赁
7	湖北增值税	49909966	04200170	47213345	2018年03月	12406490	武汉市车城	914201007	武汉经济技	中国农业银	湖北楚天高	91421202	咸宁市咸安	工行大冶城中支行1803021	*经营租赁
8	湖南增值税	49909966	04300180	01939605	2018年03月	16046901	武汉市车城	914201007	武汉经济技	中国农业银	现代投资股	914300001	湖南省长沙	中国建设银行长沙和平路支	*经营租赁
9	湖南增值税	49909978	04300180	07011955	2018年03月	08424113	武汉市车城	914201007	武汉经济技	中国农业银	湖南道岳高	914306007	湖南省岳阳	中国银行岳阳云溪支行604	*经营租赁
10	广东增值税	49909965	04400170	72411126	2018年03月	036962797	武汉市车城	914201007	武汉经济技	中国农业银	广深珠高速	91440000	广东省广州	中国银行虎门支行7250577	*经营租赁
11	广东增值税	49909889	04400170	32616467	2018年03月	166998451	武汉市车城	914201007	武汉经济技	中国农业银	广东广惠高	914400007	广州市萝岗	中国银行股份有限公司广州	*经营租赁
12	广东增值税	49909965	04400170	29414219	2018年03月	09415114	武汉市车城	914201007	武汉经济技	中国农业银	广州市北二	91440101	广州市白云	工行广州庙前直街支行360	*经营租赁

图 9-6　发票信息

Neo4j 是一个高性能的 NOSQL 图形数据库，更是一个嵌入式的、基于磁盘的、具备完全的事务特性的 Java 持久化引擎，它会将结构化数据存储在网络（从数学角度叫做图）上而不是表中。Neo4j 支持表 9-1 所示的查询语法（CQL）。

表 9-1

序号	关键字	关键字作用
1	CREATE	创建
2	MATCH	匹配
3	RETURN	加载
4	WHERE	过滤检索条件
5	DELETE	删除节点和关系

续表

序号	关键字	关键字作用
6	REMOVE	删除节点和关系的属性
7	ORDER BY	排序
8	SET	添加或更新属性

9.5.2 使用库 pandas 提取 Excel 数据

编写文件 invoice_neo4j.py，功能是通过函数 data_extraction 和函数 relation_extrantion 分别提取构建知识图谱所需要的节点数据以及联系数据，构建三元组。在提取 Excel 数据时，使用 pandas 将 Excel 数据转换成 dataframe 类型。文件 invoice_neo4j.py 的具体实现代码如下所示：

```
from dataToNeo4jClass.DataToNeo4jClass import DataToNeo4j
import pandas as pd


# 提取 excel 表格中数据，将其转换成 dateframe 类型
invoice_data = pd.read_excel('Invoice_data_Demo.xls', header=0, encoding='utf8')
print(invoice_data)


def data_extraction():
    """ 节点数据抽取 """

    # 取出发票名称到 list
    node_list_key = []
    for i in range(0, len(invoice_data)):
        node_list_key.append(invoice_data['发票名称'][i])

    # 去除重复的发票名称
    node_list_key = list(set(node_list_key))

    # 检索出 node 中的 node
    node_list_value = []
    for i in range(0, len(invoice_data)):
        for n in range(1, len(invoice_data.columns)):
            # 取出表头名称 invoice_data.columns[i]
            node_list_value.append(invoice_data[invoice_data.columns[n]][i])
    # 去重
    node_list_value = list(set(node_list_value))
    # 将 list 中浮点及整数类型全部转换成 string 类型
    node_list_value = [str(i) for i in node_list_value]

    return node_list_key, node_list_value


def relation_extraction():
    """ 联系数据抽取 """

    links_dict = {}
    name_list = []
    relation_list = []
    name2_list = []

    for i in range(0, len(invoice_data)):
        m = 0
        name_node = invoice_data[invoice_data.columns[m]][i]
```

```
        while m < len(invoice_data.columns)-1:
            relation_list.append(invoice_data.columns[m+1])
            name2_list.append(invoice_data[invoice_data.columns[m+1]][i])
            name_list.append(name_node)
            m += 1

    # 将数据中 int 类型全部转换成 string
    name_list = [str(i) for i in name_list]
    name2_list = [str(i) for i in name2_list]

    # 整合数据，将三个 list 整合成一个 dict
    links_dict['name'] = name_list
    links_dict['relation'] = relation_list
    links_dict['name2'] = name2_list
    # 将数据转换成 DataFrame
    df_data = pd.DataFrame(links_dict)
    return df_data

# 实例化对象
data_extraction()
relation_extraction()
create_data = DataToNeo4j()

create_data.create_node(data_extraction()[0], data_extraction()[1])
create_data.create_relation(relation_extraction())
```

执行后会提取 Excel 文件 Invoice_data_Demo.xls 中的数据，提取了构建知识图谱所需要的节点数据以及联系数据，并构建了一个三元组。执行后输出结果如下：

```
"C:\Program Files\Anaconda3\python.exe" H:/pythonshuju/11/9-5/tu/invoice_neo4j.py
          发票名称          机器编号         发票代码       发票号码        开票日期  \
0   山东增值税电子普通发票  499099649091  37001700112    476941  2018年03月23日
1   湖北增值税电子普通发票  499099661823  42001700112  13208805  2018年03月23日
2   湖北增值税电子普通发票  499098921515  42001700112  27908343  2018年03月23日
3   湖北增值税电子普通发票  499098892671  42001700112  47502662  2018年03月23日
4   湖北增值税电子普通发票  499098892697  42001700112  47153800  2018年03月23日
5   湖北增值税电子普通发票  499099660214  42001700112  47213345  2018年03月23日
6   湖南增值税电子普通发票  499099661882  43001800112   1939605  2018年03月23日
7   湖南增值税电子普通发票  499099785543  43001800112   7011955  2018年03月23日
8   广东增值税电子普通发票  499099656872  44001700112  72411126  2018年03月23日
9   广东增值税电子普通发票  499098893067  44001700112  32616467  2018年03月23日
10  广东增值税电子普通发票  499099653583  44001700112  29414219  2018年03月23日
11  广东增值税电子普通发票  499098894641  44001700112  83722452  2018年03月23日
12  广东增值税电子普通发票  499098891512  44001700112  75668743  2018年03月23日
13  广东增值税电子普通发票  499098892970  44001700112  36151707  2018年03月23日
14  广东增值税电子普通发票  499098890376  44001700112   5680333  2018年03月23日
15  广东增值税电子普通发票  499099655458  44001700112  31742233  2018年03月23日
16  广东增值税电子普通发票  499099653284  44001700112  59018159  2018年03月23日
17  广东增值税电子普通发票  499099655600  44001700112  58098285  2018年03月23日
18  广东增值税电子普通发票  499098892347  44001719112      6589  2018年03月23日
19  广东增值税电子普通发票  499098890819  44001709112  10622269  2018年03月23日
20  广东增值税电子普通发票  499099662033  44001700112  11218811  2018年03月23日
21  广东增值税电子普通发票  499098891951  44001709112   1454500  2018年03月23日
22  广东增值税电子普通发票  499099655503  44001711112  10691690  2018年03月23日
23  广东增值税电子普通发票  499098891686  44001711112   3981240  2018年03月23日
24  广东增值税电子普通发票  499099662201  44001711112   9932014  2018年03月23日
25  江西增值税电子普通发票  499098906112  36001700112   5393233  2018年03月23日
26  江西增值税电子普通发票  499098906112  36001700112   5393233  2018年03月23日
27  江西增值税电子普通发票  499099661153  36001700112  12547202  2018年03月23日
28  湖北增值税电子普通发票  499098892081  42001700112  11162913  2018年03月23日
29  湖北增值税电子普通发票  499099660821  42001700112   7007198  2018年03月23日
30  湖北增值税电子普通发票  499098911093  42001700112  19440150  2018年03月23日
31  湖北增值税电子普通发票  499099656098  42001700112  10077287  2018年03月23日
```

```
    32  湖北增值税电子普通发票  499098892726  42001700112  47251670  2018年03月23日
    33  广东增值税电子普通发票  499098890827  44001700112  83072331  2018年03月23日
    34  广东增值税电子普通发票  499098891627  44001721112     15382  2018年03月23日

                 校验码                  购买方名称           购买方纳税人识别号    \
    0   08385227000109856742  武汉市车城物流有限公司  91420100748306245 7
    1   10350711227340909622  武汉市车城物流有限公司  914201007483062457
    2   08110641107709826910  武汉市车城物流有限公司  914201007483062457
    3   08795327402254684486  武汉市车城物流有限公司  914201007483062457
    4   11808554414702210080  武汉市车城物流有限公司  914201007483062457
    5   12406490987068798034  武汉市车城物流有限公司  914201007483062457
    6   16046901234429288942  武汉市车城物流有限公司  914201007483062457
    7   08424113382897776596  武汉市车城物流有限公司  914201007483062457
    8   03696279796705775278  武汉市车城物流有限公司  914201007483062457
    9   16699845128671790777  武汉市车城物流有限公司  914201007483062457
    10  09415114643911023305  武汉市车城物流有限公司  914201007483062457
    11  07277542713402041348  武汉市车城物流有限公司  914201007483062457
    12  05160031932430170487  武汉市车城物流有限公司  914201007483062457
    13  13610032493225906141  武汉市车城物流有限公司  914201007483062457
    14  18152402194137637371  武汉市车城物流有限公司  914201007483062457
    15  12170444432639147551  武汉市车城物流有限公司  914201007483062457
    16  01719212105817985192  武汉市车城物流有限公司  914201007483062457
    17  08827458019817618492  武汉市车城物流有限公司  914201007483062457
    18  10381866457825785395  武汉市车城物流有限公司  914201007483062457
    19  13139987529386485607  武汉市车城物流有限公司  914201007483062457
    20  06397214072807234709  武汉市车城物流有限公司  914201007483062457
    21  11955961629009062357  武汉市车城物流有限公司  914201007483062457
    22  17866792090175297885  武汉市车城物流有限公司  914201007483062457
    23  06834913840024355085  武汉市车城物流有限公司  914201007483062457
    24  06900265555805556595  武汉市车城物流有限公司  914201007483062457
    25  08204989498324515730  武汉市车城物流有限公司  914201007483062457
    26  08204989498324515730  武汉市车城物流有限公司  914201007483062457
    27  12927690393569329018  武汉市车城物流有限公司  914201007483062457
    28  11241636206802490280  武汉市车城物流有限公司  914201007483062457
    29  12636666022332927910  武汉市车城物流有限公司  914201007483062457
    30  11753239368958298076  武汉市车城物流有限公司  914201007483062457
    31  16167883775201694087  武汉市车城物流有限公司  914201007483062457
    32  17912771164864567783  武汉市车城物流有限公司  914201007483062457
    33  17248100991537201542  武汉市车城物流有限公司  914201007483062457
    34  06775424551826339304  武汉市车城物流有限公司  914201007483062457

                   购买方地址、电话                        购买方开户行及账号 ...    \
    0   武汉经济技术开发区车城大道7号84289348   中国农业银行股份有限公司武汉开发区支行17-
071201040004598 ...
    ####后面省略好多执行效果
```

9.5.3 将数据保存到 Neo4j 数据库并构建知识图谱

在本小节中，我们将从文件 Invoice_data_Demo.xls 中提取的数据保存到 Neo4j 数据库中，然后基于 Neo4j 数据构建可视化的知识图谱。

（1）下载并搭建 Neo4j 数据库开发环境，在控制台中启动 Neo4j 数据库服务，如图 9-7 所示。

```
H:\neo4j-community-3.5.2-windows\neo4j-community-3.5.2\bin>neo4j.bat console
2019-01-24 08:37:28.211+0000 INFO  ======== Neo4j 3.5.2 ========
2019-01-24 08:37:50.385+0000 INFO  Starting...
2019-01-24 08:40:36.550+0000 INFO  Bolt enabled on 0.0.0.0:7687.
2019-01-24 08:41:49.691+0000 INFO  Started.
2019-01-24 08:42:04.665+0000 INFO  Remote interface available at http://localhost:7474/
```

图 9-7　启动 Neo4j 数据库服务

（2）编写文件 DataToNeo4jClass.py，功能是将提取的数据保存到 Neo4j 数据库，准备好建立知识图谱所需的节点和边数据。文件 DataToNeo4jClass.py 的具体实现代码如下所示：

```
    from py2neo import Node, Graph, Relationship

    class DataToNeo4j(object):
        """将excel中数据存入neo4j"""

        def __init__(self):
            """建立连接"""
                link = Graph("http://localhost//:7474", username="neo4j",
password="66688888")
            self.graph = link
            # 定义label
            self.invoice_name = '发票名称'
            self.invoice_value = '发票值'
            self.graph.delete_all()

        def create_node(self, node_list_key, node_list_value):
            """建立节点"""
            for name in node_list_key:
                name_node = Node(self.invoice_name, name=name)
                self.graph.create(name_node)
            for name in node_list_value:
                value_node = Node(self.invoice_value, name=name)
                self.graph.create(value_node)

        def create_relation(self, df_data):
            """建立联系"""

            m = 0
            for m in range(0, len(df_data)):
                try:
                    rel = Relationship(self.graph.find_one(label=self.invoice_name,
property_key='name', property_value=df_data['name'][m]),
                                    df_data['relation'][m], self.graph.find_
one(label=self.invoice_value, property_key='name',
                                property_value=df_data['name2'][m]))
                    self.graph.create(rel)
                except AttributeError as e:
                    print(e, m)
```

（3）编写文件 neo4j_to_dataframe.py，功能是建立与 neo4j 数据库服务器的连接，实现知识图谱数据接口。文件 neo4j_to_dataframe.py 的具体实现代码如下所示：

```
    from py2neo import Graph
    import re
    from pandas import DataFrame


    class Neo4jToJson(object):
        """知识图谱数据接口"""

        # 与neo4j服务器建立连接
        graph = Graph("http://localhost//:7474", username="neo4j", password="neo4j")
        links = []
        nodes = []

        def post(self):
            """与前端交互"""
            # 前端传过来的数据
            select_name = '南京审计大学'
            label_name = '单位名称'
```

```
        # 取出所有节点数据
          nodes_data_all = self.graph.run("MATCH (n:" + label_name + ") RETURN
n").data()
        # node 名存储
        nodes_list = []
        for node in nodes_data_all:
            nodes_list.append(node['n']['name'])
        # 根据前端的数据，判断搜索的关键字是否在 nodes_list 中存在，如果存在则返回相应数据，
否则返回全部数据
        if select_name in nodes_list:
            # 获取知识图谱中相关节点数据
             links_data = self.graph.run("MATCH (n:" + label_name + "{name:'" +
select_name + "'})-[r]-(b) return r").data()
        else:
            # 获取知识图谱中所有节点数据
            links_data = self.graph.run("MATCH ()-[r]->() RETURN r").data()

        data_for_df = self.get_links(links_data)

        # 将列表转换成 dataframe
        df = DataFrame(data_for_df, columns=['source', 'name', 'target'])
        return df

    def get_links(self, links_data):
        """知识图谱关系数据获取"""
        i = 1
        dict = {}

        # 匹配模式
        pattern = '^\(|\{\}\]\-\>\(|\)\-\[\:|\)$'

        for link in links_data:
            # link_data 样式：(南京审计大学) - [: 学校地址 {}]->(江苏省南京市浦口区雨山
西路 86 号)
            link_data = str(link['r'])
            # 正则，用 split 将 string 切成：['', '南京审计大学', '学校地址 ', '江苏省南
京市浦口区雨山西路 86 号', '']
            links_str = re.split(pattern, link_data)

            for data in links_str:
                if len(data) > 1:
                    if i == 1:
                        dict['source'] = data
                    elif i == 2:
                        dict['name'] = data
                    elif i == 3:
                        dict['target'] = data
                        self.links.append(dict)
                        dict = {}
                        i = 0
                    i += 1
        return self.links


if __name__ == '__main__':
    data_neo4j = Neo4jToJson()
    print(data_neo4j.post())
```

在 Neo4j 中构建的知识图谱如图 9-8 所示。

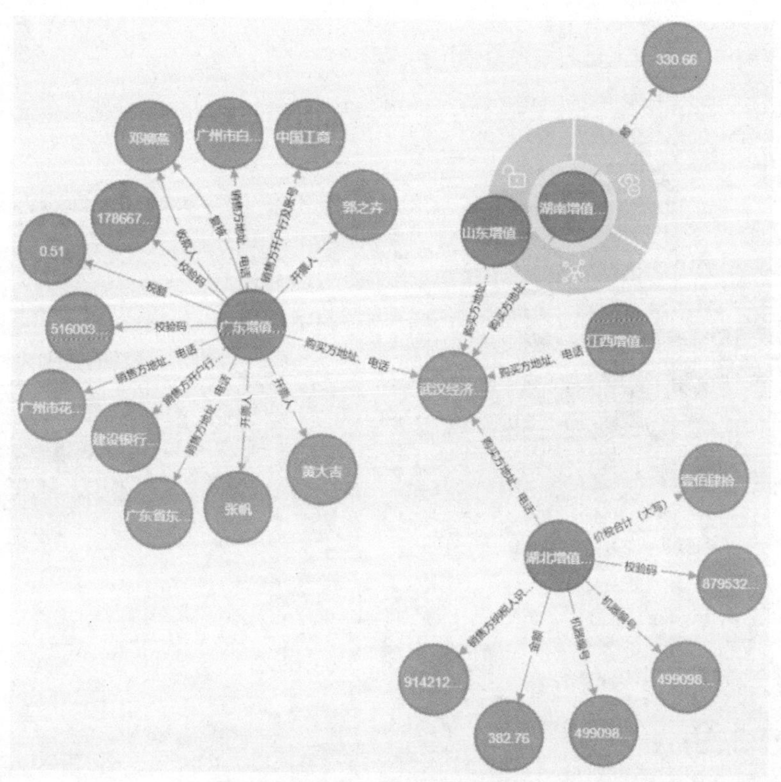

图 9-8　在 Neo4j 中构建的知识图谱

第 10 章

当 Seaborn 遇到 matplotlib

库 Seaborn 把经常用到的可视化绘图功能进行了函数封装，形成的一个“快捷方式”。Seaborn 和 matplotlib 相比代码更加简洁，只需使用几行代码就能创建出漂亮的图表。因为 Seaborn 是建立在 matplotlib 之上的，所以在开发者掌握 matplotlib 之后，学习 Seaborn 将会变得事半功倍。

10.1 搭建 Seaborn 环境

环境搭建是使用 Seaborn 的第一步；在具体实践中，Seaborn 的安装方式有 pip 安装和 conda 安装两种；这里我们选择 pip 命令安装的方式。

↑扫码看视频（本节视频课程时间：1 分 32 秒）

在使用库 Seaborn 之前，需要使用如下 pip 命令进行安装：

```
pip install seaborn
```

安装完毕后，就可以在我们的电脑中使用 Seaborn 绘制可视化图形了。如果想使用 Seaborn 官方提供的数据，需要在 Github 网站单独下载 Seaborn 提供的 Data 文件，然后将下载的文件夹“seaborn-data”保存到电脑的主目录中。

Windows 系统的主目录是 C 盘中的“User”目录，Linux 系统的主目录是“Home”文件夹。例如作者的电脑是 Windows 10 系统，User 目录是“C:\Users\apple”，所以可以将下载的文件夹“seaborn-data”保存到“C:\Users\apple”中，如图 10-1 所示。

此电脑 › BOOTCAMP (C:) › 用户 › apple

名称	修改日期	类型
.PyCharm2019.3	2020/4/0 16:40	文件夹
.tooling	2020/4/11 11:45	文件夹
.vscode	2020/5/29 22:07	文件夹
3D 对象	2020/6/27 1:56	文件夹
AppData	2020/6/27 1:54	文件夹
eclipse	2020/4/11 11:41	文件夹
eclipse-workspace	2020/6/3 22:57	文件夹
Intel	2020/3/29 8:35	文件夹
MicrosoftEdgeBackups	2020/3/29 19:07	文件夹
OneDrive	2020/5/10 19:26	文件夹
PycharmProjects	2020/4/8 15:12	文件夹
seaborn-data	2020/7/3 21:41	文件夹

图 10-1 保存下载的文件夹“seaborn-data”

10.2 绘制基本的可视化图

在使用 pip 命令安装库 Seaborn 之后，就可以在电脑中使用 Seaborn 绘制可视化图形了。在本节的内容中，将详细讲解使用库 Seaborn 绘制基本的可视化图形的知识和实践。

↑扫码看视频（本节视频课程时间：9 分 18 秒）

10.2.1 第一个 Seaborn 图形程序

Seaborn 旨在以数据可视化为中心来挖掘与理解数据，它提供的面向数据集制图函数主要是对行列索引和数组的操作，包含对整个数据集进行内部的语义映射与统计整合，以此生成富于信息的图表。请看下面的实例文件 001.py，功能是结合 matplotlib 绘制了一个简单的散点图。

```
import matplotlib.pyplot as plt
import seaborn as sns
sns.set()
tips = sns.load_dataset("tips")
sns.relplot(x="total_bill", y="tip", col="time",
            hue="smoker", style="smoker", size="size",
            data=tips);
plt.show()
```

接下来开始分析上述代码的具体含义：

（1）因为 Seaborn 实际上是调用了 matplotlib 来绘图，所以在上述代码中使用 import 分别导入了 matplotlib 和 seaborn；

（2）使用函数 sns.set() 设置并使用 seaborn 默认的主题、尺寸大小以及调色板；

（3）通过如下代码中的函数 load_dataset() 加载使用样例数据集，这些数据集被保存在我们在前面下载的文件夹“seaborn-data”中。在现实应用中，通常使用 tips 数据集来绘图，tips 数据集提供了一种“整洁”的整合数据的方式。

```
tips = sns.load_dataset("tips")
```

（4）通过如下代码绘制一个多子图散点图，并分配语义变量。在函数 relplot() 中设置了 tips 数据集中五个变量的关系，其中三个是数值变量，另外两个是类别变量。其中 total_bill 和 tip 这两个数值变量决定了轴上每个点出现的位置，另外一个 size 变量影响着出现点的大小。第一个类别变量 time 将散点图分为两个子图，第二个类别变量 smoker 决定点的形状。

```
sns.relplot(x="total_bill", y="tip", col="time",
            hue="smoker", style="smoker", size="size",
            data=tips)
```

（5）最后调用 matplotlib 函数 plt.show() 绘图，执行效果如图 10-2 所示。

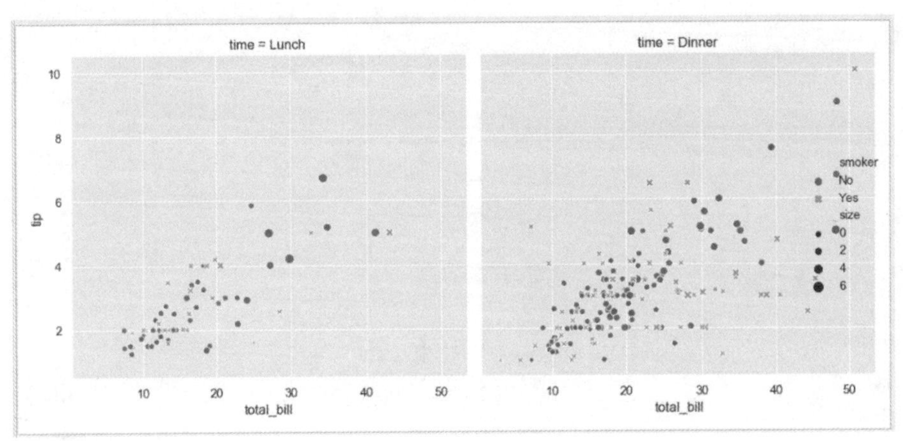

图 10-2　绘制的图

10.2.2　绘制指定样式的散点图

在 Seaborn 中使用内置函数 relplot() 绘制散点图，各个参数的具体说明如表 10-1 所示。

表 10-1

参数名称	说　明
x、*y*、data	x、y 轴，显示数据
hue	不同类别不同颜色
style	不同类别不同样式（*,+）
palette	自定义颜色（ch:r=-0.5,l=0.75）
size	点的大小对应的数值来决定
sizes	每个点的大小统一设置，例如：sizes=（500,500）
kind	line 是折线图
sort	False 禁用 x 在绘图之前按值对数据进行排序
ci	是 x 通过绘制平均值周围的平均值和 95%区间来聚合每个值的多个测量值，默认值是 None，表示不显示聚合

请看下面的实例文件 002.py，功能是绘制一个指定样式的散点图。设置 style 的属性值为“smoker”，这个值是一个默认样式。

```
import matplotlib.pyplot as plt
import seaborn as sns

# 准备数据：自带数据集
tips = sns.load_dataset("tips")
print(tips.head())

# 绘画散点图
sns.relplot(x="total_bill", y="tip", data=tips, hue="sex", style="smoker",
size="size")
sns.relplot(x="total_bill", y="tip", data=tips, hue="sex", style="smoker",
size="size", sizes=(100, 100))
# 显示
plt.show()
```

执行后的效果如图 10-3 所示。

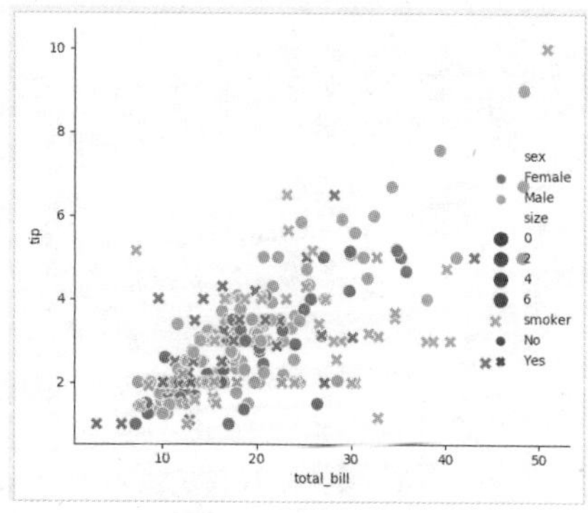

图 10-3　执行效果

10.2.3　绘制折线图

在 Seabor 中有两种绘制折线图的方法，一种是将函数 relplot() 的参数 kind 设置 line，另一种是直接使用函数 lineplot() 绘制折线图。其中使用函数 lineplot() 绘制折线图的方式比较灵活，更具针对性，所以建议大家尽量使用函数 lineplot() 绘制折线图 。

请看下面的实例文件 003.py，功能是使用函数 relplot() 绘制一个折线图。

```
import matplotlib.pyplot as plt
import seaborn as sns

# 数据集
data = sns.load_dataset("fmri")
print(data.head())
# 绘画折线图
sns.relplot(x="timepoint", y="signal", kind="line", data=data, ci=None)
# 显示
plt.show()
```

执行后的效果如图 10-4 所示。

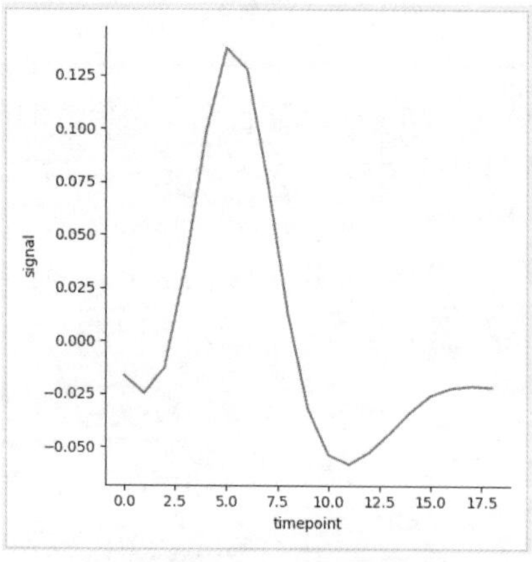

图 10-4　执行效果

请看下面的实例文件 004.py，功能是使用函数 lineplot() 绘制一个折线图。

```
import matplotlib.pyplot as plt
import seaborn as sns

# 数据集
```

```
    data = sns.load_dataset("fmri")
    print(data.head())
    # 绘画折线图:
    sns.lineplot(x="timepoint", y="signal", data=data, ci=95)
    # 显示
    plt.show()
```

执行后的效果如图 10-5 所示。

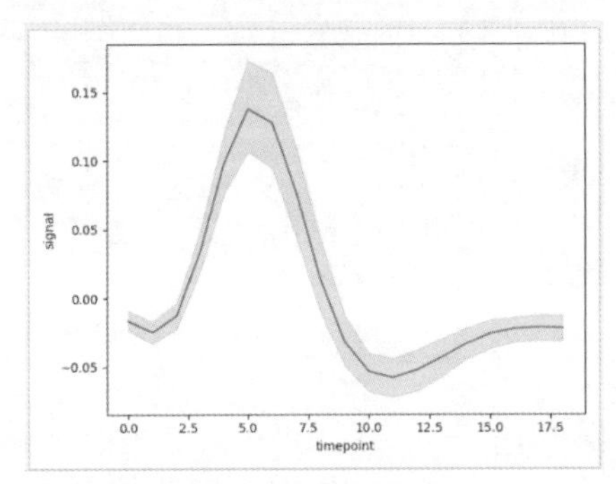

图 10-5　执行效果

请看下面的实例文件 005.py，功能是使用 sns.lineplot() 绘制了含有两个子图的折线图。在绘制第二个子图时，将 style 属性值设置为”event”。

```
    import matplotlib.pyplot as plt
    import seaborn as sns

    # 数据集
    data = sns.load_dataset("fmri")
    print(data.head())
    # 绘画折线图
    f, axes = plt.subplots(nrows=1, ncols=2, figsize=(14, 6))
    sns.lineplot(x="timepoint", y="signal", data=data, ci=95, ax=axes[0])
    sns.lineplot(x="timepoint", y="signal", hue="region", style="event", data=data, ci=None, ax=axes[1])
    plt.show()
```

执行后的效果如图 10-6 所示。

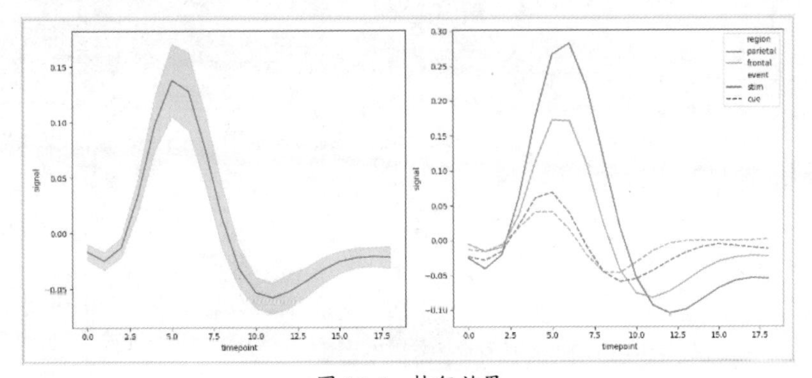

图 10-6　执行效果

10.2.4　绘制箱体图

在 Seabor 中使用 Boxplot 表示箱体图，这可能是最常见的图形类型之一。箱体图能够很好地表示数据的分布规律。箱体图方框的末尾显示了上下四分位数。极线显示最高和最低值，不包括异常值。在 Seaborn 中通常使用函数 boxplot() 箱体图。

请看下面的实例文件 006.py，功能是使用函数 boxplot() 绘制一个箱体图。

```
import matplotlib.pyplot as plt
import seaborn as sns
#调用 seaborn 自带数据集
df = sns.load_dataset('iris')
# 一个数值变量 One numerical variable only
# 如果您只有一个数字变量，则可以使用此代码获得仅包含一个组的箱线图。
# Make boxplot for one group only
# 显示花萼长度 sepal_length
sns.boxplot( y=df["sepal_length"] );

plt.show()
```

执行后的效果如图 10-7 所示。

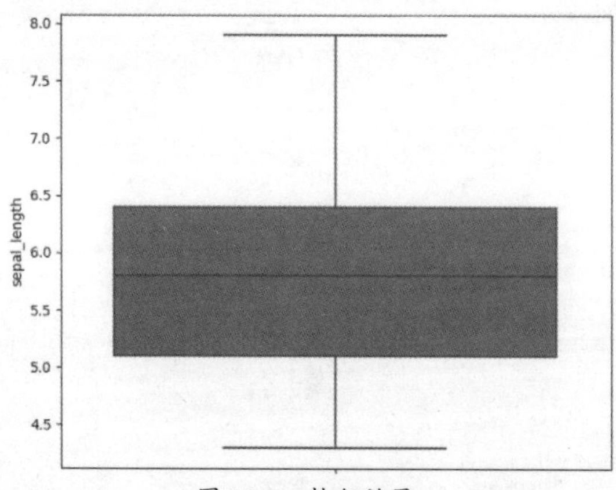

图 10-7　执行效果

请看下面的实例文件 007.py，功能是使用函数 boxplot() 绘制一个分组箱体图。

```
import matplotlib.pyplot as plt
import seaborn as sns
#调用 seaborn 自带数据集
df = sns.load_dataset('iris')
sns.boxplot( x=df["species"], y=df["sepal_length"] );

plt.show()
```

执行后的效果如图 10-8 所示。

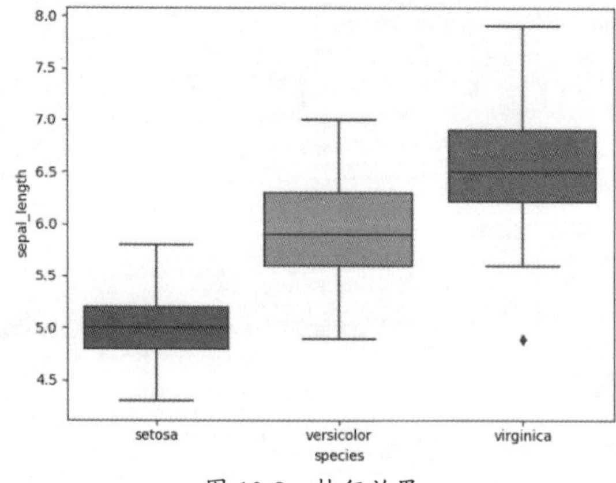

图 10-8　执行效果

请看下面的实例文件 008.py，功能是自定义设置绘制的箱体图的样式，通过属性 linewidth 设置线条宽度为 5。

```
import matplotlib.pyplot as plt
import seaborn as sns
# 调用 seaborn 自带数据集
df = sns.load_dataset('iris')
# 根据 linewidth 改变线条宽度
sns.boxplot( x=df["species"], y=df["sepal_length"], linewidth=5);
plt.show()
```

执行后的效果如图 10-9 所示。

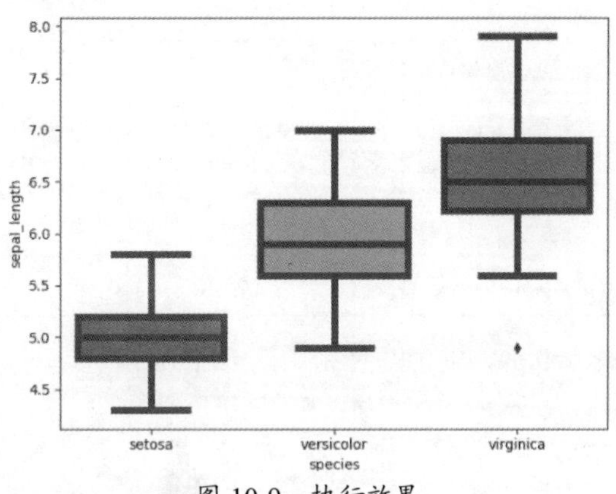

图 10-9　执行效果

10.2.5　绘制柱状图

柱状图又称为条形图，在 Seabor 中使用函数 barplot() 绘制柱状图，各个参数的具体说明如表 10-2 所示。

表 10-2

参数	说　明
x	指定 label 值，可以是一个序列
y	对应每个 label 上的数据，可以是一个序列
hue	指定分类变量
data	使用的数据集
order/hue_order	order 控制 bar 绘制的顺序，hue_order 控制一个 bra 内每个类的绘图顺序
estimator	设置每　个 label 上显示的统计量类型，默认为平均值，可修改为最大值、中位值等。注意，若修改为非平均值，那么误差线都需要做出修改，因为前面的误差线解释都是基于平均值的
ci	在 seaborn.barplot() 中误差线默认表示的是均值的置信区间，因此当 ci 为（0,100）间的值时表示置信区间的置信度，默认为 95；ci 还可以取值为 ‘sd’，此时误差线表示的是标准误差；当 ci 取值为 None 时，则不显示误差线
n_boot	计算代表置信区间的误差线时，默认会采用 bootstrap 抽样方法（在样本量较小时比较有用），该参数控制 bootstrap 抽样的次数
orient	设置柱状图水平绘制还是竖直绘制，“h”表示水平，“v”表示竖直

续表

参数	说　明
color	设置 bar 的颜色，这里似乎用于将所有的 bar 设置为同一种颜色
pattle	调色板，设置 bar 以不同颜色显示，所有的颜色选择都要是 matplotlib 能识别的颜色
saturation	设置颜色的饱和度取值为 [0,1] 之间
errcolor	设置误差线的颜色，默认为黑色
errwidth	设置误差线的显示线宽
capsize	设置误差线顶部、底端处横线的显示长度
dodge	当使用分类参数“hue”时，可以通过 dodge 参数设置将不同的类分别用一个 bar 表示，还是在一个 bar 上通过不同颜色表示，左边是 dodge=True，右边是 dodge=False，默认为 True
ax	选择将图形显示在哪个 Axes 对象上，默认为当前 Axes 对象
kwargs	matplotlib.plot.bar() 中其他的参数

请看下面的实例文件 009.py，功能是使用函数 barplot() 绘制垂直方向的柱状图。

```
import matplotlib.pyplot as plt
import seaborn as sns
sns.set(style="whitegrid")
tips = sns.load_dataset("tips")
ax = sns.barplot(x="day", y="total_bill", data=tips)
plt.show()
```

执行后的效果如图 10-10 所示。

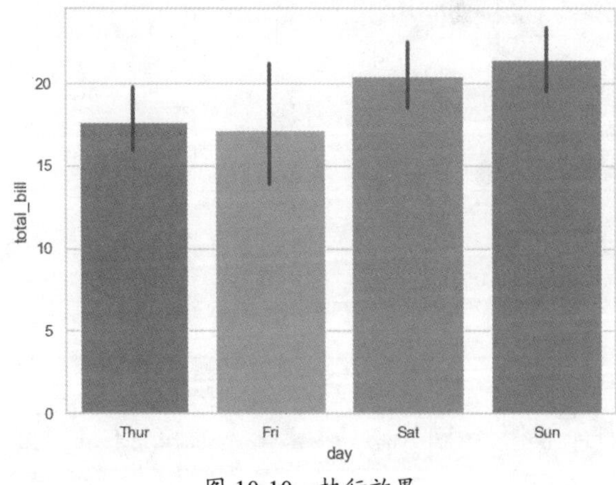

图 10-10　执行效果

请看下面的实例文件 010.py，功能是绘制一个带有图示功能的柱状图。

```
import matplotlib.pyplot as plt
import seaborn as sns
sns.set(style="whitegrid")
tips = sns.load_dataset("tips")
ax = sns.barplot(x="day", y="total_bill", hue="sex", data=tips)
plt.show()
```

执行后的效果如图 10-11 所示。

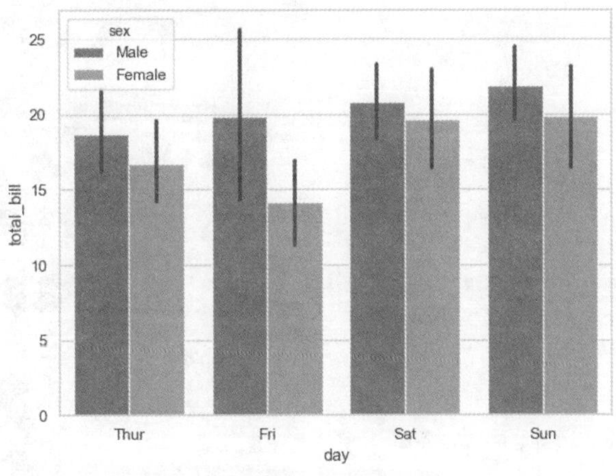

图 10-11　执行效果

请看下面的实例文件 011.py，功能是绘制一个横向显示的柱状图。

```
import matplotlib.pyplot as plt
import seaborn as sns
sns.set(style="whitegrid")
tips = sns.load_dataset("tips")
ax = sns.barplot(x="tip", y="day", data=tips)
plt.show()
```

执行后的效果如图 10-12 所示。

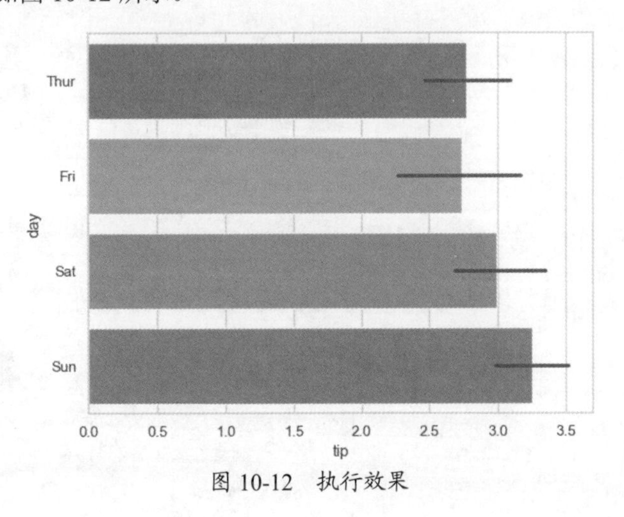

图 10-12　执行效果

10.2.6　设置显示中文

请看下面的实例文件 012.py，功能是使用 rcParams 设置在可视化绘图中显示中文的方法。

```
import numpy as np
import matplotlib.pyplot as plt
import seaborn as sns

plt.rcParams['font.sans-serif'] = ['SimHei']  # 中文字体设置 - 黑体
plt.rcParams['axes.unicode_minus'] = False  # 解决保存图像是负号 '-' 显示为方块的问题
sns.set(font='SimHei')  # 解决 Seaborn 中文显示问题

x = np.arange(-2*np.pi, 2*np.pi, 0.01)
```

```
    y1 = np.sin(x)
    y2 = np.cos(x)
    plt.figure(figsize=(10, 7))
    plt.plot(x, y1, label='$sinx$')
    plt.plot(x, y2, label='$cosx$')
    plt.legend(loc='upper right')
    plt.xlim(-2*np.pi-1, 2*np.pi+3)
    plt.xticks([-2*np.pi, -np.pi, 0, np.pi, 2*np.pi], ['$-2\pi$', '$-\pi$', '$0$',
'$\pi$', '$2\pi$'])
    plt.title('三角-函数')
    plt.xlabel('横坐标')
    plt.ylabel('纵坐标')
    plt.axhline(y=0, c='black')

    plt.show()
```

执行后的效果如图 10-13 所示。

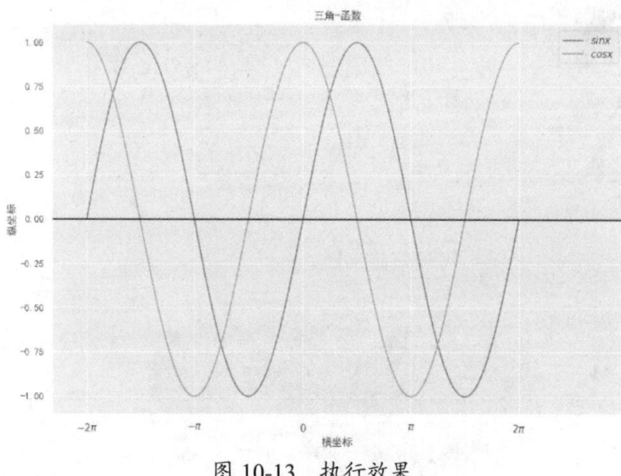

图 10-13　执行效果

10.3　实践案例：可视化分析实时疫情信息

在本节的内容中，将使用腾讯提供的 API 接口获取实时疫情信息，并使用 matplotlib 和 Seabor 绘制可视化统计图，帮助大家更直观地了解疫情信息。

↑扫码看视频（本节视频课程时间：18 分 45 秒）

10.3.1　列出统计的省和地区的名字

编写实例文件 test01-spider.py，功能是获取腾讯 API 接口中提供的 JSON 数据，使用 for 循环打印输出腾讯疫情平台中国省级行政区的信息。

```
    import time, json, requests
    # 抓取腾讯疫情实时 json 数据
    url = 'https://view.inews.qq.com/g2/getOnsInfo?name=disease_
h5&callback=&_=%d'%int(time.time()*1000)
    data = json.loads(requests.get(url=url).json()['data'])
    print(data)
```

```
    print(data.keys())

    # 统计省份信息(34个省份 湖北 广东 河南 浙江 湖南 安徽 ....)
    num = data['areaTree'][0]['children']
    print(len(num))
    for item in num:
        print(item['name'],end=" ")    # 不换行
    else:
        print("\n")                     # 换行
```

执行后会输出：

```
    34
    北京 香港 上海 四川 甘肃 河北 陕西 广东 辽宁 台湾 重庆 福建 浙江 澳门 天津 江苏 云南 湖南 海
南 吉林 江西 黑龙江 山西 河南 湖北 西藏 贵州 安徽 内蒙古 宁夏 山东 广西 新疆 青海
```

10.3.2 查询北京地区的实时数据

编写实例文件 test02-spider.py，功能是获取腾讯 API 接口中提供的 JSON 数据，使用 for 循环打印输出北京地区的疫情信息。

```
    import time, json, requests
    # 抓取腾讯疫情实时json数据
    url = 'https://view.inews.qq.com/g2/getOnsInfo?name=disease_
h5&callback=&_=%d'%int(time.time()*1000)
    data = json.loads(requests.get(url=url).json()['data'])
    print(data)
    print(data.keys())

    # 统计省份信息(34个省级行政区 湖北 广东 河南 浙江 湖南 安徽 ....)
    num = data['areaTree'][0]['children']
    print(len(num))
    for item in num:
        print(item['name'],end=" ")    # 不换行
    else:
        print("\n")                     # 换行

    # 显示北京数据
    hubei = num[0]['children']
    for data in hubei:
        print(data)
```

执行后输出结果如下：

```
    34
    北京 香港 上海 四川 甘肃 陕西 河北 广东 辽宁 台湾 重庆 福建 云南 江苏 浙江 澳门 天津 西藏 河
南 江西 山东 新疆 宁夏 贵州 湖南 黑龙江 山西 安徽 广西 湖北 吉林 青海 内蒙古 海南

    {'name': '丰台', 'today': {'confirm': 1, 'confirmCuts': 0, 'isUpdated': True},
'total': {'nowConfirm': 227, 'confirm': 270, 'suspect': 0, 'dead': 0, 'deadRate':
'0.00', 'showRate': False, 'heal': 43, 'healRate': '15.93', 'showHeal': False}}
    {'name': '大兴', 'today': {'confirm': 0, 'confirmCuts': 0, 'isUpdated': False},
'total': {'nowConfirm': 65, 'confirm': 104, 'suspect': 0, 'dead': 0, 'deadRate':
'0.00', 'showRate': False, 'heal': 39, 'healRate': '37.50', 'showHeal': False}}
    {'name': '海淀', 'today': {'confirm': 0, 'confirmCuts': 0, 'isUpdated': False},
'total': {'nowConfirm': 18, 'confirm': 82, 'suspect': 0, 'dead': 0, 'deadRate':
'0.00', 'showRate': False, 'heal': 64, 'healRate': '78.05', 'showHeal': False}}
    {'name': '西城', 'today': {'confirm': 0, 'confirmCuts': 0, 'isUpdated': False},
'total': {'nowConfirm': 6, 'confirm': 59, 'suspect': 0, 'dead': 0, 'deadRate': '0.00',
'showRate': False, 'heal': 53, 'healRate': '89.83', 'showHeal': False}}
    {'name': '东城', 'today': {'confirm': 0, 'confirmCuts': 0, 'isUpdated': False},
'total': {'nowConfirm': 5, 'confirm': 19, 'suspect': 0, 'dead': 0, 'deadRate': '0.00',
'showRate': False, 'heal': 14, 'healRate': '73.68', 'showHeal': False}}
```

```
    {'name': '房山', 'today': {'confirm': 0, 'confirmCuts': 0, 'isUpdated': False}, 'total': {'nowConfirm': 4, 'confirm': 20, 'suspect': 0, 'dead': 0, 'deadRate': '0.00', 'showRate': False, 'heal': 16, 'healRate': '80.00', 'showHeal': False}}
    {'name': '门头沟', 'today': {'confirm': 0, 'confirmCuts': 0, 'isUpdated': False}, 'total': {'nowConfirm': 2, 'confirm': 5, 'suspect': 0, 'dead': 0, 'deadRate': '0.00', 'showRate': False, 'heal': 3, 'healRate': '60.00', 'showHeal': False}}
    {'name': '朝阳', 'today': {'confirm': 0, 'confirmCuts': 0, 'isUpdated': False}, 'total': {'nowConfirm': 2, 'confirm': 77, 'suspect': 0, 'dead': 0, 'deadRate': '0.00', 'showRate': False, 'heal': 75, 'healRate': '97.40', 'showHeal': False}}
    {'name': '昌平', 'today': {'confirm': 0, 'confirmCuts': 0, 'isUpdated': False}, 'total': {'nowConfirm': 1, 'confirm': 30, 'suspect': 0, 'dead': 0, 'deadRate': '0.00', 'showRate': False, 'heal': 29, 'healRate': '96.67', 'showHeal': False}}
    {'name': '通州', 'today': {'confirm': 0, 'confirmCuts': 0, 'isUpdated': False}, 'total': {'nowConfirm': 1, 'confirm': 20, 'suspect': 0, 'dead': 9, 'deadRate': '45.00', 'showRate': False, 'heal': 10, 'healRate': '50.00', 'showHeal': False}}
    {'name': '石景山', 'today': {'confirm': 0, 'confirmCuts': 0, 'isUpdated': False}, 'total': {'nowConfirm': 1, 'confirm': 15, 'suspect': 0, 'dead': 0, 'deadRate': '0.00', 'showRate': False, 'heal': 14, 'healRate': '93.33', 'showHeal': False}}
    {'name': '延庆', 'today': {'confirm': 0, 'confirmCuts': 0, 'isUpdated': False}, 'total': {'nowConfirm': 0, 'confirm': 1, 'suspect': 0, 'dead': 0, 'deadRate': '0.00', 'showRate': False, 'heal': 1, 'healRate': '100.00', 'showHeal': False}}
    {'name': '境外输入', 'today': {'confirm': 0, 'confirmCuts': 0, 'isUpdated': False}, 'total': {'nowConfirm': 0, 'confirm': 174, 'suspect': 0, 'dead': 0, 'deadRate': '0.00', 'showRate': False, 'heal': 174, 'healRate': '100.00', 'showHeal': True}}
    {'name': '外地来京', 'today': {'confirm': 0, 'confirmCuts': 0, 'isUpdated': False}, 'total': {'nowConfirm': 0, 'confirm': 25, 'suspect': 0, 'dead': 0, 'deadRate': '0.00', 'showRate': False, 'heal': 25, 'healRate': '100.00', 'showHeal': False}}
    {'name': '密云', 'today': {'confirm': 0, 'confirmCuts': 0, 'isUpdated': False}, 'total': {'nowConfirm': 0, 'confirm': 7, 'suspect': 0, 'dead': 0, 'deadRate': '0.00', 'showRate': False, 'heal': 7, 'healRate': '100.00', 'showHeal': False}}
    {'name': '顺义', 'today': {'confirm': 0, 'confirmCuts': 0, 'isUpdated': False}, 'total': {'nowConfirm': 0, 'confirm': 10, 'suspect': 0, 'dead': 0, 'deadRate': '0.00', 'showRate': False, 'heal': 10, 'healRate': '100.00', 'showHeal': False}}
    {'name': '怀柔', 'today': {'confirm': 0, 'confirmCuts': 0, 'isUpdated': False}, 'total': {'nowConfirm': 0, 'confirm': 7, 'suspect': 0, 'dead': 0, 'deadRate': '0.00', 'showRate': False, 'heal': 7, 'healRate': '100.00', 'showHeal': False}}
    {'name': '地区待确认', 'today': {'confirm': 0, 'confirmCuts': 0, 'isUpdated': True}, 'total': {'nowConfirm': -9, 'confirm': 1, 'suspect': 0, 'dead': 0, 'deadRate': '0.00', 'showRate': False, 'heal': 10, 'healRate': '1000.00', 'showHeal': False}}
```

10.3.3 查询并显示国内各省行政区的实时数据

编写实例文件 test03-spider.py，功能是获取腾讯 API 接口中提供的 JSON 数据，使用 for 循环打印输出国内各省级行政区的实时数据。

```
import time, json, requests
# 抓取腾讯疫情实时 json 数据
url = 'https://view.inews.qq.com/g2/getOnsInfo?name=disease_h5&callback=&_=%d'%int(time.time()*1000)
data = json.loads(requests.get(url=url).json()['data'])
print(data)
print(data.keys())

# 统计省份信息(34 个省级行政区 湖北 广东 河南 浙江 湖南 安徽 ....)
num = data['areaTree'][0]['children']
print(len(num))
for item in num:
    print(item['name'],end=" ")    # 不换行
else:
    print("\n")                    # 换行
```

```
    # 解析数据（确诊 疑似 死亡 治愈）
    total_data = {}
    for item in num:
        if item['name'] not in total_data:
            total_data.update({item['name']:0})
        for city_data in item['children']:
            total_data[item['name']] +=int(city_data['total']['confirm'])
    print(total_data)
```

执行后会输出：

```
    34
    北京 香港 上海 四川 甘肃 陕西 河北 广东 辽宁 台湾 重庆 福建 云南 江苏 浙江 澳门 天津 西藏 河
南 江西 山东 新疆 宁夏 贵州 湖南 黑龙江 山西 安徽 广西 湖北 吉林 青海 内蒙古 海南

    {'北京': 926, '香港': 1247, '上海': 715, '四川': 595, '甘肃': 164, '陕西':
320, '河北': 349, '广东': 1643, '辽宁': 156, '台湾': 449, '重庆': 582, '福建': 363,
'云南': 186, '江苏': 654, '浙江': 1269, '澳门': 46, '天津': 198, '西藏': 1, '河南':
1276, '江西': 932, '山东': 792, '新疆': 76, '宁夏': 75, '贵州': 147, '湖南': 1019,
'黑龙江': 947, '山西': 198, '安徽': 991, '广西': 254, '湖北': 68135, '吉林': 155, '
青海': 18, '内蒙古': 238, '海南': 171}
```

10.3.4　绘制实时全国疫情确诊人数对比图

编写实例文件 test04-matplotlib.py，功能是获取腾讯 API 接口中提供的 JSON 数据，然后使用 matplotlib 绘制实时全国疫情确诊人数对比图。

```
    import time, json, requests
    # 抓取腾讯疫情实时 json 数据
    url = 'https://view.inews.qq.com/g2/getOnsInfo?name=disease_
h5&callback=&_=%d'%int(time.time()*1000)
    data = json.loads(requests.get(url=url).json()['data'])
    print(data)
    print(data.keys())

    # 统计省份信息(34 个省级行政区 湖北 广东 河南 浙江 湖南 安徽 ....)
    num = data['areaTree'][0]['children']
    print(len(num))
    for item in num:
        print(item['name'],end=" ")    # 不换行
    else:
        print("\n")                    # 换行

    # 显示湖北省数据
    hubei = num[0]['children']
    for item in hubei:
        print(item)
    else:
        print("\n")

    # 解析数据（确诊 疑似 死亡 治愈）
    total_data = {}
    for item in num:
        if item['name'] not in total_data:
            total_data.update({item['name']:0})
        for city_data in item['children']:
            total_data[item['name']] +=int(city_data['total']['confirm'])
    print(total_data)
    # {'湖北': 48206, '广东': 1241, '河南': 1169, '浙江': 1145, '湖南': 968, ...,
'澳门': 10, '西藏': 1}

    #-----------------------------------------------------------------------
    # 第二步：绘制柱状图
```

```
#----------------------------------------------------------------------------------
import matplotlib.pyplot as plt
import numpy as np

plt.rcParams['font.sans-serif'] = ['SimHei']  #用来正常显示中文标签
plt.rcParams['axes.unicode_minus'] = False    #用来正常显示负号

# 获取数据
names = total_data.keys()
nums = total_data.values()
print(names)
print(nums)

# 绘图
plt.figure(figsize=[10,6])
plt.bar(names, nums, width=0.3, color='green')

# 设置标题
plt.xlabel("地区", fontproperties='SimHei', size=12)
plt.ylabel("人数", fontproperties='SimHei', rotation=90, size=12)
plt.title("全国疫情确诊数对比图", fontproperties='SimHei', size=16)
plt.xticks(list(names), fontproperties='SimHei', rotation=-45, size=10)
# 显示数字
for a, b in zip(list(names), list(nums)):
    plt.text(a, b, b, ha='center', va='bottom', size=6)
plt.show()
```

执行效果如图 10-14 所示。

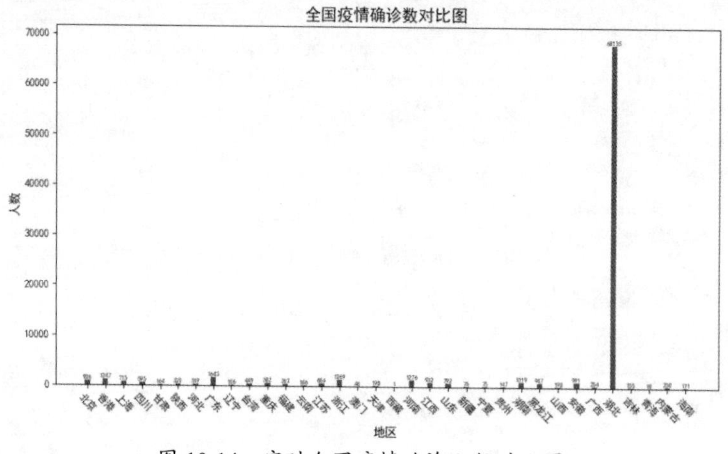

图 10-14　实时全国疫情确诊人数对比图

10.3.5　绘制实时确诊人数、新增确诊人数、死亡人数、治愈人数对比图

编写实例文件 test05-matplotlib.py，功能是获取腾讯 API 接口中提供的 JSON 数据，然后使用 matplotlib 绘制实时国内各地确诊人数、新增确诊人数、死亡人数、治愈人数对比图。

（1）首先抓取数据腾讯 API 的实时数据信息，加载显示获取的 JSON 数据，代码如下：

```
import time, json, requests
# 抓取腾讯疫情实时 json 数据
url = 'https://view.inews.qq.com/g2/getOnsInfo?name=disease_h5&callback=&_=%d'%int(time.time()*1000)
data = json.loads(requests.get(url=url).json()['data'])
print(data)
print(data.keys())
```

（2）统计国内各省级行政区的信息，代码如下：

```
num = data['areaTree'][0]['children']
print(len(num))
for item in num:
    print(item['name'],end=" ")    # 不换行
else:
    print("\n")                     # 换行
```

（3）提取国内各省级行政区的确诊人数数据，代码如下：

```
# 解析确诊数据
total_data = {}
for item in num:
    if item['name'] not in total_data:
        total_data.update({item['name']:0})
    for city_data in item['children']:
        total_data[item['name']] +=int(city_data['total']['confirm'])
print(total_data)
# {'湖北': 48206, '广东': 1241, '河南': 1169, '浙江': 1145, '湖南': 968, ...,
'澳门': 10, '西藏': 1}
```

（4）提取国内各省级行政区的疑似患者人数数据，代码如下：

```
# 解析疑似数据
total_suspect_data = {}
for item in num:
    if item['name'] not in total_suspect_data:
        total_suspect_data.update({item['name']:0})
    for city_data in item['children']:
        total_suspect_data[item['name']] +=int(city_data['total']['suspect'])
print(total_suspect_data)
```

（5）提取国内各省级行政区的死亡人数数据，代码如下：

```
# 解析死亡数据
total_dead_data = {}
for item in num:
    if item['name'] not in total_dead_data:
        total_dead_data.update({item['name']:0})
    for city_data in item['children']:
        total_dead_data[item['name']] +=int(city_data['total']['dead'])
print(total_dead_data)
```

（6）提取国内各省级行政区的治愈人数数据，代码如下：

```
# 解析治愈数据
total_heal_data = {}
for item in num:
    if item['name'] not in total_heal_data:
        total_heal_data.update({item['name']:0})
    for city_data in item['children']:
        total_heal_data[item['name']] +=int(city_data['total']['heal'])
print(total_heal_data)
```

（7）提取国内各省级行政区的新增确诊人数数据，代码如下：

```
# 解析新增确诊数据
total_new_data = {}
for item in num:
    if item['name'] not in total_new_data:
        total_new_data.update({item['name']:0})
    for city_data in item['children']:
            total_new_data[item['name']] +=int(city_data['today']['confirm']) #
today
print(total_new_data)
```

（8）使用 matplotlib 分别绘制实时确诊人数、新增确诊人数、死亡人数、治愈人数对比

图，代码如下：

```
#-------------------------------------------------------------------------------
# 第二步：绘制柱状图
#-------------------------------------------------------------------------------
import matplotlib.pyplot as plt
import numpy as np

plt.figure(figsize=[10,6])
plt.rcParams['font.sans-serif'] = ['SimHei']  #用来正常显示中文标签
plt.rcParams['axes.unicode_minus'] = False    #用来正常显示负号

#-----------------------------1.绘制确诊数据 ------------------------------------
p1 = plt.subplot(221)

# 获取数据
names = total_data.keys()
nums = total_data.values()
print(names)
print(nums)
print(total_data)
plt.bar(names, nums, width=0.3, color='green')

# 设置标题
plt.ylabel("确诊人数", rotation=90)
plt.xticks(list(names), rotation=-60, size=8)
# 显示数字
for a, b in zip(list(names), list(nums)):
    plt.text(a, b, b, ha='center', va='bottom', size=6)
plt.sca(p1)

#-----------------------------2.绘制新增确诊
数据-----------------------------------
p2 = plt.subplot(222)
names = total_new_data.keys()
nums = total_new_data.values()
print(names)
print(nums)
plt.bar(names, nums, width=0.3, color='yellow')
plt.ylabel("新增确诊人数", rotation=90)
plt.xticks(list(names), rotation=-60, size=8)
# 显示数字
for a, b in zip(list(names), list(nums)):
    plt.text(a, b, b, ha='center', va='bottom', size=6)
plt.sca(p2)

#-----------------------------3.绘制死亡数据 ------------------------------------
p3 = plt.subplot(223)
names = total_dead_data.keys()
nums = total_dead_data.values()
print(names)
print(nums)
plt.bar(names, nums, width=0.3, color='blue')
plt.xlabel("地区")
plt.ylabel("死亡人数", rotation=90)
plt.xticks(list(names), rotation=-60, size=8)
for a, b in zip(list(names), list(nums)):
    plt.text(a, b, b, ha='center', va='bottom', size=6)
plt.sca(p3)

#-----------------------------4.绘制治愈数据 ------------------------------------
p4 = plt.subplot(224)
names = total_heal_data.keys()
nums = total_heal_data.values()
```

```
    print(names)
    print(nums)
    plt.bar(names, nums, width=0.3, color='red')
    plt.xlabel("地区")
    plt.ylabel("治愈人数", rotation=90)
    plt.xticks(list(names), rotation=-60, size=8)
    for a, b in zip(list(names), list(nums)):
        plt.text(a, b, b, ha='center', va='bottom', size=6)
    plt.sca(p4)
    plt.show()
```

执行效果如图 10-15 所示。

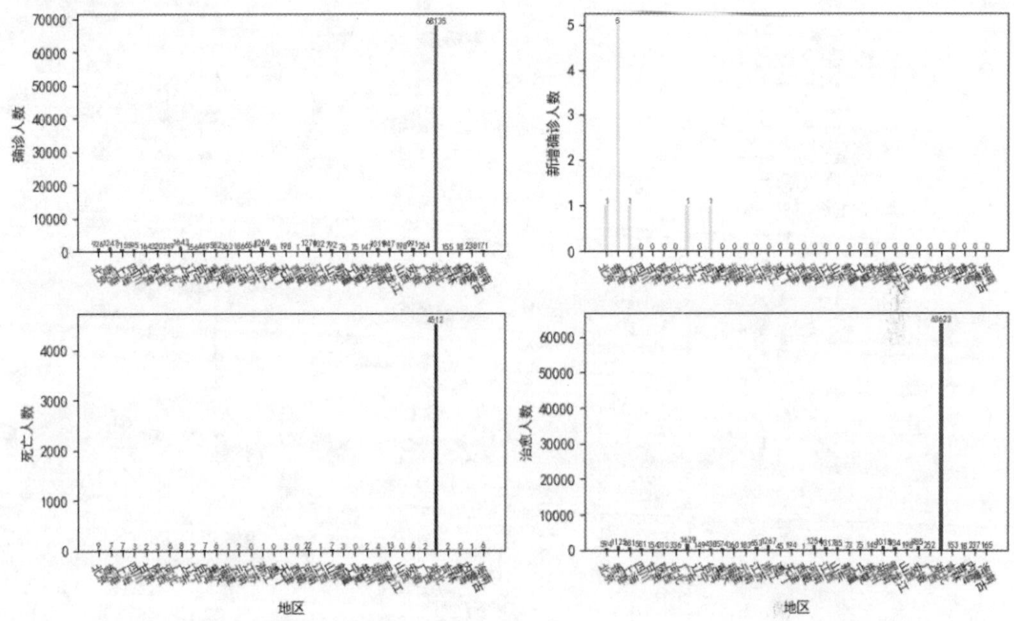

图 10-15　实时确诊人数、新增确诊人数、死亡人数、治愈人数对比图

10.3.6　将实时疫情数据保存到 CSV 文件

编写实例文件 test06-seaborn-write.py，功能是获取腾讯 API 接口中提供的 JSON 数据，然后将抓取的国内各地的实时疫情数据保存到 CSV 文件“2020-07-04-all.csv”中，其中文件名中的“2020-07-04-all.csv”不是固定的，和当前日期相对应。

```
    import time, json, requests
    # 抓取腾讯疫情实时 json 数据
    url  =  'https://view.inews.qq.com/g2/getOnsInfo?name=disease_
h5&callback=&_=%d'%int(time.time()*1000)
    data = json.loads(requests.get(url=url).json()['data'])
    print(data)
    print(data.keys())

    # 统计省级行政区信息 (34 个省级行政区 湖北 广东 河南 浙江 湖南 安徽 ....)
    num = data['areaTree'][0]['children']
    print(len(num))
    for item in num:
        print(item['name'],end=" ")    # 不换行
    else:
        print("\n")                     # 换行
```

```
    # 显示湖北省数据
    hubei = num[0]['children']
    for item in hubei:
        print(item)
    else:
        print("\n")

    # 解析确诊数据
    total_data = {}
    for item in num:
        if item['name'] not in total_data:
            total_data.update({item['name']:0})
        for city_data in item['children']:
            total_data[item['name']] +=int(city_data['total']['confirm'])
    print(total_data)
    # {'湖北': 48206, '广东': 1241, '河南': 1169, '浙江': 1145, '湖南': 968, ...,
'澳门': 10, '西藏': 1}

    # 解析疑似数据
    total_suspect_data = {}
    for item in num:
        if item['name'] not in total_suspect_data:
            total_suspect_data.update({item['name']:0})
        for city_data in item['children']:
            total_suspect_data[item['name']] +=int(city_data['total']['suspect'])
    print(total_suspect_data)

    # 解析死亡数据
    total_dead_data = {}
    for item in num:
        if item['name'] not in total_dead_data:
            total_dead_data.update({item['name']:0})
        for city_data in item['children']:
            total_dead_data[item['name']] +=int(city_data['total']['dead'])
    print(total_dead_data)

    # 解析治愈数据
    total_heal_data = {}
    for item in num:
        if item['name'] not in total_heal_data:
            total_heal_data.update({item['name']:0})
        for city_data in item['children']:
            total_heal_data[item['name']] +=int(city_data['total']['heal'])
    print(total_heal_data)

    # 解析新增确诊数据
    total_new_data = {}
    for item in num:
        if item['name'] not in total_new_data:
            total_new_data.update({item['name']:0})
        for city_data in item['children']:
              total_new_data[item['name']] +=int(city_data['today']['confirm']) #
today
    print(total_new_data)

    #-----------------------------------------------------------------------------
    # 第二步：存储数据至 CSV 文件
    #-----------------------------------------------------------------------------
    names = list(total_data.keys())           # 省份名称
    num1 = list(total_data.values())          # 确诊数据
    num2 = list(total_suspect_data.values())  # 疑似数据（全为 0）
    num3 = list(total_dead_data.values())     # 死亡数据
    num4 = list(total_heal_data.values())     # 治愈数据
```

```
    num5 = list(total_new_data.values())      # 新增确诊病例
    print(names)
    print(num1)
    print(num2)
    print(num3)
    print(num4)
    print(num5)

    # 获取当前日期命名 (2020-02-13-all.csv)
    n = time.strftime("%Y-%m-%d") + "-all.csv"
    fw = open(n, 'w', encoding='utf-8')
    fw.write('province,confirm,dead,heal,new_confirm\n')
    i = 0
    while i<len(names):
        fw.write(names[i]+','+str(num1[i])+','+str(num3[i])+','+str(num4[i])+','+str
(num5[i])+'\n')
        i = i + 1
    else:
        print("Over write file!")
        fw.close()
```

执行后会创建一个以当前日期命名的 CSV 文件，例如作者执行上述实例文件的时间是“2020-07-04”，所以会创建文件“2020-07-04-all.csv”，在文件里面保存了当前国内疫情的实时数据，如图 10-16 所示。

```
Plugins supporting *.csv files foun
1  province,confirm,dead
2  北京,926,9,594,1
3  香港,1247,7,1125,5
4  上海,715,7,681,1
5  四川,595,3,581,0
6  甘肃,164,2,154,0
7  陕西,320,3,310,0
8  河北,349,6,336,0
9  广东,1643,8,1629,1
10 辽宁,156,2,149,0
11 台湾,449,7,438,1
12 重庆,582,6,574,0
13 福建,363,1,360,0
14 云南,186,2,183,0
15 江苏,654,0,653,0
16 浙江,1269,1,1267,0
17 澳门,46,0,45,0
18 天津,198,3,194,0
19 西藏,1,0,1,0
20 河南,1276,22,1254,0
21 江西,932,1,931,0
22 山东,792,7,785,0
23 新疆,76,3,73,0
24 宁夏,75,0,75,0
25 贵州,147,2,145,0
26 湖南,1019,4,1015,0
27 黑龙江,947,13,934,0
28 山西,198,0,198,0
29 安徽,991,6,985,0
```

图 10-16　文件“2020 07 04-all.csv”

10.3.7　绘制国内实时疫情统计图

编写实例文件 test07-seaborn-write.py，功能提取刚才创建的 CSV 文件“2020-07-04-all.csv”中的数据，使用 Seaborn 绘制国内实时疫情统计图。

```
    import time
    import matplotlib
```

```
import numpy as np
import seaborn as sns
import pandas as pd
import matplotlib.pyplot as plt

# 读取数据
n = time.strftime("%Y-%m-%d") + "-all.csv"
data = pd.read_csv(n)

# 设置窗口
fig, ax = plt.subplots(1,1)
print(data['province'])

# 设置绘图风格及字体
sns.set_style("whitegrid",{'font.sans-serif':['simhei','Arial']})

# 绘制柱状图
g = sns.barplot(x="province", y="confirm", data=data, ax=ax,
            palette=sns.color_palette("hls", 8))

# 在柱状图上显示数字
i = 0
for index, b in zip(list(data['province']), list(data['confirm'])):
    g.text(i+0.05, b+0.05, b, color="black", ha="center", va='bottom', size=6)
    i = i + 1

# 设置 Axes 的标题
ax.set_title('全国疫情最新情况')

# 设置坐标轴文字方向
ax.set_xticklabels(ax.get_xticklabels(), rotation=-60)

# 设置坐标轴刻度的字体大小
ax.tick_params(axis='x',labelsize=8)
ax.tick_params(axis='y',labelsize=8)

plt.show()
```

执行效果如图 10-17 所示。

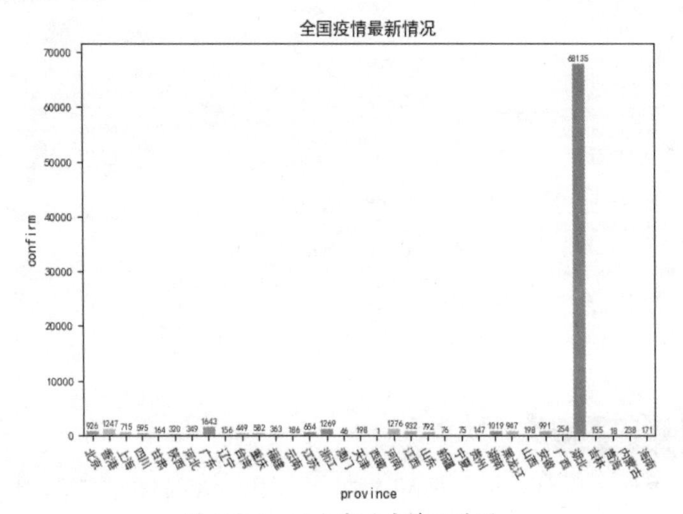

图 10-17 国内实时疫情统计图

10.3.8　可视化实时疫情的详细数据

编写实例文件 test08-seaborn-write-4db.py，功能是获取腾讯 API 接口中提供的 JSON 数据，然后将抓取的国内各地的实时疫情的详细数据保存到 CSV 文件"2020-07-04-all-4db.csv"中。其中文件名中不是固定的，和当前日期相对应。详细数据包括确诊人数、治愈人数、死亡人数和新增确诊人数，最后使用 Seaborn 绘制实时疫情详细数据的统计图。

（1）抓取腾讯网 API 中的 JSON 数据，代码如下：

```
import time, json, requests
# 抓取腾讯疫情实时 json 数据
url = 'https://view.inews.qq.com/g2/getOnsInfo?name=disease_h5&callback=&_=%d'%int(time.time()*1000)
data = json.loads(requests.get(url=url).json()['data'])
print(data)
print(data.keys())
```

（2）使用 Seaborn 绘制各地确诊人数、治愈人数、死亡人数和新增确诊人数信的统计图，代码如下：

```
import time
import matplotlib
import numpy as np
import seaborn as sns
import pandas as pd
import matplotlib.pyplot as plt

# 读取数据
n = time.strftime("%Y-%m-%d") + "-all-4db.csv"
data = pd.read_csv(n)

# 设置窗口
fig, ax = plt.subplots(1,1)
print(data['province'])

# 设置绘图风格及字体
sns.set_style("whitegrid",{'font.sans-serif':['simhei','Arial']})

# 绘制柱状图
g = sns.barplot(x="province", y="data", hue="tpye", data=data, ax=ax,
            palette=sns.color_palette("hls", 8))

# 设置 Axes 的标题
ax.set_title(' 全国疫情最新情况 ')

# 设置坐标轴文字方向
ax.set_xticklabels(ax.get_xticklabels(), rotation=-60)

# 设置坐标轴刻度的字体大小
ax.tick_params(axis='x',labelsize=8)
ax.tick_params(axis='y',labelsize=8)

plt.show()
```

执行后会创建一个以当前日期命名的 CSV 文件，例如作者执行上述实例文件的时间是"2020-07-04"，所以会创建文件"2020-07-04-all-4db.csv"，在文件里面保存了当前国内疫情的实时详细数据。如图 10-18 所示，并且还绘制了各地确诊人数、治愈人数、死亡人数和新增确诊人数信的统计图，如图 10-19 所示。

```
1  province,tpye,data
2  北京,confirm,926
3  北京,dead,9
4  北京,heal,594
5  北京,new_confirm,1
6  香港,confirm,1247
7  香港,dead,7
8  香港,heal,1125
9  香港,new_confirm,5
10 上海,confirm,715
11 上海,dead,7
12 上海,heal,681
13 上海,new_confirm,1
14 四川,confirm,595
15 四川,dead,3
16 四川,heal,581
17 四川,new_confirm,0
18 甘肃,confirm,164
19 甘肃,dead,2
20 甘肃,heal,154
21 甘肃,new_confirm,0
```

图 10-18　文件“2020-07-04-all-4db.csv”

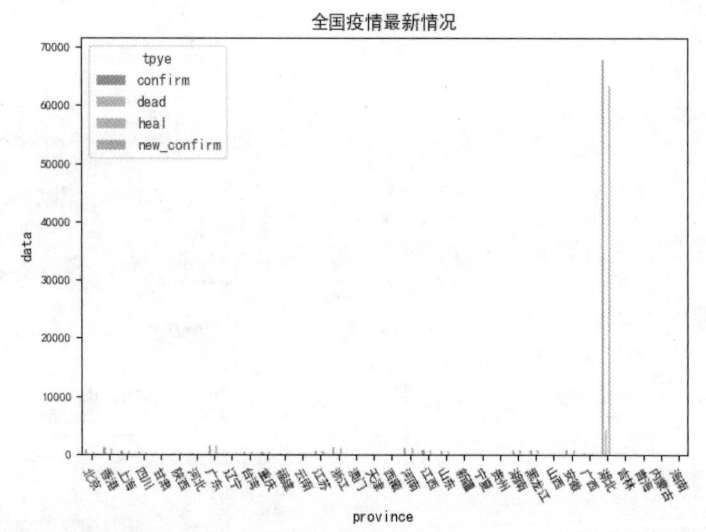

图 10-19　各地确诊人数、治愈人数、死亡人数和新增确诊人数信的统计图

10.3.9　绘制实时疫情信息统计图

编写实例文件 test09-seaborn-write.py，功能是根据 CSV 文件“2020-07-04-all-4db.csv”中的数据，使用 Seaborn 绘制国内各地实时疫情信息统计图。

```
# 读取数据
data = pd.read_csv("2020-07-04-all-4db.csv")

# 设置窗口
fig, ax = plt.subplots(1,1)
print(data['province'])

# 设置绘图风格及字体
sns.set_style("whitegrid",{'font.sans-serif':['simhei','Arial']})

# 绘制柱状图
```

```
    g = sns.barplot(x="province", y="data", hue="tpye", data=data, ax=ax,
                palette=sns.color_palette("hls", 8))

    # 设置 Axes 的标题
    ax.set_title('全国疫情最新情况')

    # 设置坐标轴文字方向
    ax.set_xticklabels(ax.get_xticklabels())

    # 设置坐标轴刻度的字体大小
    ax.tick_params(axis='x',labelsize=8)
    ax.tick_params(axis='y',labelsize=8)

    plt.show()
```

执行效果如图 10-20 所示。

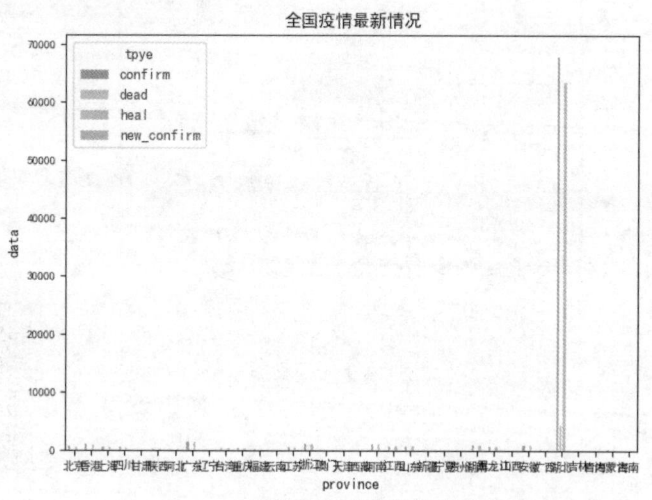

图 10-20　国内各地实时疫情信息统计图

10.3.10　绘制本年度国内疫情曲线图

编写实例文件 test10-qushi.py，功能是获取腾讯 API 接口中提供的 JSON 数据，然后使用 matplotlib 绘制本年度国内疫情曲线图，时间在 1 月开始，到当前时间结束，最后将绘制的图保存为图片文“nCoV 疫情曲线 .png”。

```
    # 抓取腾讯疫情实时 json 数据
    def catch_daily():
         url = 'https://view.inews.qq.com/g2/getOnsInfo?name=wuwei_ww_cn_day_
counts&callback=&_=%d'%int(time.time()*1000)
        data = json.loads(requests.get(url=url).json()['data'])
        data.sort(key=lambda x:x['date'])

        date_list = list() # 日期
        confirm_list = list() # 确诊
        suspect_list = list() # 疑似
        dead_list = list() # 死亡
        heal_list = list() # 治愈
        for item in data:
            month, day = item['date'].split('/')
                date_list.append(datetime.strptime('2020-%s-%s'%(month, day),
'%Y-%m-%d'))
```

```
        confirm_list.append(int(item['confirm']))
        suspect_list.append(int(item['suspect']))
        dead_list.append(int(item['dead']))
        heal_list.append(int(item['heal']))
    return date_list, confirm_list, suspect_list, dead_list, heal_list

# 绘制每日确诊和死亡数据
def plot_daily():


     date_list, confirm_list, suspect_list, dead_list, heal_list = catch_daily()
# 获取数据

    plt.figure('疫情统计图表', facecolor='#f4f4f4', figsize=(10, 8))
    plt.title('nCoV疫情曲线', fontsize=20)

    plt.rcParams['font.sans-serif'] = ['SimHei']  #用来正常显示中文标签
    plt.rcParams['axes.unicode_minus'] = False    #用来正常显示负号

    plt.plot(date_list, confirm_list, 'r-', label='确诊')
    plt.plot(date_list, confirm_list, 'rs')
    plt.plot(date_list, suspect_list, 'b-',label='疑似')
    plt.plot(date_list, suspect_list, 'b*')
    plt.plot(date_list, dead_list, 'y-', label='死亡')
    plt.plot(date_list, dead_list, 'y+')
    plt.plot(date_list, heal_list, 'g-', label='治愈')
    plt.plot(date_list, heal_list, 'gd')

    plt.gca().xaxis.set_major_formatter(mdates.DateFormatter('%m-%d')) # 格式化时
间轴标注
    plt.gcf().autofmt_xdate() # 优化标注（自动倾斜）
    plt.grid(linestyle=':') # 显示网格
    plt.legend(loc='best') # 显示图例
    plt.savefig('nCoV疫情曲线.png') # 保存为文件
    plt.show()

if __name__ == '__main__':
    plot_daily()
```

执行效果如图 10-21 所示。

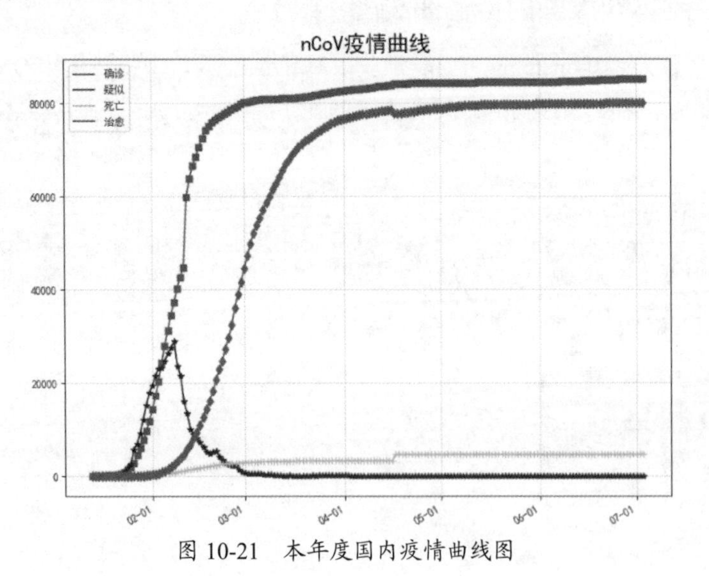

图 10-21　本年度国内疫情曲线图

10.3.11　绘制山东省实时疫情数据统计图

编写实例文件 test14-spider-shandong.py，功能是获取腾讯 API 接口中提供的 JSON 数据，然后将抓取的山东省各地的实时疫情的详细数据保存到 CSV 文件“2020-07-04-all-4db.csv”中。其中文件名中不是固定的，和当前日期相对应。详细数据包括确诊人数、治愈人数、死亡人数和新增确诊人数，最后使用 Seaborn 绘制山东省实时疫情详细数据的统计图。

（1）抓取腾讯网 API 中的 JSON 数据，代码如下：

```
import time, json, requests
# 抓取腾讯疫情实时 json 数据
url = 'https://view.inews.qq.com/g2/getOnsInfo?name=disease_
h5&callback=&_=%d'%int(time.time()*1000)
data = json.loads(requests.get(url=url).json()['data'])
print(data)
print(data.keys())
```

（2）提取 JSON 数据中心省份为“山东”的数据，代码如下：

```
# 统计省份信息(34 个省份 湖北 广东 河南 浙江 湖南 安徽 ....)
num = data['areaTree'][0]['children']
print(len(num))

# 获取山东下标
k = 0
for item in num:
    print(item['name'],end=" ")   # 不换行
    if item['name'] in "山东":
        print("")
        print(item['name'], k)
        break
    k = k + 1
print("") # 换行

# 显示山东省数据
gz = num[k]['children']
for item in gz:
    print(item)
else:
    print("\n")
```

（3）分别获取山东省各地的确诊数据、疑似数据、死亡数据、治愈数据和新增确诊数据，代码如下：

```
total_data = {}
for item in gz:
    if item['name'] not in total_data:
        total_data.update({item['name']:0})
    total_data[item['name']] = item['total']['confirm']
print('确诊人数')
print(total_data)


# 解析疑似数据
total_suspect_data = {}
for item in gz:
    if item['name'] not in total_suspect_data:
        total_suspect_data.update({item['name']:0})
    total_suspect_data[item['name']] = item['total']['suspect']
print('疑似人数')
print(total_suspect_data)
```

```
# 解析死亡数据
total_dead_data = {}
for item in gz:
    if item['name'] not in total_dead_data:
        total_dead_data.update({item['name']:0})
    total_dead_data[item['name']] = item['total']['dead']
print('死亡人数')
print(total_dead_data)

# 解析治愈数据
total_heal_data = {}
for item in gz:
    if item['name'] not in total_heal_data:
        total_heal_data.update({item['name']:0})
    total_heal_data[item['name']] = item['total']['heal']
print('治愈人数')
print(total_heal_data)

# 解析新增确诊数据
total_new_data = {}
for item in gz:
    if item['name'] not in total_new_data:
        total_new_data.update({item['name']:0})
    total_new_data[item['name']] = item['today']['confirm'] # today
print('新增确诊人数')
print(total_new_data)

#--------------------------------------------------------------------------------
```

（4）将山东省各地的实时疫情数据信息保存到 CSV 文件中，代码如下：

```
names = list(total_data.keys())              # 省份名称
num1 = list(total_data.values())             # 确诊数据
num2 = list(total_suspect_data.values()) # 疑似数据（全为 0)
num3 = list(total_dead_data.values())        # 死亡数据
num4 = list(total_heal_data.values())        # 治愈数据
num5 = list(total_new_data.values())         # 新增确诊病例
print(names)
print(num1)
print(num2)
print(num3)
print(num4)
print(num5)

# 获取当前日期命名(2020-02-13-gz.csv)
n = time.strftime("%Y-%m-%d") + "-sd-4db.csv"
fw = open(n, 'w', encoding='utf-8')
fw.write('province,type,data\n')
i = 0
while i<len(names):
    fw.write(names[i]+',confirm,'+str(num1[i])+'\n')
    fw.write(names[i]+',dead,'+str(num3[i])+'\n')
    fw.write(names[i]+',heal,'+str(num4[i])+'\n')
    fw.write(names[i]+',new_confirm,'+str(num5[i])+'\n')
    i = i + 1
else:
    print("Over write file!")
    fw.close()
```

（5）调用 Seaborn 绘制山东省实时疫情数据统计图，代码如下：

```
import time
import matplotlib
import numpy as np
```

```
import seaborn as sns
import pandas as pd
import matplotlib.pyplot as plt

# 读取数据
n = time.strftime("%Y-%m-%d") + "-sd-4db.csv"
data = pd.read_csv(n)

# 设置窗口
fig, ax = plt.subplots(1,1)
print(data['province'])

# 设置绘图风格及字体
sns.set_style("whitegrid",{'font.sans-serif':['simhei','Arial']})

# 绘制柱状图
g = sns.barplot(x="province", y="data", hue="type", data=data, ax=ax,
            palette=sns.color_palette("hls", 8))

# 设置 Axes 的标题
ax.set_title(' 山东疫情最新情况 ')

# 设置坐标轴文字方向
ax.set_xticklabels(ax.get_xticklabels(), rotation=-60)

# 设置坐标轴刻度的字体大小
ax.tick_params(axis='x',labelsize=8)
ax.tick_params(axis='y',labelsize=8)

plt.show()
```

执行后会创建一个以当前日期命名的 CSV 文件，例如作者执行上述实例文件的时间是“2020-07-04”，所以会创建文件“2020-07-04-all-4db.csv”，在文件里面保存了当前山东省各地疫情的实时数据。如图 10-22 所示，并且还绘制了山东省各地确诊人数、治愈人数、死亡人数和新增确诊人数信的统计图，如图 10-23 所示。

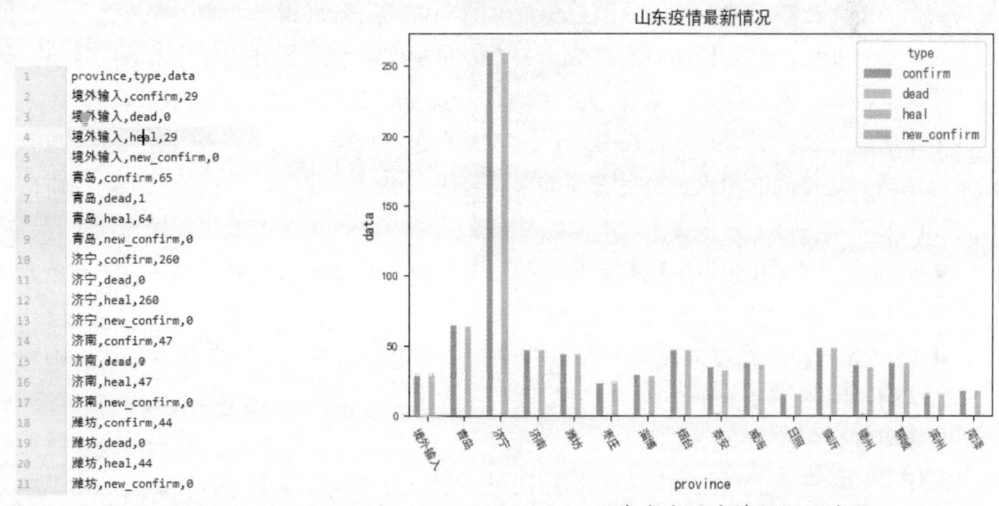

```
1	province,type,data
2	境外输入,confirm,29
3	境外输入,dead,0
4	境外输入,heal,29
5	境外输入,new_confirm,0
6	青岛,confirm,65
7	青岛,dead,1
8	青岛,heal,64
9	青岛,new_confirm,0
10	济宁,confirm,260
11	济宁,dead,0
12	济宁,heal,260
13	济宁,new_confirm,0
14	济南,confirm,47
15	济南,dead,0
16	济南,heal,47
17	济南,new_confirm,0
18	潍坊,confirm,44
19	潍坊,dead,0
20	潍坊,heal,44
21	潍坊,new_confirm,0
```


图 10-22　文件“2020-07-04-all-4db.csv”　　　图 10-23　山东省实时疫情数据统计图

第 11 章 综合实战：招聘信息可视化

在众多招纳人才的途径中，招聘网站成为最重要的渠道之一。无论对于用人单位还是应聘者，数据分析招聘网站的招聘信息是十分重要的。在本章的内容中，将通过一个综合实例的实现过程，详细讲解爬虫抓取某知名招聘网站中招聘信息的方法，并讲解可视化分析招聘信息的过程。

11.1 系统背景介绍

在当今的社会环境下，招聘已然成为人力资源管理的热点，猎头公司、招聘网站、人才测评等配套服务机构应运而生，其核心在于为企业提供人才信息渠道。这些专业机构为企业提供专业服务，猎头公司、招聘网站解决的是“符合企业要求的人才在哪里”的问题，人才测评公司解决的是“这个人到底有何素质、适合做什么”的问题。随着企业用人需求的弹性化和动态变化，需要企业内部专业经理们与外部专业机构解决“企业到底需要什么样的人”这一问题。

↑扫码看视频（本节视频课程时间：3 分 31 秒）

随着时代的发展，很多公司在招聘时都会收到成千上万的简历，如何挑选合适的应聘者成为公司 HR 比较棘手的事情，这给招聘单位的人事部门带来相当大的工作压力。与其他传统的人才中介相比，网上招聘具有低成本、大容量、速度快和强调个性化服务的优势。伴随着新增岗位源源不断地涌现，即使广泛存在的岗位对人的要求也变得模糊起来。对于用人单位的 HR 来说，及时了解招聘行情是自己最基本的业务范畴。而对于应聘者来说，根据自己的情况选择合适待遇的用人单位是首要的应聘目的。在这个时候，将招聘网中的招聘信息进行可视化处理就变得十分重要了。下面以某开发公司招聘 Python 开发工程师为例进行说明：

- 开发公司的 HR 可以可视化分析招聘网中的和 Python 相关的招聘信息，了解不同学历和不同工作经验对应的薪资水平。
- Python 应聘者通过可视化分析招聘网中的和 Python 相关的招聘信息，了解不同学历和不同工作经验对应的薪资水平。

目前可视化招聘信息已经成为人力资源经理关注的焦点，在特定的发展阶段、特定的文化背景下，面对变动的市场环境和弹性的岗位要求，企业到底需要什么样的人，为不同层次的人才提供什么样的待遇是他们格外关注的问题。基于目前招聘信息可视化需求分析的重要性，很多专业招聘网和猎头机构热衷于用可视化图表展示人才的供求状况，可视化分析招聘信息大有可为。

11.2　系统架构分析

本项目将首先使用爬虫抓取某知名招聘网站中的招聘信息，然后可视化分析抓取到招聘信息，为用人单位 HR 和应聘者提供强有力的数据支撑。本项目的功能模块如图 11-1 所示。

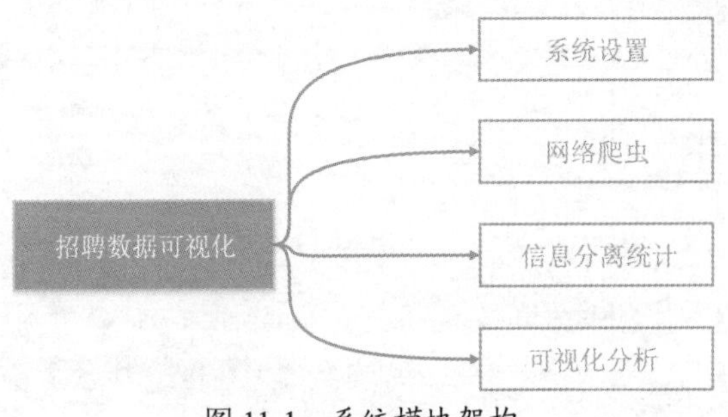

图 11-1　系统模块架构

在图 11-1 中各个模块的具体说明如下所示。

（1）系统设置

设置使用 MySQL 数据库保存用爬虫抓取到的数据，然后用 Flask Web 框架提取数据库中的数据，用网页的形式可视化展示招聘信息。

（2）网络爬虫

根据用户输入的关键字，使用网络爬虫技术抓取招聘网中的招聘信息，将抓取到的招聘信息添加到 MySQL 数据库中。

（3）信息分离统计

提取在 MySQL 数据库中保存的爬虫数据，分别根据“工作地区”“工作经验”“薪资水平”和“学历水平”提取并分离招聘信息。

（4）可视化分析

提取在 MySQL 数据库中保存的爬虫数据，然后使用开源框架 Highcharts 绘制柱状图和饼状图，可视化展示招聘数据信息。

11.3　系统设置

本项目使用 MySQL 数据库保存用爬虫抓取到的数据，编写程序文件 config.py，设置连接 MySQL 数据库的参数，参数说明如表 11-1 所示。

表 11-1

参数名称	说明
HOST	数据库服务器的地址，本地服务器的 IP 默认是 '127.0.0.1'
PORT	端口号，默认为 3306
USERNAME	MySQL 数据库的用户名
PASSWORD	MySQL 数据库的密码
DATABASE	MySQL 数据库的名字
mysql+pymysql	连接 MySQL 数据库的语句，在语句中调用上面的参数

具体实现代码如下所示：

```
    HOST = '127.0.0.1'
    PORT = '3306'
    USERNAME = 'root'
    PASSWORD = '66688888'
    DATABASE = 'u1'
    DB_URI  =  'mysql+pymysql://{username}:{password}@{host}:{port}/{db}?charset=utf8mb4'.format(username=USERNAME, password=PASSWORD, host=HOST, port=PORT, db=DATABASE)
```

11.4 网络爬虫

在本项目网络爬虫模块的实现文件是 data.py，功能是根据输入的关键字抓取招聘网站中的招聘信息，将抓取到的招聘信息添加到 MySQL 数据库中。

↑扫码看视频（本节视频课程时间：5 分 28 秒）

在本实例中，首先建立和 MySQL 数据库的连接，然后编写函数 def get_url(url) 和 def get_data(urls) 分别抓取目标 URL 中的每个招聘信息中的职位名称、公司、城市地区、经验、发布日期、学历、薪资、招聘详情等信息；最后将抓取到的信息保存到 MySQL 数据库中。

11.4.1 建立和数据库的连接

因为需要将爬取的数据添加到 MySQL 数据库中，所以需要导入在前面配置文件 config.py 中设置的数据库连接参数。对应的实现代码如下所示：

```
    host = config.HOST
    post = config.PORT
    username = config.USERNAME
    password = config.PASSWORD
    database = config.DATABASE

    etree = html.etree
    tlock=threading.Lock()


    # 拿到游标
    cursor = db.cursor()
```

11.4.2 设置 HTTP 请求头 User-Agent

User-Agent 会告诉网站服务器，访问者是通过什么工具来请求的。如果是爬虫请求，一般会拒绝，如果是用户浏览器就会应答。通过设置 HTTP 请求头 User-Agent，可以确保能够爬取到数据。对应的实现代码如下所示：

```
    user_agent = [

        # Firefox
        "Mozilla/5.0 (Windows NT 6.1; WOW64; rv:34.0) Gecko/20100101 Firefox/34.0",
         "Mozilla/5.0 (X11; U; Linux x86_64; zh-CN; rv:1.10.2.10) Gecko/20100922 Ubuntu/10.10 (maverick) Firefox/3.6.10",
        # Safari
```

```
        "Mozilla/5.0 (Windows NT 6.1; WOW64) AppleWebKit/534.57.2 (KHTML, like
Gecko) Version/5.1.7 Safari/534.57.2",
        # chrome
        "Mozilla/5.0 (Windows NT 6.1; WOW64) AppleWebKit/537.36 (KHTML, like Gecko)
Chrome/310.0.2171.71 Safari/537.36",
        "Mozilla/5.0 (X11; Linux x86_64) AppleWebKit/537.11 (KHTML, like Gecko)
Chrome/23.0.1271.64 Safari/537.11",
        "Mozilla/5.0 (Windows; U; Windows NT 6.1; en-US) AppleWebKit/534.16 (KHTML,
like Gecko) Chrome/10.0.648.133 Safari/534.16",
        # 360
        "Mozilla/5.0 (Windows NT 6.1; WOW64) AppleWebKit/537.36 (KHTML, like Gecko)
Chrome/30.0.15910.101 Safari/537.36",
        "Mozilla/5.0 (Windows NT 6.1; WOW64; Trident/7.0; rv:11.0) like Gecko",
        # 猎豹浏览器
        "Mozilla/5.0 (Windows NT 6.1; WOW64) AppleWebKit/537.1 (KHTML, like Gecko)
Chrome/21.0.1180.71 Safari/537.1 LBBROWSER",
        "Mozilla/5.0 (compatible; MSIE 10.0; Windows NT 6.1; WOW64; Trident/5.0;
SLCC2; .NET CLR 2.0.50727; .NET CLR 3.5.30729; .NET CLR 3.0.30729; Media Center PC
6.0; .NET4.0C; .NET4.0E; LBBROWSER)",
        "Mozilla/4.0 (compatible; MSIE 6.0; Windows NT 5.1; SV1; QQDownload 732;
.NET4.0C; .NET4.0E; LBBROWSER)",
        # QQ浏览器
        "Mozilla/5.0 (compatible; MSIE 10.0; Windows NT 6.1; WOW64; Trident/5.0;
SLCC2; .NET CLR 2.0.50727; .NET CLR 3.5.30729; .NET CLR 3.0.30729; Media Center PC
6.0; .NET4.0C; .NET4.0E; QQBrowser/7.0.3698.400)",
        "Mozilla/4.0 (compatible; MSIE 6.0; Windows NT 5.1; SV1; QQDownload 732;
.NET4.0C; .NET4.0E)"
    ]

    def get_user_agent():
        """随机获取一个请求头"""
        return {'User-Agent': random.choice(user_agent)}


    def requst(url):
        """requests到url的HTML"""
        html = requests.get(url,headers=get_user_agent())
        html.encoding = 'gbk'
        return etree.HTML(html.text)
```

11.4.3 抓取信息

分别编写函数 def get_url(url) 和 def get_data(urls)，根据设置的要抓取的 URL 地址，然后分别抓取目标 URL 中的每个招聘信息中的职位名称、公司、城市地区、经验、发布日期、学历、薪资、招聘详情等信息。对应的实现代码如下所示：

```
    def get_url(url):
        data = requst(url)
        href = data.xpath('//*[@id="resultList"]/div/p/span/a/@href')
        print(len(href))
        return href

    def get_data(urls):
        """获取xxjob职位信息，并存入数据库"""
        list_all = []
        for url in urls:
            regjob = re.compile(r'https://(.*?)51job.com', re.S)
            it = re.findall(regjob, url)
            if it != ['jobs.']:
                print('不匹配')
                continue
            try:
```

```
            data = requst(url)
            # 职位名称
              titles = data.xpath('/html/body/div[3]/div[2]/div[2]/div/div[1]/
h1/@title')[0]
            # 公司
             company = data.xpath('/html/body/div[3]/div[2]/div[2]/div/div[1]/
p[1]/a[1]/@title')[0]

              ltype = data.xpath('/html/body/div[3]/div[2]/div[2]/div/div[1]/
p[2]/@title')[0]
            ltype_str = "".join(ltype.split())
            # print(ltype_str)
            ltype_list = ltype_str.split('|')
            # 城市地区
            addres = ltype_list[0]
            # 经验
            exper = ltype_list[1]
            # 发布日期
            if len(ltype_list)>=5:
                # 学历
                edu = ltype_list[2]
                dateT = ltype_list[4]
            else:
                # 学历
                edu = "没有要求"
                dateT = ltype_list[-1]
            # 薪资
              salary = data.xpath('/html/body/div[3]/div[2]/div[2]/div/div[1]/
strong/text()')
            if len(salary) == 0:
                salary_list = [0,0]
            else:
                salary_list = salary_alter(salary)[0]
            # 招聘详情
             contents = data.xpath('/html/body/div[3]/div[2]/div[3]/div[1]/div')
[0]
            content = contents.xpath('string(.)')
            # content = content.replace(' ','')
            content = "".join(content.split())

             list_all.append([titles,company,addres,salary_list[0],salary_list[1
],dateT,edu,exper,content])
             item = [titles, company, addres, salary_list[0], salary_list[1],
dateT, edu, exper, content]
            write_db(item)
        except:
            print('爬取失败')
```

11.4.4 将抓取的信息添加到数据库

编写函数 write_db(data)，功能是将抓取到的招聘信息添加到 MySQL 数据库中。对应的实现代码如下所示：

```
def write_db(data):
    """ 写入数据库 """
    print(data)
    try:
        tlock.acquire()
        # rows 变量得到数据库中被影响的数据行数。
        rescoun = cursor.execute(
                "insert into data (post,company,address,salary_min,salary_
```

```
max,dateT,edu,exper,content) values(%s,%s,%s,%s,%s,%s,%s,%s,%s)", data)
            # 向数据库提交
            db.commit()
            tlock.release()
            # 如果没有 commit()，库中字段已经向下移位但内容没有写进，自动生成的 ID 会自动增加。
            print('成功')
            global db_item
            db_item = db_item + 1

        except:
            # 发生错误时回滚
            db.rollback()
            tlock.release()
            print('插入失败')
```

11.4.5 处理薪资数据

为了便于在可视化图标中展示薪资水平，向数据库中添加的是单位为元的数据，在图表中展示的薪资单位是千元，如果招聘信息没有工资数据则向数据库添加 0。如果在招聘信息中提供的是年薪，则向数据库中添加除以 12 的月薪。对应的实现代码如下所示：

```
def salary_alter(salarys):    #[]
    salary_list = []
    for salary in salarys:
        # print(salary)
        if salary == '':
            a = [0,0]
        re_salary = re.findall('[\d+\.\d]*', salary)  # 提取数值 -- 是文本值
        salary_min = float(re_salary[0])  # 将文本转化成数值型，带有小数，用 float()

        wan = lambda x,y : [x*10000,y*10000]
        qian = lambda x, y: [x * 1000, y * 1000]
        wanqian = lambda x,y,s :wan(x,y) if '万' in salary else qian(x,y)
        tian = lambda x, y, s: [x,y] if '元' in salary else qian(x, y)

        if '年' in salary:
            salary_max = float(re_salary[2])
            a = wanqian(salary_min,salary_max,salary)
            a[0] = round(a[0] / 12, 2)
            a[1] = round(a[1] / 12, 2)
        elif '月' in salary:
            salary_max = float(re_salary[2])
            a = wanqian(salary_min, salary_max, salary)
        elif '天' in salary:
            salary_max = float(re_salary[0])
            a = tian(salary_min, salary_max, salary)
            a[0] *= 31
            a[1] *= 31
        salary_list.append(a)
    return salary_list
```

11.4.6 清空数据库数据

为了保证可视化分析的数据具备时效性和准确性，在每一次新的爬取之前清空之前抓取的数据。对应的实现代码如下所示：

```
def data_clr():
    # 清空 data
    try:
        tlock.acquire()
        query = "truncate table `data`"
```

```
        cursor.execute(query)
        db.commit()
        tlock.release()
        print('data 原表已清空 ')
    except Exception as aa:
        print(aa)
        print(' 无 data 表！ ')
```

11.4.7 执行爬虫程序

设置要爬取的 URL 地址，开始执行爬虫程序。对应的实现代码如下所示：

```
def main(kw, city,startpage):
    print(kw)
    print(city)
    city_id = get_cityid(city)
    print(city_id)
    data_clr()
    page = startpage
    global db_item
    db_item = 0
    while(db_item <= 200):
        url = "https://search.51job.com/list/{},000000,0000,00,9,99,{},2,{}.
html".format(city_id,kw,page)
        print(url)
        # url = 'https://search.51job.com/jobsearch/search_result.php'
        a =get_url(url)
        print(a)
        get_data(a)
        page = page + 1
        print(db_item)
        # time.sleep(0.2)
```

11.5 信息分离统计

提取在 MySQL 数据库中保存的爬虫数据，分别根据“工作地区”“工作经验”“薪资水平”和“学历水平”提取招聘信息并进行统计，最终的统计结果将作为后面可视化图表的素材数据。

↑扫码看视频（本节视频课程时间：2 分 38 秒）

11.5.1 根据“工作经验”分析数据

编写程序文件 jinyan.py，根据用人单位的“工作经验”要求提取并统计招聘信息，具体实现代码如下所示：

```
def get_edu():
    try:
        cursor.execute("select exper from data ")
        salary = cursor.fetchall()
        # 向数据库提交
        db.commit()
        return salary
    except:
        # 发生错误时回滚
        db.rollback()
        print(" 查询失败 ")
```

```
        return 0

def jinyanfun():
    edu = get_edu()
    data = []
    for i in edu:
        # print(type(i))
        year = re.findall(r"\d+", i[0])
        if len(year)==1:
            data.append(year[0]+'年工作经验')
        elif len(year)==2:
            data.append(year[0]+'-'+year[1]+'年工作经验')
        elif len(year)==0:
            data.append('无工作经验')

    data = DataFrame(data)
    da = data[0].value_counts()
    # print(da)
    list_all = []
    for (i, j) in zip(da.index, da):
        print(j,i)
        list_all.append([i,j])
    # print(type(da))
    return list_all


if __name__ == '__main__':
    a = jinyanfun()
    print(a)
```

执行后会基于当前抓取到的招聘信息提取“工作经验”列的数据，并进行数据统计。

```
76 3~4 年工作经验
56 2 年工作经验
37 1 年工作经验
33 无工作经验
14 5~7 年工作经验
1 3 年工作经验
[['3~4 年工作经验', 76], ['2 年工作经验', 56], ['1 年工作经验', 37], ['无工作经验',
33], ['5~7 年工作经验', 14], ['3 年工作经验', 1]]
```

11.5.2　根据“工作地区”分析数据

编写程序文件 map.py，根据用人单位的“工作地区”要求提取并统计招聘信息，具体实现代码如下所示：

```
def get_xinzi():
    try:
        cursor.execute("select address from data ")
        salary = cursor.fetchall()
        # 向数据库提交
        db.commit()
        return salary
    except:
        # 发生错误时回滚
        db.rollback()
        print("查询失败")
        return 0
a = get_xinzi()

list = []
for i in a:
    test = i[0].split('-')
```

```
    list.append(test[0])
    # print(test)

data = DataFrame(list)
da = data[0].value_counts()


for (i,j) in zip(da.index,da):
    pass
    print(j,i)
print(type(da))
```

执行后会基于当前抓取到的招聘信息提取“工作地区”列的数据，并进行数据统计。

```
169   广州
14    佛山
12    东莞
8     珠海
4     中山
3     深圳
1     南昌
1     汕头
1     上海
1     惠州
1     澄迈
1     韶关
1     广东省
```

11.5.3 根据“薪资水平”分析数据

编写程序文件 xinzi.py，根据用人单位的“薪资水平”要求提取并统计招聘信息。在本项目中将工资分为如下 8 个档次：

- 小于 5k
- 5k~8k
- 8k~11k
- 11k~14k
- 14k~17k
- 17k~20k
- 20k~23k
- 高于 23k

文件 xinzi.py 的具体实现代码如下所示：

```
def get_xinzi():
    try:
        cursor.execute("select salary_min,salary_max from data ")
        salary = cursor.fetchall()
        # 向数据库提交
        db.commit()
        return salary
    except:
        # 发生错误时回滚
        db.rollback()
        print(" 查询失败 ")
        return 0

def xinzi():
    a = get_xinzi()
```

```
        data = []
        for i in a:
            data.append((int(i[0])+int(i[1]))/2)

        fenzu=pd.cut(data,[0,5000,8000,11000,14000,17000,20000,23000,9000000000],ri
ght=False)
        pinshu=fenzu.value_counts()
        # print(pinshu)
        list = []
        for i in pinshu:
            # print(i)
            list.append(i)

        list_all = [
            ['小于 5k', list[0]],
            ['5k~8k', list[1]],
            ['8k~11k',list[2]],
            ['11k~14k',list[3]],
            ['14k-17k', list[4]],
            ['17k-20k', list[5]],
            ['20k-23k', list[6]],
            ['23K~', list[7]]
        ]
        return list

    if __name__ == '__main__':
        a = xinzi()
        print(a)
```

执行后会基于当前抓取到的招聘信息提取“薪资水平”列的数据，并进行数据统计。

```
[11, 34, 47, 53, 31, 14, 15, 12]
```

11.5.4　根据“学历水平”分析数据

编写程序文件 xueli.py，根据用人单位的“学历水平”要求提取并统计招聘信息，在本项目中学历水平要求分为：没有要求、中专、大专、本科及硕士。具体实现代码如下所示：

```
def get_edu():
    try:
        cursor.execute("select edu from data ")
        salary = cursor.fetchall()
        # 向数据库提交
        db.commit()
        return salary
    except:
        # 发生错误时回滚
        db.rollback()
        print("查询失败")
        return 0

def xuelifun():
    edu = get_edu()
    data = []
    for i in edu:
       data.append(i)
       # print(i)

    data = DataFrame(data)
    da = data[0].value_counts()
    # print(da)
    list_all = []
    for (i, j) in zip(da.index, da):
```

```
        print(j, i)
        if '招' in i:
            # print(i)
            continue
        list_all.append([i, j])
    # print(type(da))
    return list_all
if __name__ == '__main__':
    a = xuelifun()
    print(a)
    print(type(a))
```

执行后会基于当前抓取到的招聘信息提取“学历水平”列的数据，并进行数据统计。

```
107 本科
73  大专
26  没有要求
5   硕士
2   中专
2   招若干人
1   招4人
1   招2人
[['本科', 107], ['大专', 73], ['没有要求', 26], ['硕士', 5], ['中专', 2]]
```

11.6　实现数据可视化

提取在 MySQL 数据库中保存的爬虫数据，并使用在前面分析统计步骤中得到的统计结果，使用开源框架 Highcharts 绘制柱状图和饼状图，使用 Flask 框架可视化展示招聘数据信息。

↑扫码看视频（本节视频课程时间：5 分 33 秒）

11.6.1　Flask Web 架构

编写程序文件 app.py，使用 Flask 框架创建一个 Web 项目，设置不同的 URL 参数对应的 HTML 模板文件。文件 app.py 的具体实现代码如下所示：

```
from flask import Flask  # 导入Flask模块
from flask import render_template #导入模板函数
import _thread,time

from flask import request
from servers.data import main
from servers import xinzi,xueli,jinyan
from flask_sqlalchemy import SQLAlchemy# 导入SQLAlchemy模块，连接数据库
from sqlalchemy import or_
import config# 导入配置文件
app = Flask(__name__)#Flask初始化

app.jinja_env.auto_reload = True
app.config['TEMPLATES_AUTO_RELOAD'] = True
app.config.from_object(config)# 初始化配置文件
db = SQLAlchemy(app)# 获取配置参数，将和数据库相关的配置加载到SQLAlchemy对象中

from models.models import Data    # 导入user模块
# 创建表和字段

@app.route('/')
```

```
    def first():
        return render_template("input.html")

    @app.route('/list')# 定义路由
    def list():# 定义 hello_world 函数
        # data = Data.query.all()
        page = int(request.args.get('page', 1))
        per_page = int(request.args.get('per_page', 2))
        key = request.args.get('key', '')
        # Data.address.like("%{}%".format('java'))
        data = Data.query.filter(or_(Data.post.like("%{}%".format(key)),
                                     Data.company.like("%{}%".format(key)),
                                     Data.address.like("%{}%".format(key)),
                                     Data.salary_max.like("{}".format(key)),
                                              Data.salary_min.like("{}".format(key))
)).paginate(page, 12, error_out=False)
        return render_template("data.html", datas=data,key=key)


    @app.route('/search')
    def search():
        kw =request.args.get("kw")
        city =request.args.get("city")

        # main(kw, city)
        # 创建两个线程
        try:
            _thread.start_new_thread(main, (kw, city, 1,))
            _thread.start_new_thread(main, (kw, city, 51,))
        except:
            print("Error: 无法启动线程 ")

        time.sleep(50)
        xz = xinzi.xinzi()
        xl = xueli.xuelifun()
        jy = jinyan.jinyanfun()
        return render_template('h.html', **locals())


    # 通过 url 传递信息
    @app.route('/chart')
    def charts():
        xz =xinzi.xinzi()
        xl = xueli.xuelifun()
        jy = jinyan.jinyanfun()
        return render_template('h.html',**locals())


    # 通过 url 传递信息
    @app.route('/xinzi')
    def xinzitest():
        a = xinzi.xinzi()
        print(a)
        return str(a)

    @app.route('/xueli')
    def xuelitest():
        data = xueli.xuelifun()
        print(data)

        return str(data)

    if __name__ == '__main__':
        app.run()
```

11.6.2 Web 主页

在 Flask Web 模块中，Web 主页对应的 HTML 模板文件是 input.html，功能是提供一个表单供用户分别输入“岗位名称”和“搜索的省份”，单击“搜索”按钮后会根据用户输入的职位条件爬虫抓取目标网站中的招聘信息。文件 input.html 的具体实现代码如下所示：

```
{% extends "base.html" %}
{% block content %}

    <style>
    body{
        background: url('{{ url_for('static',filename='img/backgroun.jpg') }}');
    }
    #sousuo{
        height: 500px;
    }
    </style>
    <div id="sousuo"style="text-align: center ; padding: 5px">
                <form method="get" action="/search">
                        <div class="form">
                                <p>岗位名称</p>
                                <input class="form-name"placeholder="java 软件工
程师" name="kw" type="text" autofocus>
                        </div><br>
                        <div class="form" style="margin-top: 30px;">
                                <p>搜索的省份</p>
                                <input class="form-name"placeholder="例如：全国，
江苏省" name="city" type="text" autofocus>
                        </div><br>
                        <input type="submit" value="搜索" class="btn" />
                </form>
    </div>

{% endblock %}
```

执行后的 Web 主页效果如图 11-2 所示。

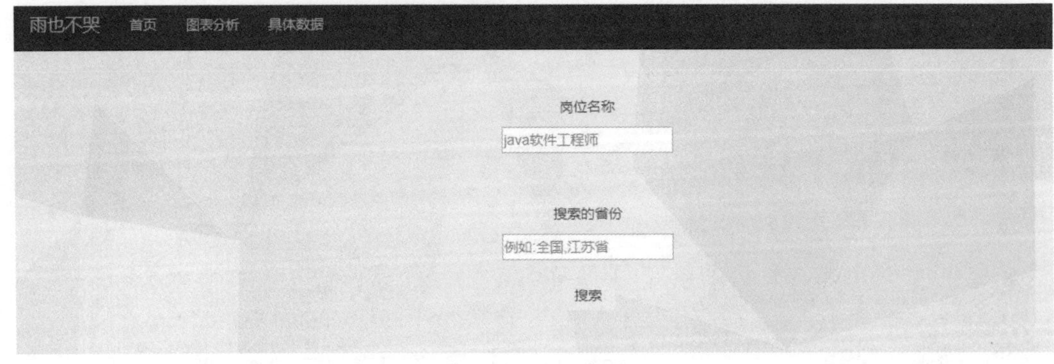

图 11-2 Web 主页

11.6.3 数据展示页面

在 Flask Web 模块中，数据展示页面对应的 HTML 模板文件是 data.html，功能是获取在 MySQL 数据库中保存的招聘信息，并通过表格分页的形式展示这些招聘信息。文件 data.html 的具体实现代码如下所示：

```
    {% extends "base.html" %}
    {% block content %}
        <form method="get" action="/list">
              <div class="input-group col-md-3" style="margin-top:0px; margin-left:
75%; positon:relative">
                <input type="text" class="form-control" placeholder="请输入要搜索的内
容" name="key"/>
                <span class="input-group-btn">
                    <button class="btn btn-info btn-search" style="margin-left:3px">
搜索</button>
                </span>
            </div>
        </form>


        <table class="table table-bordered">
        <tr>
            <th>职位</th>
            <th>公司</th>
            <th>城市</th>
            <th>最低薪资</th>
            <th>最高薪资</th>
            <th>发布日期</th>
        </tr>
            {% for a in datas.items %}
                <tr>
                    <td>{{ a.post }}</td>
                    <td>{{ a.company }}</td>
                    <td>{{ a.address }}</td>
                    <td>{{ a.salary_min }}</td>
                    <td>{{ a.salary_max }}</td>
                    <td>{{ a.dateT }}</td>
                </tr>
        {% endfor %}
        </table>


        <div style="text-align: center">
            <nav aria-label="Page navigation example">
            <ul class="pagination justify-content-center">

                {% if datas.has_prev %}
                    <li class="page-item">
                  <a class="page-link" href="/list?page={{ datas.prev_num }}" aria-
label="Previous">
                    <span aria-hidden="true">&laquo;</span>
                    <span class="sr-only">Previous</span>
                </a>
            </li>
                {% endif %}

                {% for i in datas.iter_pages() %}
                    {% if i == None %}
                                  <li class="page-item"><a class="page-link" href="#">
...</a></li>
                    {% else %}
                                <li class="page-item"><a class="page-link" href="/
list?page={{ i }}&key={{ key }}">{{ i }}</a></li>
                    {% endif %}
                {% endfor %}

                {% if datas.has_next %}
                    <li class="page-item">
```

```
                <a class="page-link" href="/list?page={{ datas.next_num }}&key={{
key }}" aria-label="Next">
                    <span aria-hidden="true">&raquo;</span>
                    <span class="sr-only">Next</span>
                </a>
            </li>
                {% endif %}
            </ul>
            </nav>
            当前页数：{{ datas.page }}
            总页数：{{ datas.pages }}
            一共有 {{ datas.total }} 条数据
            <br>
        </div>
    {% endblock %}
```

执行后的数据展示页面效果如图 11-3 所示。

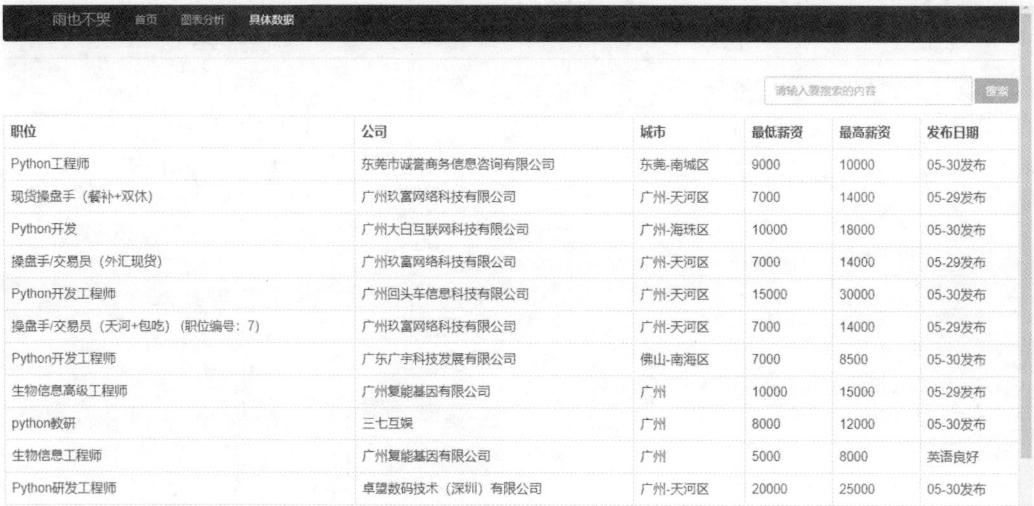

雨也不哭 首页 图表分析 具体数据

请输入要搜索的内容 搜索

职位	公司	城市	最低薪资	最高薪资	发布日期
Python工程师	东莞市诚誉商务信息咨询有限公司	东莞-南城区	9000	10000	05-30发布
现货操盘手（餐补+双休）	广州玖富网络科技有限公司	广州-天河区	7000	14000	05-29发布
Python开发	广州大白互联网科技有限公司	广州-海珠区	10000	18000	05-30发布
操盘手/交易员（外汇现货）	广州玖富网络科技有限公司	广州-天河区	7000	14000	05-29发布
Python开发工程师	广州回头车信息科技有限公司	广州-天河区	15000	30000	05-30发布
操盘手/交易员（天河+包吃）（职位编号：7）	广州玖富网络科技有限公司	广州-天河区	7000	14000	05-29发布
Python开发工程师	广东广宇科技发展有限公司	佛山-南海区	7000	8500	05-30发布
生物信息高级工程师	广州复能基因有限公司	广州	10000	15000	05-29发布
python教研	三七互娱	广州	8000	12000	05-30发布
生物信息工程师	广州复能基因有限公司	广州	5000	8000	英语良好
Python研发工程师	卓望数码技术（深圳）有限公司	广州-天河区	20000	25000	05-30发布

图 11-3　数据展示页面

11.6.4　数据可视化页面

在 Flask Web 模块中，数据可视化页面对应的 HTML 模板文件是 h.html，功能是根据对 MySQL 数据库中保存的招聘信息的统计结果，使用 Highcharts 绘制统计图表。文件 h.html 的具体实现代码如下所示：

```
    {% extends "base.html" %}
    {% block content %}
        <style>
        body{
            background: url('{{ url_for('static',filename='img/backgroun.jpg') }}');
        }
        </style>

        <div id="xinzi"></div>
        <table align="center">
        <tr>
            <td>
                <div id="xueli"></div>
            </td>
            <td>
```

```
            <div id="jinyan"></div>
        </td>
    </tr>
    </table>



    <script type="text/javascript">
    $(document).ready(function() {
        var chart = {
            type: 'column',
            backgroundColor: 'rgba(0,0,0,0)'
        };
        var title = {
            useHTML: true,
            style: {
                color: '#000',              //字体颜色
                "fontSize": "29px",         //字体大小
                fontWeight: 'bold'
            },
            text: '工资分布图'
        };
        var subtitle = {
            text: '51job.com'
        };
        var xAxis = {
             categories: ['0~5k','5~8k','8k~11k','11k~14k','14k~17k','17k~20k','
20k~23k','23K以上'],
            crosshair: true
        };
        var yAxis = {
            min: 0,
            title: {
                text: '岗位数'
            }
        };
        var tooltip = {
              headerFormat: '<span style="font-size:10px">{point.key}</
span><table>',
                    pointFormat: '<tr><td style="color:{series.
color};padding:0">{series.name}: </td>' +
                '<td style="padding:0"><b>{point.y:.1f} 个</b></td></tr>',
            footerFormat: '</table>',
            shared: true,
            useHTML: true
        };
        var plotOptions = {
            column: {
                pointPadding: 0.2,
                borderWidth: 0
            }
        };
        var credits = {
            enabled: false
        };

        var series= [{
            name: '岗位数',
            data: {{ xz|tojson }}
        }];

        var json = {};
        json.chart = chart;
```

```
        json.title = title;
        json.subtitle = subtitle;
        json.tooltip = tooltip;
        json.xAxis = xAxis;
        json.yAxis = yAxis;
        json.series = series;
        json.plotOptions = plotOptions;
        json.credits = credits;
        $('#xinzi').highcharts(json);

    });
    </script>
    <script type="text/javascript">
        $(document).ready(function() {
       let chart = {
           plotBackgroundColor: null,
           plotBorderWidth: null,
           plotShadow: false,
           backgroundColor: 'rgba(0,0,0,0)'
       };
       let title = {
           useHTML: true,
            style: {
                color: '#000',              // 字体颜色
                "fontSize": "29px",         // 字体大小
                fontWeight: 'bold'
            },
          text: '学历占比情况'
       };
       let tooltip = {
          pointFormat: '{series.name}: <b>{point.percentage:.1f}%</b>'
       };
       let plotOptions = {
          pie: {
             allowPointSelect: true,
             cursor: 'pointer',
             dataLabels: {
                enabled: true
             },
             showInLegend: true
          }
       };
       let series= [{
          type: 'pie',
          name: '学历',
          data: {{xl|tojson}}
       }];

       let json = {};
       json.chart = chart;
       json.title = title;
       json.tooltip = tooltip;
       json.series = series;
       json.plotOptions = plotOptions;
       $('#xueli').highcharts(json);
    });
    </script>
    <script type="text/javascript">
    $(document).ready(function() {
       let chart = {
           plotBackgroundColor: null,
           plotBorderWidth: null,
           plotShadow: false,
```

```
            backgroundColor: 'rgba(0,0,0,0)'
        };
        let title = {
            useHTML: true,
             style: {
                 color: '#000',            // 字体颜色
                 "fontSize": "29px",       // 字体大小
                 fontWeight: 'bold'
             },
           text: '工作年限要求'
        };
        let tooltip = {
           pointFormat: '{series.name}: <b>{point.percentage:.1f}%</b>'
        };
        let plotOptions = {
           pie: {
              allowPointSelect: true,
              cursor: 'pointer',
              dataLabels: {
                 enabled: true
              },
              showInLegend: true
           }
        };
        let series= [{
           type: 'pie',
           name: '工作年限要求',
           data: {{jy|tojson}}
        }];

        let json = {};
        json.chart = chart;
        json.title = title;
        json.tooltip = tooltip;
        json.series = series;
        json.plotOptions = plotOptions;
        $('#jinyan').highcharts(json);
    });
    </script>
{% endblock %}
```

执行后的数据可视化页面效果如图 11-4 所示。

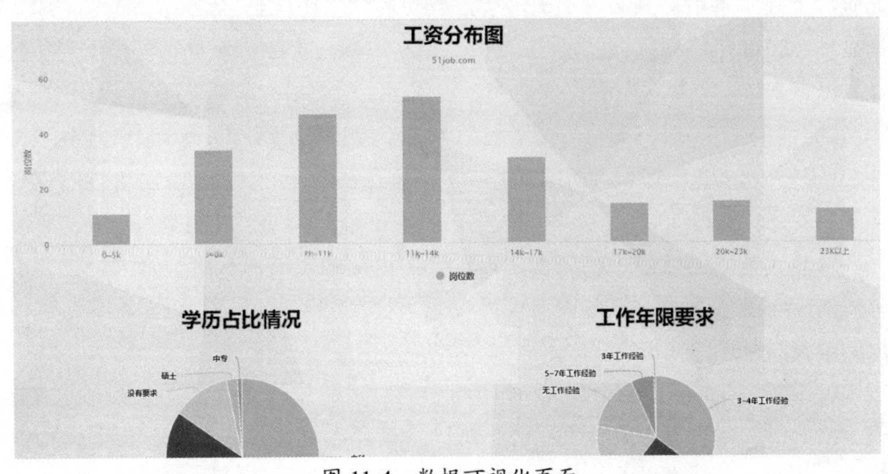

图 11-4　数据可视化页面

第12章

综合实战：民宿信息可视化

很多工作累了想出来放松一下的游客，如果不住酒店，一般可以通过民宿这个平台，感受一下当地的风土人情，体验一下不同的生活方式，基于越来越多的人们喜欢住在民宿的市场需求下，打造‘民宿 + 当地文化’的个性民宿。分析民宿市场的发展和市场定位变得愈发重要。在本章的内容中，将通过一个综合实例的实现过程，详细讲解爬虫抓取民宿信息的方法，并讲解可视化分析这些民宿信息的过程。

12.1 系统背景介绍

近年来，随着国家建设美丽乡村政策的实施，各地纷纷加大对特色小镇的建设力度，相继出台对民宿的补贴扶持方案，之所以选择从事民宿行业，大部分人是因为自己喜欢旅行，也有“隐于野”的诗意情结，他们或是放弃了稳定的工作，或是远离了大都市的生活，希望能通过民宿传递自己的生活理念。

↑扫码看视频（本节视频课程时间：3 分 26 秒）

作为一种新兴的非标准住宿业态，民宿对传统标准酒店住宿业起到明显的补充作用。目前美团民宿交易额占美团酒店交易额的比例约为 4.8%，且整体呈现上升趋势。从各省份民宿的交易额看，广东省民宿交易份额占据首位，交易额占全国市场的 11.6%，交易额排在前 10 位的依次为广东省、北京市、四川省、江苏省、山东省、陕西省、重庆市、上海市、浙江省、湖北省，上述十省市交易额占全国民宿市场交易额的比例超过 65%。

数据显示，2019 年民宿预订以女性消费者为主，占比 55.7%。从民宿产品用户年龄层分布来看，40 岁以下人群占整体消费者比例达到 86.2%，可见国内民宿产品受众偏向年轻化。其中，90 后是民宿消费的主力军，90 后消费者的订单量占比约为 58.9%，80 后占比约为 27.3%。从消费品类偏好看，用户在住民宿期间，同时消费餐饮品类的比例约占 30.8%，同时消费非餐饮品类的比例约占 28.2%。这说明民宿消费对其他品类的消费也具有一定的带动作用。

在民宿市场大发展的前提下，可视化分析民宿市场的发展现状对商家来说具有重要的意义。另外，对于消费者来说，也可以通过可视化系统及时了解民宿行情，帮助自己取得更加物美价廉的服务。

在本项目中，将使用爬虫技术抓取 X 团网中的民宿信息，然后将抓取到的民宿信息持久化保存到 MySQL 数据库中，最后可视化分析民宿信息，并通过大数据分析民宿行业的现状和发展趋势。需要大家注意的是，本项目用到了反爬技术，读者可以掌握反爬技术的原理和具体用法。

12.2 爬虫抓取信息

本项目将使用 Scrapy 作为爬虫框架，使用代理 IP 爬取业内知名民宿网站中的数据信息，然后将爬取的信息保存到 MySQL 数据库中。最后使用 Django 可视化展示在数据库中保存的民宿数据信息。在本节的内容中，将首先讲解爬虫功能的具体实现过程。

↑扫码看视频（本节视频课程时间：12 分 09 秒）

12.2.1 系统配置

在 Django 模块中设置整个项目的配置信息，在文件 settings.py 中设置数据库和缓存等配置信息，主要代码如下所示：

```
# 配置mysql数据库的参数
DATABASES = {
    'default': {
        'ENGINE': 'django.db.backends.mysql',
        'NAME': "scrapy_django",
        'USER': 'root',
        'PASSWORD': '66688888',
        'OPTIONS': {
                'charset':'utf8mb4',
                 # "init_command": "SET foreign_key_checks = 0;",
                },      # 都改成这种编码，避免emoji无法存储
        'HOST': "127.0.0.1",   # 要不要都改成远程的地址
        'PORT': '3306'
    }
}
CACHES = {   # redis做缓存
    'default': {
        'BACKEND': 'django_redis.cache.RedisCache',
        "LOCATION": "redis://127.0.0.1:6379/3",   # 本机django的redis缓存路径
        # 'LOCATION':"redis://127.0.0.1:6378/3",
        'OPTIONS':{
            "CLIENT_CLASS":"django_redis.client.DefaultClient",
        }
    }
}
```

12.2.2 Item 处理

Scrapy 为我们提供了 Item 类，这些 Item 类可以让我们自己来指定字段。比方说在某个 Scrapy 爬虫项目中定义了一个 Item 类，在这个 Item 里面包含了 title、release_date、url 等，这样的话，通过各种爬取方法爬取过来的字段，再通过 Item 类进行实例化，这样就不容易出错，因为我们在一个地方统一定义过了字段，而且这个字段具有唯一性。在本项目实例文件 items.py 中设置了 4 个 ORM 对象，这 4 个对象和本项目数据库字段是一一对应的。文件 items.py 的具体实现代码如下所示：

```
class HouseItem(DjangoItem):
    django_model = House
    jsonString = scrapy.Field()    # 此处需增加临时字段，用于把多个其他对象的属性一次性传过来

class HostItem(DjangoItem):
    django_model = Host
```

```
class LabelsItem(DjangoItem):
    django_model = Facility

class FacilityItem(DjangoItem):
    django_model = Labels

class CityItem(DjangoItem):
    django_model = City

class urlItem(scrapy.Item):  # master 专用 item
    # define the fields for your item here like:
    url = scrapy.Field()
```

12.2.3 具体爬虫功能实现

编写文件hotel.py，功能是实现具体的网络爬虫功能，根据设置的目标URL进行爬虫操作，并将爬取的民宿信息保存到数据库中。

（1）创建类 HotwordspiderSpider，设置爬虫项目的名字是“hotel”，然后分别设置爬虫的并发请求数、延时、最大的并发请求数量、保存数据管道和使用代理等信息。对应的实现代码如下所示：

```
class HotwordspiderSpider(scrapy.Spider):
    def __init__(self):
        self.ua = UserAgent()
        # for i in range(10):
        #     print(ua.random)

    name = 'hotel'
    allowed_domains = ['*']
    # start_urls = ['https://minsu.meituan.com/guangzhou/']

    custom_settings = {  # 每个爬虫使用各自的自定义设置
        #### Scrapy downloader(下载器) 处理的最大的并发请求数量。 默认 : 16
        'CONCURRENT_REQUESTS' : 2,
        #### 下载延迟的秒数，用来限制访问的频率，默认为 0，没有延时
        # 'DOWNLOAD_DELAY' : 1,
        #### 每个域名下能够被执行的最大的并发请求的数量，默认 8 个
        'CONCURRENT_REQUESTS_PER_DOMAIN' : 2,
        'ROBOTSTXT_OBEY':False,
        'COOKIES_ENABLED':False,
        #### 设置某个 IP 最大并发请求数量，默认 0 个
        'ONCURRENT_REQUESTS_PER_IP' : 2,
        'RETRY_ENABLED' :True,   # 打开重试开关
        'RETRY_TIMES': 20 , # 重试次数，没办法
        'DOWNLOAD_TIMEOUT': 60,
        'DOWNLOAD_DELAY': 2,  # 慢慢爬呗   ，这里写了下载延迟，  403IP 就被封了 todo
        'RETRY_HTTP_CODES': [404,403,406],   # 重试
        'HTTPERROR_ALLOWED_CODES': [403],   # 上面报的是 403，就把 403 加入。
        "ITEM_PIPELINES": {
            'myscrapy.pipelines.houseItemPipeline': 300,   # 启用这个管道来保存数据

        },
        "DOWNLOADER_MIDDLEWARES":{    # 这样就可以单独使用每个不同的配置
        'myscrapy.middlewares.RandomUserAgent': 100,    # 使用代理
        'myscrapy.middlewares.proxyMiddleware': 301,      # 暂时不用代理来进行爬取，测
试一下最多多少个

        },
```

（2）设置爬虫网页 HTTP 请求协议的请求报文（Request Headers），对应的实现代码如下所示：

```
        "DEFAULT_REQUEST_HEADERS": {
            'Accept': 'application/json',
            'Accept-Language': 'zh-CN,zh;q=0.9',
            'Referer': 'https://www.meituan.com/',
            'X-Requested-With': "XMLHttpRequest",
            # "cookie":"lastCity=101020100; JSESSIONID=""; Hm_lvt_194df3105ad7
148dcf2b98a91b5e727a=1532401467,1532435274,1532511047,1532534098; __c=1532534098;
__g=-; __l=l=%2Fwww.zhipin.com%2F&r=; toUrl=https%3A%2F%2Fwww.zhipin.
com%2Fc101020100-p100103%2F; Hm_lpvt_194df3105ad7148dcf2b98a91b5e727a=1532581213;
__a=4090516.1532500938.1532516360.1532534098.11.3.7.11"
            # 'Accept': 'application/json',
              # 'User-Agent': 'Mozilla/6.0 (Linux; Android 8.0; Pixel 2 Build/
OPD3.170816.012) AppleWebKit/537.36 (KHTML, like Gecko) Chrome/66.0.3359.117 Mobile
Safari/537.36',
            # 'User-Agent':  self.ua.random  # 随机  好像还是需要通过中间件来 todo
            # 'cookie':cookie self.ua

        }
    }
```

（3）编写函数 regexMaxNum()，功能是使用正则表达式返回匹配到的最大的数字就是页数，代码如下：

```
    def regexMaxNum(self,reg,text):
        temp = re.findall(reg,text)
        return max([int(num) for num in temp if num != ""])
```

（4）编写函数 start_requests()，功能设置爬虫启动时要爬取的城市列表。对应的实现代码如下所示：

```
    def start_requests(self):
        guangdong = '''广州市、韶关市、深圳市、珠海市、汕头市、佛山市、江门市、湛江市、茂名市、
肇庆市、惠州市、梅州市、汕尾市、河源市、阳江市、清远市、东莞市、中山市、潮州市、揭阳市、雷州市、陆丰市、
普宁市'''
        topcity = '北京、南京、上海、杭州、昆明市、大连市、厦门市、合肥市、福州市、哈尔滨市、
济南市、温州市、长春市、石家庄市、常州市、泉州市、南宁市、贵阳市、南昌市、南通市、金华市、徐州市、太原市、
嘉兴市、烟台市、保定市、台州市、绍兴市、乌鲁木齐市、潍坊市、兰州市。'
        pin = Pinyin()
        print("启动")
          guangdonglist = [pin.get_pinyin(i.replace("市", ""), "") for i in
guangdong.split("、")]
        for onecity in City.objects.all():  # todo 城市这里先设置为0：2 直接读取数据
库中的
            # print(i.city_
            # print(i.city_pynm)
            # 先抓取广东省内的
            # if onecity.city_pynm in guangdonglist: # 广东省内的才爬
            #     print("在广州")
            for i in range(1, 18):  # 这里最大（1, 18） 1~ 17的意思
                yield  scrapy.Request(url=f'https://minsu.meituan.com/{onecity.
city_pynm}/pn{i}',
                dont_filter=True,
                callback=self.parse)   # 暂时还不是很懂发生了什么

    def parse(self, response):
        tempPageUrl = response.xpath("//a[@target='_blank']/@href").extract()
          tempPageUrl2 = [urljoin(response.url,url) for url in tempPageUrl if
url.find("housing")!=-1]
        for url in tempPageUrl2:  # 一页里面的所有房源的链接
            print(url)
                  # yield scrapy.Request(url="https://minsu.meituan.com/
housing/9969914/",callback=self.detail,dont_filter=True) # 直接转到详情上
              yield scrapy.Request(url=url,callback=self.detail,dont_filter=True)
# 直接转到详情上
```

（5）编写函数 getRSXFPrice()，功能是提取爬虫数据中的价格信息。对应的实现代码如下所示：

```
        def getRSXFPrice(self,RSXF_TOKEN):  # 提取
            import datetime
            import requests
            import time
                url  =  'https://minsu.meituan.com/gw/corder/api/v1/order/
productPricePreview'
            data  ={
                "currentTimeMillis":int(round(time.time() * 1000)),    # 获得当前时间的
毫秒
                    "sourceType":7,
                    "checkinGuests":1,
                     "checkinDate":datetime.datetime.now().strftime("%Y%m%d"),  # 获
得今天
                        "checkoutDate":(datetime.datetime.now()+datetime.
timedelta(days=1)).strftime("%Y%m%d"), # 获得明天
                    "productId":2645048,
                    "autoChooseDiscount":'true',
                    "avgMoneyFormat":'true',
                    "deviceInfoByWeb":{
                                "ua":"Mozilla/5.0 (Windows NT 10.0; Win64; x64)
AppleWebKit/537.36 (KHTML, like Gecko) Chrome/79.0.3945.130 Safari/537.36",
                        "touchPoint":"",
                         "browserPlugins":"Microsoft Edge PDF Plugin,Microsoft Edge
PDF Viewer,Native Client",
                         "colorDepth":24,"pixelDepth":24,"screenWith":1280,"screenHe
ight":720,"browserPageWidth":653,"browserPageHeight":615}}
                header  ={'User-Agent':'Mozilla/5.0 (Windows NT 10.0; Win64; x64)
AppleWebKit/537.36 (KHTML, like Gecko) Chrome/79.0.3945.130 Safari/537.36',
                    'Referer':'https://minsu.meituan.com/housing/2645048/'}
            cookies = {"XSRF-TOKEN":RSXF_TOKEN}
             result = requests.post(url,headers=header,data=data,cookies = cookies).
content
            return result
```

（6）编写函数 detail()，功能是获取每个民宿的详细信息，包括面积、标签、标题、地址、房型、位置、城市、留言数量等信息。对应的实现代码如下所示：

```
        def detail(self, response):
            # 新增加提取预览图 todo
            try:
                house_img = response.xpath('//*[@class="item"]/img/@src').extract()
[0]
                print(house_img)
            except Exception as e:
                print(e)
                print("预览图提取失败")

            print("in detail")
            print(response.url)
            # print(response.text)
            # from pprint import pprint
            item = HouseItem()
            print("设施")
            tempDic = []  # 这个出 bug 就全出问题了，   todo fix
            facility = ""
             # facility = response.xpath("//*[@class='page-card']/*[@id='r-props-J-
facility']/text()").extract()[0]
              facility01 = response.xpath("//*[@id='r-props-J-facility']/text()").
extract()
            try:
```

```
                facility = facility01[0]
            except Exception as e:
                print("facility 错误 ")
                print(e)
                print(facility01)
                print(response.text)
            # print(facility.strip("<!--").strip("-->"))
            all = facility.replace("<!--", "").replace("-->", "")
            # print(all)

             # all = BeautifulSoup(facility[0],'lxml').find("script",attrs={"id":"r-
props-J-facility"}).get_text()
            facilityDic = json.loads(all) # 设施可以提取录入
            for i in facilityDic:
                for j in facilityDic[i]:
                    try:
                        for x in j['group']:
                            tempDic.append(x)
                    except Exception as e:
                        break

            print(" 获得第一步发布的时间 ")
            text = all
            # print(text)
              firstOnSaleTime = text[text.find('"firstOnSaleTime":')+18:18+13+text.
find('"firstOnSaleTime":')]
            print(firstOnSaleTime)
            try:
                # print(" 怎么回事 ")
                tempInt = int(firstOnSaleTime)  # 这个是房源信息第一次发布的时间
                # print(tempInt)
                      firstOnSale  =  time.strftime('%Y-%m-%d  %H:%M:%S',time.
localtime(tempInt/1000))
                print(firstOnSale)
                print(" 正常 ")
            except Exception as e:
                print(e)
                tempInt = 0 # 默认 1970 年 1 月 1 日的毫秒
                      firstOnSale  =  time.strftime('%Y-%m-%d  %H:%M:%S',time.
localtime(tempInt/1000))
                print(firstOnSale)
                print(" 元年，发布时间提取出错 ")

            print(" 获取房主数据 ")
               tempUserScript  =  response.xpath("//*[@id='r-props-J-gallery']").
xpath("text()")
            UserJson = tempUserScript.extract()[0]
```

（7）在函数 detail() 中，关于价格的计算比较麻烦，因为民宿网的价格进行了数据加密，所以需要专门的逻辑来破解这个反扒机制。对应的实现代码如下所示：

```
            UserJson = UserJson.replace('"',"'").replace(" ","")  # str 这里替换是为了
让下面统一以单引号来进行提取操作。
            # print(UserJson)
            # 这里又变成了双引号，而且自动把间隙的空格去掉了
            house_id = re.findall("(?<=housing\\/)[0-9]*?(?=\\/)",response.url)[0]
            HostId = re.findall("(?<=hostId\\'\\:).*?(?=\\,)",UserJson)
            price = re.findall("(?<=price\\'\\:)[0-9]*?(?=\\,)",UserJson)   # 提取出来
的结构都是 ['999']
            print(" 提取未加密 price 转化前 ")
            print(price)
            try:
```

```
            price = [float(price[0])/100]
        except Exception as e:
            print("price 出错")
            print(price)
            price = [0.00]


        ## 使用 UserJson
        tempjson = UserJson.strip("<!--").strip("-->").replace("'",'"')
        tempjson2 = None

        try:
            tempjson2 = json.loads(tempjson)
        except Exception as e:
            print("提取折扣价格出错")
            print(e)
            print("输出 tempjson")
            print(tempjson2)

        if price == 0.00 and tempjson2 != None:  # 补充用的
            price = [tempjson2['product']['price']]


             discountprice = re.findall("(?<=discountPrice\\'\\:)
[0-9]*?(?=\\,)",UserJson)  # 这里这个可能找不到的
        if len(discountprice) == 0:  #
            discountprice = [0.00]

        # print(tempjson2['product']['discountPrice'])
        if discountprice == [0.00] and tempjson2 != None:
            try:   #  可能真的没有
                 discountprice = [tempjson2['product']['discountPrice']]  # 折扣
价，默认为 0
                print("这里的 disprice")
                print(discountprice)    # 折扣价
            except Exception as e:
                print(e)
                print("如果没有的话")
                discountprice = [0.00]

        try:   # 单位变成元
            discountprice = [float(discountprice[0])/100]
        except Exception as e:
            print(e)
            print("discountprice 整除出错")
            print(discountprice)
            discountprice = [0.00]

        # 再提取一次价格，使用新的方法，处理字体加密
        if price == [0.00] or discountprice == [0.00]:
            try:
                # print("字体解密成功")
                price,discountprice = parsePriceMain(UserJson)
                price = [price]
                discountprice = [discountprice]
                # price = [0.00]# 这里是原价
            except Exception as e:
                print(price)
                print(discountprice)
                print(e)
                traceback.print_exc()
                print("字体加密解除失败")
```

```
        print("检查价格 price/discountprice")
        print(price)
        print(discountprice)


        title = response.xpath("//head/title").xpath("string(.)").extract()
        fullAddress = re.findall("(?<=fullAddress\\'\\:\\').*?(?=\\'\\,)",UserJson)
        layoutRoom =  re.findall("(?<=layoutRoom\\'\\:).*?(?=\\,)",UserJson)
        layoutKitchen =  re.findall("(?<=layoutKitchen\\'\\:).*?(?=\\,)",UserJson)
        layoutHall =  re.findall("(?<=layoutHall\\'\\:).*?(?=\\,)",UserJson)
        layoutWc =  re.findall("(?<=layoutWc\\'\\:).*?(?=\\,)",UserJson)
        maxGuestNumber =  re.findall("(?<=maxGuestNumber\\'\\:).*?(?=\\,)",UserJson)
        bedCount =  re.findall("(?<=bedCount\\'\\:).*?(?=\\,)",UserJson)
        roomArea =  re.findall("(?<=usableArea\\'\\:\\').*?(?=\\'\\,)",UserJson)
        longitude = re.findall("(?<=longitude\\'\\:).*?(?=\\,)",UserJson)
        latitude = re.findall("(?<=latitude\\'\\:).*?(?=\\}\\,)",UserJson)
        cityName = re.findall("(?<=cityName\\'\\:\\').*?(?=\\'\\,)",UserJson)
         earliestCheckinTime = re.findall("(?<=earliestCheckinTime\\'\\:\\').*?(
?=\\')",UserJson)

         house_type = response.xpath("//*[@class='spec-item spec-room'][1]/div/
div[@class='value']/text()").extract()[0]
        if house_type==None:
            house_type = "未分类"
        house_commentNum = re.findall("(?<=count\\'\\:)[0-9]*?(?=\\,)",UserJson)
        print(house_commentNum)
```

（8）在函数 detail() 中提取留言回复信息，对应的实现代码如下所示：

```
        if house_commentNum == None or len(house_commentNum)==0:
            try:
                          house_commentNum  =  re.findall("(?<=commentNumber\\'\\:)
[0-9]*?(?=\\,)",UserJson)
            except Exception as e:
                pass
            if house_commentNum == None or len(house_commentNum)==0:
                house_commentNum = [0,]
        print("house_commentNum")
        print(house_commentNum)

        print("房子面积开始")
        print(roomArea)
        print(house_id)
        print(price)
        # print(HostId)
        # 折扣价格和原价这个有区别。
        print(discountprice)  # 这个确实没有了，因为找不到现在的价格。
        # print(title)
        # print(fullAddress)
        # print(longitude)
        # print(latitude)
        # print(layoutWc)
        # print(layoutHall)
        # print(layoutRoom)    # 卧室数量
        # print(layoutKitchen)
        # print(maxGuestNumber)
        # print(bedCount)
        # print(house_commentNum)
```

（9）在函数 detail() 中提取促销和普通标签的信息，对应的实现代码如下所示：

```
        print("标签来了")
        try:
            # tempJson = json.loads("".join(tagList).replace("'",'"').strip().
```

```
strip(":")
                    dictString = json.loads(UserJson.strip("<!--").strip("-->").replace
("'", '"'))
                    temp = dictString['product']['productTagInfoList']  # 头疼，重新来提取
这个 label，标签    todo
                    discountList = {"1": [], "0": []}
                    for i in temp:
                        #     print(i['tagName']," ",i['tagType']," ",i['styleType']," ",
i['tagDesc'])
                        #     print()
                        if i['tagType'] == 1:
                            discountList['0'].append([i['tagName'], i['tagDesc']])  # 这
里管道需要先检查后添加进来
                        else:
                            discountList['1'].append([i['tagName'], i['tagDesc']])  # 这
里管道需要先检查后添加进来
                    print("提取折扣")
                    print(discountList)  # 这里是提取折扣

                ################# 这部分是用来提取折扣的

                except Exception as e:
                    print("提取折扣出错")
                    print(e)

                favCount =  re.findall("(?<=favCount\\'\\:).*?(?=\\,)",UserJson)
                # 这个也是可能为 0，就是新房子
                if favCount==None:
                    favCount = 0
                print("喜爱数量")
                print(favCount)
```

（10）在函数 detail() 中提取评价信息和回复率信息，对应的实现代码如下所示：

```
                print()
                print("下面是房主的信息（我估计很多都是二次房主）")
                    host_name = response.xpath("//a[@class='nick-name S--host-link']/
text()").extract()
                host_name = host_name[0].replace(" ","").replace("\n","")
                host_main = response.xpath("//ul[@class='host-score-board']")
                host_infos = []  # 评价数，回复率，房源数
                for div in host_main.xpath("li"):
                    temp = div.xpath("*[@class='value']/span/text()").extract()[0]
                    if temp!=None or temp .find("%") != -1:
                        temp = temp.replace("%","")
                        print(div.xpath("*[@class='value']/span/text()").extract())
                    else:
                        temp = 0  # 如果没有评价，或者没有回复（新房主）那就是 0
                    host_infos.append(temp)


                # print(host_infos)
                print("下面开始赋值给 item")
                # print(HostId)  # 这些是房主的信息
                # print(host_infos[0])
                # print(host_infos[1])
                # print(host_infos[2])

                host_commentNum = host_infos[0]
                if not isinstance(host_infos[1],str):
                    host_replayRate =  host_infos[1]
                else:
                    host_replayRate = 0
                host_RoomNum = host_infos[2]
```

（11）在函数 detail() 中提取好评平均分信息，对应的实现代码如下所示：

```
            # pointer_num = response.xpath("//span[@class='zg-price']/text()").
extract()
        # print(pointer_num[0])
        # print(pointer_num[1])
        # print(pointer_num[2])

        # print("这个也不是每个都有的。")
        # print("这个是平均分")
         avarageScore = response.xpath("//*[@class='sum-score-circle']/text()").
extract()
        # 新房子
        if avarageScore==None or len(avarageScore)==0:
            print(f"新房子，检查一下{response.url}")
            avarageScore =[0,]
        # print(avarageScore)


        fourScore = []    # 描述，沟通，卫生，位置的评分
            for score in response.xpath("//ul[@class='score-chart']").xpath("li"):
                tempscore = score.xpath("div/div[@class='score']/text()").extract()
            print(tempscore)
            if tempscore==None:
                fourScore.append([0])
            else:
                fourScore.append(tempscore)
        # print("评分")
        print(fourScore)
        if fourScore == []:  # 找不到的时候评分就是空的
            fourScore = [[0],[0],[0],[0]]
        # print(fourScore)    # 评分可能是没有的

        house_descScore = fourScore[0][0]
        house_talkScore = fourScore[1][0]
        house_hygieneScore = fourScore[2][0]
        house_positionScore = fourScore[3][0]
```

12.2.4 破解反扒字体加密

在网站中的价格信息是加密的，为了获取每个民宿的价格信息，需要对“.woff”格式的加密字体进行破解。编写文件 parseTool.py，功能是破解“.woff”格式的价格信息，主要实现代码如下所示：

```
    # 获得j-gallery这段字符串
    def getFontUrl(UserJson):
        j_gallery_text = UserJson  # 这里再处理一遍去掉可能出现的东西
        UserJson = j_gallery_text.replace("'",'"').replace("\\","")   # 这样才可以去掉一个杠
        test = re.findall('(?<=cssPath\\"\\:\\").*?(?=\\}\\,)',UserJson)[0]

        print()
         wofflist = re.findall('(?<=\\(\\").*?(?=\\)\\;)',test)
    #     print(wofflist)
        print()
        font_url = ''
        for woffurl in wofflist:
            if woffurl.find("woff")!=-1:
                tempwoff = re.findall('(?<=\\").*?(?=\\")',woffurl)
    #           print(tempwoff)
                for j in tempwoff:
                    if j.find("woff")!=-1:
                        print("https:"+j)
```

```
                        font_url = "https:"+ j

        # 提取字体成功
    #     print(font_url)
        return font_url

    def download_font(img_url,imgName,path=None):
          headers = {'User-Agent':"Mozilla/5.0 (Windows NT 6.1; WOW64)
AppleWebKit/537.1 (KHTML, like Gecko) Chrome/22.0.1207.1 Safari/537.1",
                    }   ## 浏览器请求头（大部分网站没有这个请求头会报错、请务必加上）
        try:
            img = requests.get(img_url, headers=headers)
            dPath = os.path.join("woff",imgName)  # imgName 传进来不需要带时间
            # print(dPath)
            print(" 字体的文件名 "+dPath)
            f = open(dPath, 'ab')
            f.write(img.content)
            f.close()
            print(" 下载成功 ")
            return dPath
        except Exception as e:
            print(e)


    # 从字体文件中获得字形数据备用待对比
    def getGlyphCoordinates(filename):
        """
        获取字体轮廓坐标，手动修改 key 值为对应数字
        """
        font = TTFont("woff/"+f'{filename}')  # 自动带上了 woff 文件夹
        # font.saveXML("bd10f635.xml")
        glyfList = list(font['glyf'].keys())
        data = dict()
        for key in glyfList:
            # 剔除非数字的字体
            if key[0:3] == 'uni':
                data[key] = list(font['glyf'][key].coordinates)
        return data


    def getFontData(font_url):
        # 合并两个操作，如果有的话就不用下载
        filename = os.path.basename(font_url)
        font_data = None
        if os.path.exists("woff/"+filename):
            # 直接读取
            font_data = getGlyphCoordinates(filename)  # 读取的时候自带 woff 文件夹
        else:
            # 先下载再读取
            download_font(font_url, filename, path=None)
            font_data = getGlyphCoordinates(filename)
        if font_data == None:
            print(" 字题文件读取出错，请检查 ")
        else:
            #          print(font_data)
            return font_data

    # 自动分割并且大写，这两个要连着来调用，那么全部封装成一个对象即可
    def splitABC(price_unicode):
        raw_price = price_unicode.split("&amp;")
        temp_price_unicode = []
        for x in raw_price:
            if x != "":
```

```
                    temp_price_unicode.append(x.upper().replace("#X", "").
replace(";", ""))
        return temp_price_unicode  # 提取出简化大写的  4。0  这个是原价 ，折扣价才是 280 所以

    def getBothSplit(UserJson):
        UserJson = UserJson.replace("\\", "").replace("'", '"')
        result_price = []
        result_discountprice = []
        try:
            price_unicode = re.findall('(?<=price\\"\\:\\").*?(?=\\"\\,)', UserJson)
[0]  # 原价数字 400
            result_price = splitABC(price_unicode)
        except Exception as e:
            print("没有找到价格")
            print(e)

        try:  # 可能没有找到，那就会有
                discountprice_unicode = re.findall('(?<=discountPri
ce\\"\\:\\").*?(?=\\"\\,)', UserJson)[0]  # 原价数字 400
            result_discountprice = splitABC(discountprice_unicode)
        except Exception as e:
            print("没有找到折扣价")
            print(e)
        if result_discountprice == [] and result_price != []:
            result_discountprice = result_price      # 如果折扣价为 0 的话，那么就等于原价好了
        return result_price,result_discountprice      # 如果没有折扣价，就只返回处理后的价格编码

    def pickdict(dict):  # 序列化这个字典
        with open(os.path.join(os.path.abspath('.'),"label_dict.pickle"), "wb") as f:
            pickle.dump(dict, f)
```

12.2.5　下载器中间件

下载器中间件是在引擎及下载器之间的特定钩子（specific hook），处理 Downloader 传递给引擎的 response（也包括引擎传递给下载器的 Request）。其提供了一个简便的机制，通过插入自定义代码来扩展 Scrapy 功能。在本项目中的下载器中间件文件 middlewares.py 中，主要实现了在线代理 IP 功能。

（1）先创建类 EnvironmentIP 和 EnvironmentFlag，对应实现代码如下所示：

```
class EnvironmentIP:                                     # 这里设置一个全局变量，单例模式
    _env = None

    def __init__(self):
        self.IP = 0                                      # 用那个计数的来操作就可以

    @classmethod
    def get_instance(cls):
        """
        返回单例 Environment 对象
        """
        if EnvironmentIP._env is None:
            cls._env == cls()
        return cls._env

    def set_flag(self, IP): # 里面放的是数字
        self.IP = IP

    def get_flag(self):
        return self.IP
```

```
envVarIP = EnvironmentIP()  # 这个变量看看是否切换使用代理的

class EnvironmentFlag:  # 这里设置一个全局变量，单例模式
    _env = None
    def __init__(self):
        self.flag = False   # 默认不使用代理

    @classmethod
    def get_instance(cls):
        """
        返回单例 Environment 对象
        """
        if EnvironmentFlag._env is None:
            cls._env == cls()
        return cls._env

    def set_flag(self, flag):
        self.flag = flag

    def get_flag(self):
        return self.flag

envVarFlag = EnvironmentFlag()  # 这个变量看看是否切换使用代理的

class Environment:  # 这里设置一个全局变量，单例模式
    _env = None
    def __init__(self):
        self.countTime = datetime.datetime.now()

    @classmethod
    def get_instance(cls):
        """
        返回单例 Environment 对象
        """
        if Environment._env is None:
            cls._env == cls()
        return cls._env

    def set_countTime(self, time):
        self.countTime = time

    def get_countTime(self):
        return self.countTime

envVar = Environment()
```

（2）定义类 RandomUserAgent 实现随机生成 IP 功能，通过函数 process_request() 和 process_response() 及时获取响应信息，用于判断这个 IP 是否可用。对应实现代码如下所示：

```
class RandomUserAgent(object):  # ua 中间件
    # def __init__(self):

    @classmethod
    def from_crawler(cls, crawler):
        s = cls()
        crawler.signals.connect(s.spider_opened, signal=signals.spider_opened)
        return s

    def process_request(self, request, spider):
        ua = UserAgent()
        print(ua.random)
        request.headers['User-Agent'] = ua.random
        return None
```

```
    def process_response(self, request, response, spider):
        # Called with the response returned from the downloader.
        print(f"请求的状态码是  {response.status}")
        print("调试ing")
        print(request.url)
        HTML = response.body.decode("utf-8")
        # print(HTML)
        print(HTML[:200])
        try:
            # print('进来中间件调试')
            if HTML.find("code")!=-1:
                if re.findall('(?<=code\\"\\:).*?(?=\\,)',HTML)[0]=='406':
                  # 自动会是双引号
                    print("正在重新请求（网络不好）")
                    return request
        except Exception as e:
            print(request.url)
            print(e)

        try:
            temp = json.loads(HTML)
            if temp['code'] == 406:  #
                print("正在重新请求（网络不好）状态码406")
                request.meta["code"] = 406
                return request     # 重新发给调度器，重新请求
        except Exception as e:
            print(e)
        return response


    def process_exception(self, request, exception, spider):
        pass

    def spider_opened(self, spider):
        spider.logger.info('Spider opened: %s' % spider.name)
```

（3）定义类 proxyMiddleware 实现在线代理 IP 功能，创建了 redis 代理连接池，用列表 remote_iplist 中的 IP 轮询访问，并打印输出对应的响应信息。对应实现代码如下所示：

```
class proxyMiddleware(object):  # 代理中间件
    # 这里是使用代理 ip
    # MYTIME = 0   # 类变量用来设定切换代理的频率

    def __init__(self):
        # self.count = 0
        from redis import StrictRedis, ConnectionPool
        # 使用默认方式连接到数据库
            pool = ConnectionPool(host='localhost', port=6378,
db=0,password='Zz123zxc')
        self.redis = StrictRedis(connection_pool=pool)

    @classmethod
    def from_crawler(cls, crawler):
        # This method is used by Scrapy to create your spiders.
        s = cls()
        crawler.signals.connect(s.spider_opened, signal=signals.spider_opened)
        return s

    def get_proxy_address(self):
        proxyTempList = list(self.redis.hgetall("useful_proxy"))
        # proxyTempList = list(redis.hgetall("useful_proxy"))
        return str(random.choice(list(proxyTempList)), encoding="utf-8")
```

```
        def process_request(self, request, spider):
            # 这里是用来代理的
                remote_iplist = ['125.105.70.77:4376', '58.241.203.162:4386',
'119.5.181.109:4358', '14.134.186.95:4372', '125.111.150.25:4305',
'122.246.193.161:4375']

            print()
            print("proxyMiddleware")
            now = datetime.datetime.now()
            print("flag")
            print("time")
            print(f" 现在时间 {now}")
            print(f" 变量内时间 {envVar.get_countTime()}")
            print(" 变量状态 {True} 才使用代理 ")
            print(envVarFlag.get_flag())
            print(" 相减之后的结果 ")
            print((now-envVar.get_countTime()).seconds / 40)
             if envVarFlag.get_flag() ==True:  #envVarFlag.get_flag() == True:  # 好像
还不如不用这个代理呢
                if (now-envVar.get_countTime()).seconds / 20 >= 1:
                    envVarFlag.set_flag(not envVarFlag.get_flag())  # 切换为使用代理
                    envVar.set_countTime(now)
                print(" 使用代理中池中的 ip")
                proxy_address = None
                try:
                    proxy_address = self.get_proxy_address()
                    if proxy_address is not None:
                        print(f' 代理 IP --  {proxy_address}')
                         request.meta['proxy'] = f"http://{proxy_address}"   # 如果出
现了 302 错误，则可能是因为代理的类型不对
                    else:
                        print(" 代理池中没有代理 ip 存在 ")
                except Exception as e:
                    print(" 检查到代理池里面已经没有 ip 了，使用本地 ")
            else:  # 不使用代理，这里轮流使用本地 ip 和外面的 ip
                if (now-envVar.get_countTime()).seconds / 40 >= 1:  # 这个进来是切换状态的
                    envVarFlag.set_flag(not envVarFlag.get_flag())  # 切换为使用代理
                    envVar.set_countTime(now)

                if envVarIP.get_flag() <= len(remote_iplist)-1:   ## 直接本地，也用下代理的吧
                    remoteip = remote_iplist[envVarIP.get_flag()]
                    print(f' 使用远程 ip --  {remoteip}')
                    request.meta['proxy'] = f"http://{remoteip}"
                    envVarIP.set_flag(envVarIP.get_flag()+1)
                else:
                    envVarIP.set_flag(0)  # 把这个 ip 设置为 0，这个是使用本地 ip
                    print(" 使用到本地 ip")
                pass
```

12.2.6 保存爬虫信息

编写实例文件 pipelines.py，功能是将爬取的民宿房源信息保存到本地数据库中。为了提高程序的扩展性，我们分别使用不同类保存不同的字段信息。

（1）编写类 urlItemPipeline，保存房源的 URL 信息，对应实现代码如下所示：

```
    class urlItemPipeline(object):                          # master 专用管道
        def __init__(self):
            self.redis_url = "redis://Zz123zxc:@localhost:6379/"  # master 端属于本地 redis
            self.r = redis.Redis.from_url(self.redis_url,decode_response=True)

        def process_item(self, item, spider):
            if isinstance(item, urlItem):
```

```
            print("urlItem item")
            try :
                # item.save()
                self.r.lpush("Meituan:start_urls",item['url'])
            except Exception as e:
                print(e)
        return item
```

（2）编写类 cityItemPipeline，保存房源的城市信息，对应实现代码如下所示：

```
class cityItemPipeline(object):
    def process_item(self, item, spider):
        if isinstance(item, CityItem):
            print("CityItem item")
            try :
                item.save()
            except Exception as e:
                print(e)
        return item
```

（3）编写类 houseItemPipeline，保存房源的详细信息，主要包括 Labels、Facility 和 Host 等信息。对应实现代码如下所示：

```
class houseItemPipeline(object):
    def __init__(self):
        pass

    def process_item(self, item, spider):
        if isinstance(item, HouseItem):
            print("HouseItem item")
            house = None
            try:
                house = item.save()  # 最后才可以 save 这个
                print("hosuse 保存成功 ")

            except Exception as e:
                print("hosuse 保存失败，后面的跳过保存 ")
                print(e)
                print(item)
                return item

            jsonString = item.get("jsonString")
            labelsList = jsonString['Labels']
            facilityList = jsonString['Facility']
            hostInfos = jsonString['Host']

                # house = House.objects.filter(**{'house_id':item.get("house_
id"),"house_date":item.get("house_date")}).first()
            # 查询一次就可以，多条件查询，查询两个联合的主键
            # 这要是标签的多对多写入
            for onetype in labelsList:  # 1.添加所有标签的
                for one in labelsList[onetype]:  # one 都是一个标签
                    try:   # 先找找有没有，然后把已经有的添加进来
                                        label = Labels.objects.filter(**{'label_
name':one[0],"label_desc":one[1]})  # 找到的话就直接加入另一个 Meiju 对象中
                        # print(" 长度 ")
                        if len(label) == 0:
                            # print(" 需要创建后添加 ")
                            l = Labels()
                            l.label_name = one[0]
                            l.label_desc = one[1]
                            # print(" 检查 label")
                            # print(f"onetype:{onetype}")
                            # print(one)
```

```
                    if onetype == "1":    # 优惠标签
                        l.label_type = 1  # 数字也行把，应该，这里没改
                    else:
                        l.label_type = 0
                    l.save()
                         # label = Labels.objects.filter(**{'label_
name':one[0],"label_desc":one[1]})  # 找到的话就直接加入另一个Meiju对象中
                    house.house_labels.add(l)  # 直接把刚才的添加进来
                else:
                    # print("找到就直接添加")
                    # print(label)
                    house.house_labels.add(label.first())  # 这样添加进来的
                    # print("添加成功")
            except Exception as e:
                print(e)
                print("label已存在，跳过插入")
                # print(e)


        # 这里开始是Facility的写入
        # print(facilityList)
        # 先全部写入一遍，然后再把有的添加进来就好，这个效率不太高的样子
        for facility in facilityList:
            if 'metaValue' in facility:
                print()  # 执行先检查后添加
                try:   # 先找找有没有，然后把已经有的添加进来
                           fac = Facility.objects.filter(**{'facility_
name':facility['value']})  # 找到的话就直接加入另一个Meiju对象中
                    # print("长度")
                    if len(fac) == 0:
                        # print("需要创建后添加")
                        l = Facility()
                        l.facility_name = facility['value']
                        l.save()
                             # fac = Facility.objects.filter(**{'label_
name':facility['value']})  # 找到的话就直接加入另一个Meiju对象中
                        # print(l)
                        house.house_facility.add(l)
                    else:
                        # print("找到就直接添加")
                        house.house_facility.add(fac.first())  # 这样添加进来的
                        # print("添加成功")
                except Exception as e:
                    print("facility已经存在，跳过插入")
                    # print(e)

        print("下面开始host信息的添加")
        '''{'hostId': '36438164',
            'host_RoomNum': '51',
            'host_commentNum': '991',
            'host_name': '【塔宿】醉美小蛮腰',
            'host_replayRate': '100'}'''
        print(hostInfos)

        try:   # 先找找有没有，然后把已经有的添加进来,fixing
            hosts = Host.objects.filter(**{
                'host_id':hostInfos['hostId'],
                'host_updateDate':house.house_date})  # 找到直接加入另一个Meiju对象中
            print("长度")
            print("输出查找到的结果")
            print(hosts)
            # if len(hosts) == 0:   # 都可以创建，可以看房东的评价变化
            try:
```

```
                    # print("需要创建后添加")
                    l = Host()
                    l.host_name = hostInfos['host_name']
                    l.host_id = hostInfos['hostId']
                    l.host_RoomNum = hostInfos['host_RoomNum']
                    l.host_commentNum = hostInfos['host_commentNum']
                    l.host_replayRate = hostInfos['host_replayRate']
                    l.save()  # 保存不成功会自然进行处理
                            # fac = Facility.objects.filter(**{'label_
name':facility['value']})  # 找到的话就直接加入另一个Meiju对象中
                    # print("__label_")
                    # print(l)
                    house.house_host.add(l)
                except Exception as e:
                    print(e)
                # else:
                    # print("不为空找到了")
                    # print("__label_")
                    # print(fac)
                    house.house_host.add(hosts.first())  # 这样添加进来的
                    print("已有的情况下添加成功")
            except Exception as e:
                print("以有这个host跳过插入")
                print(e)

            return item  # 这个暂时是无所谓的，因为只有一个管道
        return item
```

通过如下命令运行爬虫程序：

```
scrapy crawl hotel
```

爬虫数据被保存在 MySQL 数据库中，如图 12-1 所示。

图 12-1　数据库中的爬虫数据

12.3 实现民宿信息数据可视化

本项目使用Django框架实现可视化功能，提取在MySQL数据库中保存的民宿数据信息，然后使用Echarts实现数据可视化功能。在本节的内容中，将详细讲解实现数据可视化功能的具体过程。

↑扫码看视频（本节视频课程时间：6分47秒）

为了提高系统的可扩展性，本项目使用Django框架开发一个Web程序，在Dango Web程序中展示可视化结果。

12.3.1 数据库设计

编写文件models.py实现数据库模型设计功能，文件中每个类和MySQL数据库中的表一一对应，每个变量和数据库表中的字段一一对应。文件models.py主要实现代码如下所示：

```
class City(models.Model):
    city_nm = models.CharField(max_length=50,unique=True)    # 城市名字
    city_pynm = models.CharField(max_length=50,unique=True)  # 减少冗余的代价是时间
代价
    city_statas = models.BooleanField(default=False)
    # 这个是让爬虫选择是否进行爬取的城市（缩小爬取范围才可以全部爬下来）

    def __str__(self):
        return self.city_nm + "  " + self.city_pynm


class Labels(models.Model):
    TYPE_CHOICE = (
        (0, "普通标签"),
        (1, "优惠标签"),
        )
    label_type = models.IntegerField(choices=TYPE_CHOICE)  # 类型1为营销， 0为默认标签
    label_name = models.CharField(max_length=191,unique=True)
    label_desc = models.CharField(max_length=191,unique=True)  # 减少冗余的代价是时间

    def __str__(self):
        return str(self.label_type)+"  "+self.label_name+"  "+self.label_desc

    class Meta:
        # 联合约束 独立性约束
        unique_together = ('label_name',"label_desc")


class Host(models.Model):
    '''自己会自动创建一个id的'''
    host_name = models.CharField(max_length=191)  # 房东名字
    host_id = models.IntegerField()  # 房东id
    host_replayRate = models.IntegerField(default=0)  # 回复率
    host_commentNum = models.IntegerField(default=0)  # 评价总数会变也要保存，这样可
以看变化率
    host_RoomNum = models.IntegerField(default=0)  # 不同时间段的房子含有数量
    host_updateDate = models.DateField(default=timezone.now)  # 自动创建时间，不可修改

    def __str__(self):
        return str(self.host_id)+ " "+self.host_name
```

```
        class Meta:
                # 每一个房子一天最多一个数据（一个价格）
            unique_together = ('host_id',"host_updateDate")

    class Facility(models.Model):
        ''' 设施类型 85 个设施
            todo django 如何设置不用外键约束，这样可以更高效，暂时还是使用外键约束。
        '''
        facility_name = models.CharField(max_length=50,unique=True)    # 设施名字

        def __str__(self):
            return self.facility_name


    # 这个是酒店公寓的 model 类
    class House(models.Model):
        '''
        house_id 为主键，然后 unique_together(house_id,date) 为联合唯一约束，一天只能插入一次，然后这里每天的房子都不一定一样，那么要设置好一个爬取的时间，例如每天凌晨的 12 点。
        '''

        house_img = models.CharField(max_length=191,default="static/media/default.jpg")  # 这个是预览图，默认设置

        house_id =  models.IntegerField(default=0)    # 用来标识唯一房子的，可以
        house_cityName = models.CharField(max_length=50,default=" 未知城市 ")
        house_title = models.CharField(max_length=191)    # 我确实想存 YMR
        house_url = models.CharField(max_length=191)
        house_date = models.DateField(default=timezone.now)    # 爬取时间
        house_firstOnSale = models.DateTimeField(default=datetime.datetime(1970, 1, 1, 1, 1, 1, 499454)) # 发布时间
        # 用户评价
        house_favcount = models.IntegerField(default=0)  # 房子页面的点赞数
        house_commentNum = models.IntegerField(default=0)  # 评分人数（也是评论人数）

        house_descScore = models.FloatField(default=0)  # 房子四个分数
        house_talkScore = models.FloatField(default=0)
        house_hygieneScore = models.FloatField(default=0)
        house_positionScore = models.FloatField(default=0)
        house_avarageScore = models.FloatField(default=0)  # 总的平均分 5.0 满分

        # 房子的具体内容信息
        house_type = models.CharField(max_length=50,default=" 未分类 ")  # 整套 / 单间 / 合住
        house_area = models.IntegerField(default=0)    # 房子的面积单位 m²
        house_kitchen = models.IntegerField(default=0)  # 厨房数量 0 就是没有
        house_living_room =  models.IntegerField(default=0)  # 客厅数量、
        house_toilet = models.IntegerField(default=0)  # 卫生间数量
        house_bedroom = models.IntegerField(default=0)  # 卧室数量
        house_capacity = models.IntegerField()  # 可以容纳的人数
        house_bed = models.IntegerField(default=1)  # 床的数量

        # 房子的价格信息
        house_oriprice = models.DecimalField(max_digits=16,decimal_places=2)  # 刚发布价格
        house_discountprice = models.DecimalField(max_digits=16,decimal_places=2,default=0.00)
        # 现在价格，这个可能搞不定，如果这个 discountPrice 没有的话，那就等于现价

        # 房源位置
        house_location_text = models.CharField(max_length=191) # 因为使用 utf8mb4 格式，char 最长为 191，四个字节为一个字符
        house_location_lat = models.DecimalField(max_digits=16,decimal_places=6)
        # 纬度，小数点后 6 位
        house_location_lng = models.DecimalField(max_digits=16,decimal_places=6)
```

```
        # 经度

        # 房源设施
         house_facility = models.ManyToManyField(Facility)   # 一对多外键，但是这样好像增
加保存的难度了，不过可以一次性进行显示等。

        # 房东信息
        house_host = models.ManyToManyField(Host)  # 多对多

        # 普通标签和优惠标签
        house_labels = models.ManyToManyField(Labels)

        earliestCheckinTime = models.TimeField(default="00:00")  # 这个时间有没有

        def __str__(self):
              return  str(self.house_id)+":" +f"{str(self.house_cityName)}" + ":"+
f"{str(self.house_title[0:15])}..." + ":"+str(self.house_oriprice) + " ¥/晚"

        class Meta:
               # 联合约束 独立性约束
            unique_together = ('house_id',"house_date")  # 每个房子一天最多一个数据

            # 联合索引，优化的东西先不弄，索引先不建立
            # index_together = ["user", "good"]

    # class TestUser(models.Model):
    #     account = models.IntegerField(default=0)
    #     password = models.CharField(max_length=191)


    class Favourite(models.Model):  # 收藏夹
        user =  models.OneToOneField(User,unique=True,on_delete=models.CASCADE)
    #     # fav_house = models.CharField(max_length=50,unique=True)   # 城市名字

         fav_city = models.ForeignKey(City, on_delete=models.CASCADE)   # 偏好城市，前
端处理，默认都是广州
        fav_houses = models.ManyToManyField(House)

        def __str__(self):
            return str(self.user.username) + ":" + str(self.fav_city)
```

12.3.2 登录验证表单

为了提高系统的安全性，只有合法用户才能查看可视化结果。编写文件 forms.py 实现登录表单验证功能，主要实现代码如下所示：

```
    from django import forms
    class LoginForm(forms.Form):
          username = forms.CharField(widget=forms.TextInput(attrs={'class': 'mdui-
textfield-input', 'placeholder':"用户名"}))
        password = forms.CharField(widget=forms.PasswordInput(attrs={'class': 'mdui-
textfield-input', 'placeholder':"密码"}))  # 秘密啊

    class RegistrationForm(forms.ModelForm):   # 这个是注册表单的，表单继承 model 就是
modelform
        username =  forms.CharField(label="用户名",widget=forms.TextInput(attrs= {'class':
'mdui-textfield-input', 'placeholder':"用户名"}))
        password = forms.CharField(label="密码",widget=forms.PasswordInput(attrs={'class':
'mdui-textfield-input',

'pattern':"^.*(?=.{8,})(?=.*[a-z])(?=.*[A-Z]).*$",'placeholder':"密码"}))
        password2 = forms.CharField(label="再次输入密码",widget=forms.PasswordInput(attrs=
```

```
{'class': 'mdui-textfield-input',

'pattern':"^.*(?=.{8,})(?=.*[a-z])(?=.*[A-Z]).*$",'placeholder':" 密码 "}))
        email = forms.CharField(label=' 邮箱 ',widget=forms.EmailInput(attrs={'type':
"email",'class': 'mdui-textfield-input', 'placeholder':" 密码 "})
                                ,error_messages={'required': " 邮箱不能为空 "})

        class Meta:
            model = User
            fields = ("username","email")

            def clean_password2(self):
                cd = self.cleaned_data
                if cd['password'] != cd['password2']:
                    raise forms.ValidationError(" 两次输入的密码不相同 ")
                return cd["password2"]
```

12.3.3 视图显示

在本项目中，数据可视化功能是通过 View 视图文件和模板文件实现的。本项目的 View 视图文件是 drawviews.py，具体实现流程可分为如下 7 个步骤。

（1）编写函数 bar_base() 获取系统内的爬虫数量，分别显示房源总数和城市数量，具体代码如下：

```
    def bar_base() -> Bar:  # 返回给前端用来显示图的 json 设置，按城市分组来统计数量
        nowdate = time.strftime('%Y-%m-%d', time.localtime(time.time()))
         count_total_city = House.objects.filter(house_date=nowdate).values("house_
cityName").annotate(
            count=Count("house_cityName")).order_by("-count")
        # for i in count_total_city:
        #     print(i['house_cityName']," ",str(i['count']))
        c = (
            Bar(init_opts=opts.InitOpts(theme=ThemeType.WONDERLAND))
                .add_xaxis([city['house_cityName'] for city in count_total_city])
                .add_yaxis(" 房源数量 ", [city['count'] for city in count_total_city])
                  .set_global_opts(title_opts=opts.TitleOpts(title=" 今天城市房源数量 ",
subtitle=" 如图 "),
                                   xaxis_opts=opts.AxisOpts(axislabel_opts=opts.
LabelOpts(rotate=-90)),
                                 )
                .set_global_opts(
                  datazoom_opts={'max_': 2, 'orient': "horizontal", 'range_start':
10, 'range_end': 20, 'type_': "inside"})
                .dump_options_with_quotes()
        )
        return c
```

（2）编写类 PieView 统计数据库中的房型数据并绘制饼图，具体代码如下：

```
    class PieView(APIView):
        def get(self, request, *args, **kwargs):
            result = fetchall_sql(
                  "select house_type,count(house_type) from (select distinct house_
id ,house_type from hotelapp_house  group by house_id,house_type ) hello group by
house_type")
            c = (
                Pie()
                    .add("", [z for z in zip([i[0] for i in result], [i[1] for i in result])])
                    # .add("",[list(z) for z in zip([x['house_type'] for x in house_
type_count],[x['count'] for x in house_type_count])])
                    .set_global_opts(title_opts=opts.TitleOpts(title=" 总房屋类型 "))
                    .set_series_opts(label_opts=opts.LabelOpts(
```

```
                formatter="{b}: {c} | {d}%",
            ))
                .dump_options_with_quotes()
        )
        return JsonResponse(json.loads(c))
```

（3）编写类 getMonthPostTime、getMonthPostTime2 和 timeLineView，获取数据库中保存的发布时间信息，并绘制发布时间折线图和最近 7 天的折线图，具体代码如下：

```
    # 这个是按月查询的
    class getMonthPostTime(APIView):  # 理论上这个按年份来进行统计
        def get(self, request, *args, **kwargs):
            result = fetchall_sql('''select DATE_FORMAT(house_firstOnSale,'%Y-%m')
              as mydate,count(DATE_FORMAT(house_firstOnSale,'%Y-%m'))as
             mydate_count from hotelapp_house group by mydate ORDER BY mydate'''
                                  )
            context = {"result": result}
            return JsonResponse(context)


    class getMonthPostTime2(APIView):  # 按各个月份来进行统计
        def get(self, request, *args, **kwargs):
            result = fetchall_sql(
                      '''select DATE_FORMAT(house_firstOnSale,'%m') as mydate,count
(DATE_FORMAT(house_firstOnSale,'%Y-%m'))as mydate_count from hotelapp_house group by
mydate ORDER BY mydate'''
                  )
            context = {"result": result}
            # for i in result:
            #     print(i)
            return JsonResponse(context)


    class timeLineView(APIView):
        def get(self, request, *args, **kwargs):
            # week_name_list = getLatestSevenDay()  # 获得最近七天的日期时间列折线图
            # 七天前的那个日期
            today = datetime.datetime.now()
            # // 计算偏移量
            offset = datetime.timedelta(days=-6)
            # // 获取想要的日期的时间
            re_date = (today + offset).strftime('%Y-%m-%d')
               house_sevenday  =  House.objects.filter(house_date__gte=re_date).
values("house_date"). \
              annotate(count=Count("house_date")).order_by("house_date")
            week_name_list = [day['house_date'] for day in house_sevenday]
            date_count = [day['count'] for day in house_sevenday]
            c = (
                Line(init_opts=opts.InitOpts(width="1600px", height="800px"))
                    .add_xaxis(xaxis_data=week_name_list)
                    .add_yaxis(
                    series_name=" 抓取的数量 ",
                    # y_axis=high_temperature,
                    y_axis=date_count,
                    markpoint_opts=opts.MarkPointOpts(
                        data=[
                            opts.MarkPointItem(type_="max", name=" 最大值 "),
                            opts.MarkPointItem(type_="min", name=" 最小值 "),
                        ]
                    ),
                    markline_opts=opts.MarkLineOpts(
                        data=[opts.MarkLineItem(type_="average", name=" 平均值 ")]
                    ),
```

```
        )
            .set_global_opts(
            title_opts=opts.TitleOpts(title="最近七天抓取情况", subtitle=""),
            xaxis_opts=opts.AxisOpts(type_="category", boundary_gap=False),
        )
            .dump_options_with_quotes()
    )
    return JsonResponse(json.loads(c))
```

（4）编写类 drawMap 绘制房源分布热力图，具体代码如下：

```
class drawMap(APIView):  # 要加 apiview # 美团房源数量热力图
    def get(self, request, *args, **kwargs):
        from pyecharts import options as opts
        from pyecharts.charts import Map
        from pyecharts.faker import Faker

        result = cache.get('house_city', None)  # 使用缓存，可以共享，真好。
        if result is None:  # 如果无，则向数据库查询数据
            print("使用缓存房源城市统计")
            result = fetchall_sql(
                """select house_cityName,count(house_cityName) as count from
(SELECT distinct(house_id),house_cityName FROM  hotelapp_house) hello group by
house_cityName""")
            cache.set('house_city', result, 3600 * 12)  # 设置缓存

        else:
            pass
        c = (
            Geo()
                .add_schema(maptype="china")
                .add(
                "房源",
                    [z for z in zip([i[0] for i in result], [i[1] for i in result])],
                type_=ChartType.HEATMAP,
            )
                .set_series_opts(label_opts=opts.LabelOpts(is_show=False))
                .set_global_opts(
                visualmap_opts=opts.VisualMapOpts(),
                title_opts=opts.TitleOpts(title="美团民宿房源热力图"),
            )
                .dump_options_with_quotes()
        )

        return JsonResponse(json.loads(c))  # f 安徽这个
```

（5）编写函数 houseScoreLine() 绘制详情页评分折线图，具体代码如下：

```
class houseScoreLine(APIView):
    def get(self, request):
        house_id = request.GET.get("house_id")
        result = fetchall_sql(
                  f'select house_date,house_avarageScore,house_img,house_
descScore,house_hygieneScore,house_positionScore,house_talkScore from hotelapp_house
where house_id="{house_id}"  '
              f'and house_oriprice !=0.00 and house_discountprice!=0.00  order by
house_date limit 0,7')  # todo 这个很有意思
        week_name_list = [house[0] for house in result]
        avarageScore = [house[1] for house in result]
        # discountprice = [house[2] for house in result]
        descScore = [house[3] for house in result]
        hygieneScore = [house[4] for house in result]
        positionScore = [house[5] for house in result]
        talkScore = [house[6] for house in result]
        c = (
```

```
                    Line(init_opts=opts.InitOpts(width="1600px", height="800px",
theme=ThemeType.LIGHT))
                    .add_xaxis(xaxis_data=week_name_list)
                    # .add_xaxis(xaxis_data=date_count)
                    .add_yaxis(
                    series_name="平均分",
                    # y_axis=high_temperature,
                    y_axis=avarageScore,
                    markpoint_opts=opts.MarkPointOpts(
                        data=[
                            opts.MarkPointItem(type_="max", name="最高分"),
                            #     opts.MarkPointItem(type_="min", name="最高价"),
                        ]
                    )
                )
                    .add_yaxis(
                    series_name="描述得分",
                    is_connect_nones=True,
                    y_axis=descScore,
                )
                    .add_yaxis(
                    series_name="沟通得分",
                    is_connect_nones=True,
                    y_axis=talkScore,
                ).add_yaxis(
                    series_name="卫生得分",
                    is_connect_nones=True,
                    y_axis=hygieneScore,
                )
                    .add_yaxis(
                    series_name="位置得分",
                    is_connect_nones=True,
                    y_axis=positionScore,
                )

                    .set_global_opts(
                    title_opts=opts.TitleOpts(title="最近房源评价", subtitle="满分5分"),
                    # tooltip_opts=opts.TooltipOpts(trigger="axis"),
                    # toolbox_opts=opts.ToolboxOpts(is_show=True),
                        xaxis_opts=opts.AxisOpts(type_="category", boundary_gap=False),
                )
                    .dump_options_with_quotes()  # 这个是序列化为json，最后必须加上参数
            )
            return JsonResponse(json.loads(c))
```

（6）编写函数 get_twoLatestYear() 提取每个民宿信息的发布时间，获得最近两年的发布数量的情况，具体代码如下：

```
    class get_twoLatestYear(APIView):  # 按各个月份来进行统
        # @cache_response(timeout=60 * 60*3, cache='default')
        def get(self, request, *args, **kwargs):
            yearRange = request.GET.get("yearRange")
            oneYear = yearRange.split("-")[0]
            twoYear = yearRange.split("-")[1]
            print(yearRange)

            temp_df = cache.get('house_firstOnSale_df', None)  # 使用缓存，可以共享，真好。
            if temp_df is None:  # 如果无，则向数据库查询数据
                print("没有缓存，重新查询 ,house_firstOnSale_df")
                 result = fetchall_sql_dict('''SELECT distinct(id),house_firstOnSale
FROM hotelapp_house ''')
                temp_df = pd.DataFrame(result)
                temp_df.index = pd.to_datetime(temp_df.house_firstOnSale)
```

```
            # df.resample("Q-DEC").count()  # 季度
            cache.set('house_firstOnSale_df', temp_df, 3600 * 12)  # 设置缓存

        dff = temp_df.id.resample("QS-JAN").count().to_period("Q")

        import pyecharts.options as opts
        from pyecharts.charts import Line
        c = (
            Line(init_opts=opts.InitOpts(width="1600px", height="800px"))
                .add_xaxis(
                # xaxis_data=[str(j) for j in dff[oneYear].index],
                xaxis_data=[str(twoYear) + "Q" + str(j) for j in range(1, 5)],
            )
                .extend_axis(
                # xaxis_data=[str(j) for j in dff[twoYear].index],
                xaxis_data=[str(oneYear) + "Q" + str(jx) for jx in range(1, 5)],

                xaxis=opts.AxisOpts(
                    type_="category",
                    axistick_opts=opts.AxisTickOpts(is_align_with_label=True),
                    axisline_opts=opts.AxisLineOpts(
                                    is_on_zero=False, linestyle_opts=opts.
LineStyleOpts(color="#6e9ef1")
                    ),
                    axispointer_opts=opts.AxisPointerOpts(
                        is_show=True,
                        # label=opts.LabelOpts(formatter=JsCode(js_formatter)
                        #                        )
                    ),
                ),
            )
                .add_yaxis(
                series_name=f"{oneYear}",
                is_smooth=True,
                symbol="emptyCircle",
                is_symbol_show=False,
                # xaxis_index=1,
                color="#d14a61",
                 # y_axis=[2.6, 5.9, 9.0, 26.4, 28.7, 70.7, 175.6, 182.2, 48.7,
18.8, 6.0, 2.3],
                y_axis=[int(x) for x in dff[oneYear].values],

                label_opts=opts.LabelOpts(is_show=False),
                linestyle_opts=opts.LineStyleOpts(width=2),
            )
                .add_yaxis(
                series_name=f"{twoYear}",
                is_smooth=True,
                symbol="emptyCircle",
                is_symbol_show=False,
                color="#6e9ef1",
                 # y_axis=[3.9, 5.9, 11.1, 18.7, 48.3, 69.2, 231.6, 46.6, 55.4,
18.4, 10.3, 0.7],
                y_axis=[int(x) for x in dff[twoYear].values],

                label_opts=opts.LabelOpts(is_show=False),
                linestyle_opts=opts.LineStyleOpts(width=2),
            )

                .set_global_opts(
                legend_opts=opts.LegendOpts(),
                title_opts=opts.TitleOpts(title=f"{yearRange}期间发布房源的数量",
subtitle="单位个"),
```

```
                        tooltip_opts=opts.TooltipOpts(trigger="none", axis_pointer_
type="cross"),
                    xaxis_opts=opts.AxisOpts(
                        type_="category",
                        axistick_opts=opts.AxisTickOpts(is_align_with_label=True),
                        axisline_opts=opts.AxisLineOpts(
                                        is_on_zero=False, linestyle_opts=opts.
LineStyleOpts(color="#d14a61")
                        ),
                        axispointer_opts=opts.AxisPointerOpts(
                                              # is_show=True, label=opts.
LabelOpts(formatter=JsCode(js_formatter))
                        ),
                    ),
                    yaxis_opts=opts.AxisOpts(
                        type_="value",
                        splitline_opts=opts.SplitLineOpts(
                                          is_show=True, linestyle_opts=opts.
LineStyleOpts(opacity=1)
                        ),
                    ),
                )
                .dump_options_with_quotes()  # 这个是序列化为 json，最后必须加上参数
            )

            return JsonResponse(json.loads(c))
```

（7）编写函数 get_postTimeLine()，根据封装好的按月分或者按年分绘制时间折线图，具体代码如下：

```
    class get_postTimeLine(APIView):  # 按月份分，或者按年分
        # @cache_response(timeout=60 * 60*3, cache='default')
        def get(self, request, *args, **kwargs):
            timeFreq = request.GET.get("timeFreq")
            # print(timeFreq)
            temp_df = cache.get('house_firstOnSale_df', None)  # 使用缓存，可以共享，真好。
            if temp_df is None:  # 如果无，则向数据库查询数据
                print("没有缓存，重新查询")
                result = fetchall_sql_dict('''SELECT distinct(id),house_firstOnSale FROM
hotelapp_house ''')
                temp_df = pd.DataFrame(result)
                temp_df.index = pd.to_datetime(temp_df.house_firstOnSale)
                # type(df.id.resample("1M").count())  # huode 按月份来
                # df.id.resample("Y").count()
                # df.id.resample("M").count()
                # df.id.resample("D").count()  # 忽略为 0 的值来处理
                # df.resample("Q-DEC").count()  # 季度
                cache.set('house_firstOnSale_df', temp_df, 3600 * 12)  # 设置缓存
            else:
                pass
            # if timeFreq!="" or timeFreq!=None:
            label = ""
            title = ""
            if timeFreq == "month":
                timeFreq = "M"
                label = "月新增房源"
            if timeFreq == "year":
                timeFreq = 'A'
                label = "年新增房源"
            if timeFreq == "season":
                timeFreq = 'Q-DEC'
                label = "季度新增房源"
              dff = temp_df.house_firstOnSale.resample(f"{timeFreq}").count().to_
```

```
period(
                f"{timeFreq}")  # 加上这个就好了 .to_period('Q')  # 季度

            # dff.resample('A').mean().to_period('A')

            x_data = [str(y) for y in dff.index]
            y_data = [str(x) for x in dff.values]

            c = (
                Line()
                    .add_xaxis(xaxis_data=x_data)
                    .add_yaxis(
                    series_name=label,
                    stack="新增房源数量",
                    y_axis=y_data,
                    label_opts=opts.LabelOpts(is_show=False),
                )
                    .set_global_opts(
                    title_opts=opts.TitleOpts(title="房源发布时间序列分析"),
                    tooltip_opts=opts.TooltipOpts(trigger="axis"),
                    yaxis_opts=opts.AxisOpts(
                        type_="value",
                        axistick_opts=opts.AxisTickOpts(is_show=True),
                        splitline_opts=opts.SplitLineOpts(is_show=True),
                    ),
                     xaxis_opts=opts.AxisOpts(type_="category", boundary_gap=False),
                )
                    .dump_options_with_quotes()  # 这个是序列化为json，最后必须加上参数
            )

            return JsonResponse(json.loads(c))
```

为了节省本书篇幅，本项目介绍到此为止，有关本项目的具体实现流程，请参考本书的配套源码。执行本项目可视化模块后的部分效果如图 12-2 所示。

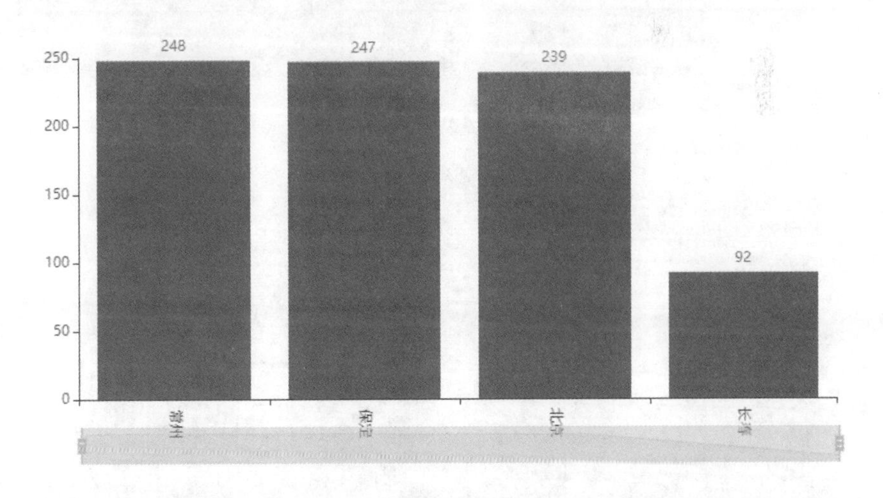

（a）数据概览

图 12-2　可视化执行效果

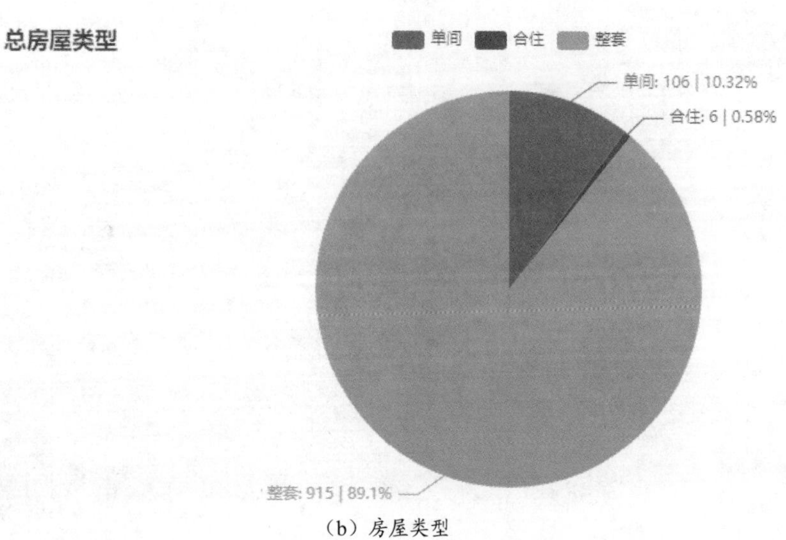

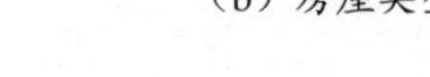

（b）房屋类型

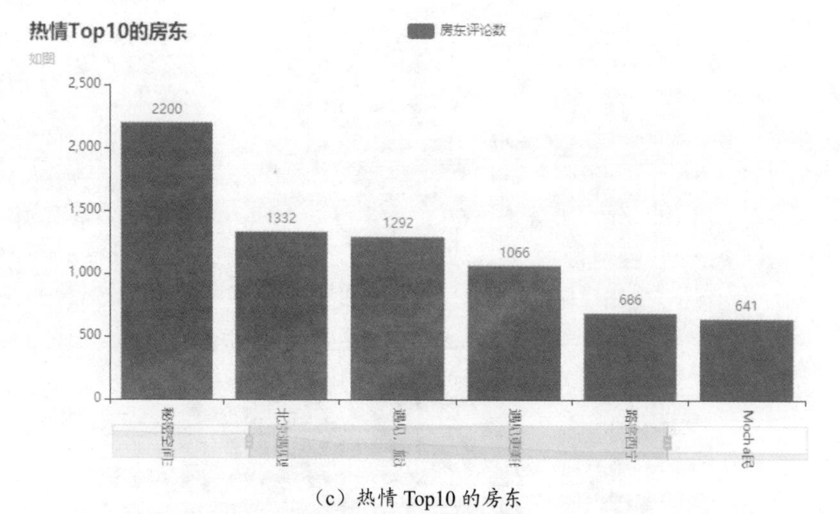

（c）热情 Top10 的房东

（d）房屋设施分析

图 12-2　可视化执行效果（续）